Thermal Analysis

Thermal Analysis

From Introductory Fundamentals to Advanced Applications

El-Zeiny M. Ebeid
Professor of Physical Chemistry, Department of Chemistry, Faculty of Science, Tanta University, Tanta, Egypt
Misr University for Science and Technology (MUST), 6th of October City, Egypt

Mohamed B. Zakaria
Lecturer of Physical Chemistry, Department of Chemistry, Faculty of Science, Tanta University, Tanta, Egypt
Humboldt Fellow, Inorganic Chemistry 1, Ruhr University of Bochum, Bochum, Germany

ELSEVIER

Elsevier
Radarweg 29, PO Box 211, 1000 AE Amsterdam, Netherlands
The Boulevard, Langford Lane, Kidlington, Oxford OX5 1GB, United Kingdom
50 Hampshire Street, 5th Floor, Cambridge, MA 02139, United States

Library of Congress Cataloging-in-Publication Data
A catalog record for this book is available from the Library of Congress

British Library Cataloguing-in-Publication Data
A catalogue record for this book is available from the British Library

ISBN: 978-0-323-90191-8

For information on all Elsevier publications
visit our website at https://www.elsevier.com/books-and-journals

Publisher: Susan Dennis
Acquisitions Editor: Kathryn Eryilmaz
Editorial Project Manager: Lena Sparks
Production Project Manager: Bharatwaj Varatharajan
Cover Designer: Victoria Pearson

Typeset by SPi Global, India

Contents

2. Temperature-modulated DSC (TMDSC)

3. Thermal analysis in materials science

4. Applications of TA

5. Thermal analysis in studying the phase problem

6. Thermal analysis in recycling and waste management

About the authors

El-Zeiny M. Ebeid is a professor of physical chemistry at Tanta University, Egypt. He received his PhD under joint supervision between Tanta University and University Centre Weston in the United Kingdom in 1980. He was a visiting professor to the University of Kent at Canterbury and Konstanz Universities. He won the Egyptian National Academy of Science Award. He coauthored several published books including the book "Photophysical and Laser-Based Techniques in Chemistry, Biology and Medicine", BookSurge (2006), USA. He is a member of the American Nano Society. He has more than 90 publications in international scientific journals. His areas of interest are the fields of laser chemistry, solid-state chemistry, and nanomaterials.

Mohamed B. Zakaria, a lecturer of physical chemistry at Tanta University, Egypt, received his PhD in engineering (2016) from Waseda University (Tokyo) supervised by Professor Yusuke Yamauchi. He then joined National Institute for Materials Science in Tsukuba (2016–2018) as a JSPS postdoctoral fellow supervised by Professor Toyohiro Chikyow. From 2018 to 2020 he moved to Australian Institute for Bioengineering and Nanotechnology (AIBN), The University of Queensland (Australia). In 2020, he joined Professor's Ulf-Peter Apfel group at Ruhr University of Bochum (RUB) and Fraunhofer UMSICHT in Germany as a Humboldt Fellow. He is a member of the American Chemical Society (ACS). He is the recipient of many awards, including JSPS Fellowship/Japan (2016), Postdoctoral International Exchange Program Award/China (2018), IAAM Young Scientist Prestigious Medal/Stockholm (2018), Humboldt Research Fellowship/Germany (2020), IRC Postdoctoral Research Award/Ireland (2020), and more. He has published more than 52 articles in leading peer-reviewed journals and has participated in more than 27 international conferences. His area of interest is the field of production and properties of functional nanomaterials for electrocatalysis and nanodevices.

Preface

Thermal analysis (TA) is a technique for characterizing materials by measuring and analyzing changes in physical or chemical properties resulting from controlled and measured changes in temperature. Meanwhile, according to the International Confederation for Thermal Analysis and Calorimetry (ICTAC 2014), TA is defined with the study of the relationship between a sample property and its temperature as the sample is heated or cooled in a controlled manner [1]. Modern thermal analytical instruments have progressed from homemade devices used by a relatively small number of scientists to instrument systems that take advantage of the most up-to-date advances in thermal measurement, electronics, computer technology, and materials of construction. Today, these sophisticated instrument systems are used routinely by physical scientists, engineers, material scientists, and biological scientists in research, product development, quality control, and production.

TA has contributed significantly to the development of materials, resulting in improved product design. By optimizing material processing and production, and by improving product quality, manufacturing economies can be realized. The return on investment can be measured in terms of cost saving, improved product performance, and growing markets. Payback usually exceeds many times for the investment in thermal analysis instruments and personnel.

In academic and research institutions, TA is an important tool for theoretical and applied research in thermodynamics, kinetics, materials science, and many other physical science disciplines. Moreover, TA is useful in materials science to study the thermal conversion of materials into new products, determine the contents of elements and moisture, examine the phase purity, estimate the molar masses of elements, track polymer degradation, sol-gel science, and more.

Differential thermal analysis (DTA) appears to have been performed first by Austen in 1899 using the equipment developed by Kurnakov and Saladin [2]. Then, modular thermal analysis was introduced commercially in 1962 [3] and since then, it has been applied extensively for studying many chemical and physical characteristics such as melting points and percent crystallinity by DSC, crystallization temperature, glass transition temperature by TMA, cure rate, degree of cure, composition, thermal stability, expansion coefficient, and chemical kinetics. Several TA techniques are currently used, such as differential scanning calorimetry (DSC), temperature modulated DSC (TMDSC), differential thermal analysis (DTA), thermogravimetry (TG) or thermogravimetric analysis (TGA), thermomechanical analysis (TMA), differential photocalorimetry (DPC), dynamic mechanical analysis (DMA), thermodilatometry (TD), dielectric thermal analysis (DEA), thermally stimulated current (TSC), emanation thermal analysis (ETA), thermoluminescence (TL), fast scanning calorimetry (FSC), and microcalorimetry [1–3].

Our book includes state of the art and definitions of various TA techniques since its appearance in 1899 by Austen until today, as described in Chapter 1. Chapter 2 reports the temperature-modulated DSC (TMDSC) method and its applications, especially its use for studying the thermodynamic properties of polymers and pharmaceuticals. We talked in this chapter about all kinds of TMDSC and the most advanced TOPEM® method and their applications. Chapter 3 focuses on the potential of TA in materials science with applications in chemistry and engineering, where it is used for studying the thermal conversion of materials which yield new functional candidates for the recent applications and determining contents of elements and moisture. Chapter 4 demonstrates in detail the various applications of TA in food, electronic industries, solid-state reactions, chemistry of polymers and large directing agents, kinetic studies, and more. Chapter 5 demonstrates the crystal structure and phase changes occurring upon heating by TA. Finally, Chapter 6 demonstrates the potential of TA in recycling and waste management.

To the best of our knowledge, this is the first time that the up-to-date thermal analysis methods that form fundamental to recent applications in materials science, industry, electronics, academics, recycling and waste management, phase changes, and food industry are combined in one book, which will be useful in academics, industry, and research institutions.

We thank Ahmed El-Zeiny Ebeid for copy editing this book.

El-Zeiny M. Ebeid, Tanta University, Tanta
Misr University for Science and Technology (MUST), 6th October City, Egypt
Mohamed B. Zakaria, Tanta University, 31527 Tanta, Egypt
Ruhr University of Bochum, 44801 Bochum, Germany
Fraunhofer UMSICHT, 46047 Oberhausen, Germany

References

[1] T. Lever, P. Haines, J. Rouquerol, E.L. Charsley, P.V. Eckeren, D.J. Burlett, ICTAC nomenclature of thermal analysis (IUPAC recommendations 2014), Pure Appl. Chem. 86 (2014) 545–553.

[2] W.C. Roberts-Austen, Metallographist 2 (1899) 186. N.S. Kurnakov, Z. Anorg, Chemie 42 (1904) 184; E. Saladin, Iron and steel metallurgy, Metallography 7 (1904) 237.

[3] D. Furniss, A.B. Seddon, in: P. Gabbott (Ed.), Principles and Applications of Thermal Analysis, Blackwell Publishing Ltd, 2008, pp. 410–449. Chapter 10; M. Wagner, Thermal Analysis in Practice, Carl Hanser Verlag, Munich (2017), pp. 158–161.

Chapter 1

State of the art and definitions of various thermal analysis techniques

1 Differential thermal analysis (DTA)

DTA is a technique in which the temperature difference between a sample and a reference is measured as a function of temperature while the sample and the reference material are subjected to a controlled temperature program [1–3].

In a DTA curve, the temperature difference (ΔT) is plotted on an ordinate with endothermic reactions downward or upward, based on the machine, and temperature (T) or time (t) on the abscissa (Fig. 1.1A).

The main components of DTA machine are sample and reference pans or crucibles (e.g., metallic or ceramic) and holders, furnace, thermocouples, a temperature programmer, and a recording system. One thermocouple is placed in the reference and another one is placed in the sample, which is connected to a voltmeter. The reference body is a material that has no phase changes during heating. Moreover, DTA can be performed with or without the reference body. One more thermocouple of the furnace is responsible for recording at temperature T (in continuous line or periodic pulse form) and for controlling linearity of heating according to the preestablished program. The phase transition of the sample by heating causes a small deflection of the voltmeter [4,5].

2 Differential scanning calorimetry (DSC)

In DSC unlike DTA, the temperature difference (ΔT) between the sample and the reference is always zero since automatic arrangements are made to supply or withhold power to the sample during the thermal scanning. Hence, the rate of change of enthalpy (ΔH) during a reaction is recorded directly as a function of temperature or time [4,5]. Moreover, the DSC measures are made in a calorimeter where C_p (thermal capacity, J g^{-1} K^{-1}) is known, i.e., the equipment gives a ΔH. The heat capacity at constant pressure is given by the equation: $C_p = (dH/dT)_p$, where H is the enthalpy of the system and T is the temperature [5].

Thermal Analysis. https://doi.org/10.1016/B978-0-323-90191-8.00001-4

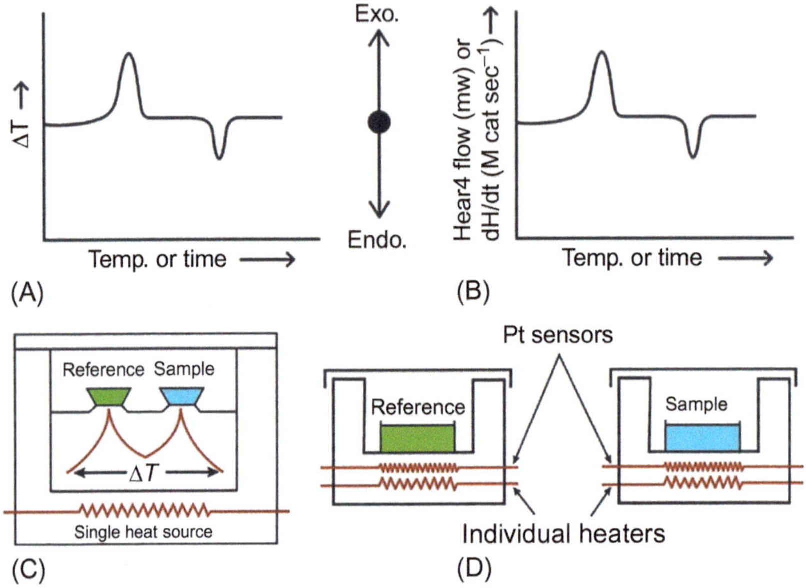

FIG. 1.1 (A) DTA and (B) DSC curves and the schematic illustrations of (C) heat flux DSC and (D) power compensated DSC systems.

Two modes: power compensation differential scanning calorimetry (power compensation DSC) and heat-flow differential scanning calorimetry (heat flux DSC), can be distinguished depending on the method of measurement used. DSC is sometimes designated as differential enthalpic analysis [5].

DSC by heat flow is a technique in which the difference between the heat-flow rates into a sample and a reference material is measured on the ordinate and temperature (*T*) or time (*t*) is plotted on the abscissa (Fig. 1.1B and C) [1].

DSC by power compensation is a technique in which the difference between the electrical powers into a sample and a reference material is measured (Fig. 1.1D) [1].

The reference used is usually an inert solid such as aluminum oxide (Al_2O_3 powder). DTA has a drawback that the emissivity of heat is different from the sample and the reference due to differences in their temperatures, which is given by Eq. (1.1):

$$\frac{dQ}{dt} = c\,(T_s - T_e) \tag{1.1}$$

where $\frac{dQ}{dt}$ is the rate of heat emissivity, c is a constant, T_s and T_e are the temperature of the sample and the environment, respectively. DSC avoids the different emissivity drawback in DTA that complicates the thermal bases of DTA. For instance, high uncertainty levels are associated with activation parameters extracted from DTA data (vide infra) compared with the more reliable DSC data. Moreover, the DSC works as a calorimeter in which the heat involved in the reaction is not dissipated into the ambient.

3 Instrumentation of DSC and DTA

Fig. 1.2 illustrates the circuitry of DSC. There are two separate heating circuits: the average heating controller and the differential heating circuit.

3.1 Heating controller

The average temperature controller

The average temperature controller is the temperature driving unit that raises the temperature of the sample and the reference (which could be the same) linearly with time. Since there could be a little mismatch between the temperatures of the sample and the reference, their average temperature is considered by the average [6].

The differential temperature controller

The differential temperature controller monitors the difference in temperature between the sample and the reference and automatically adjusts the power to

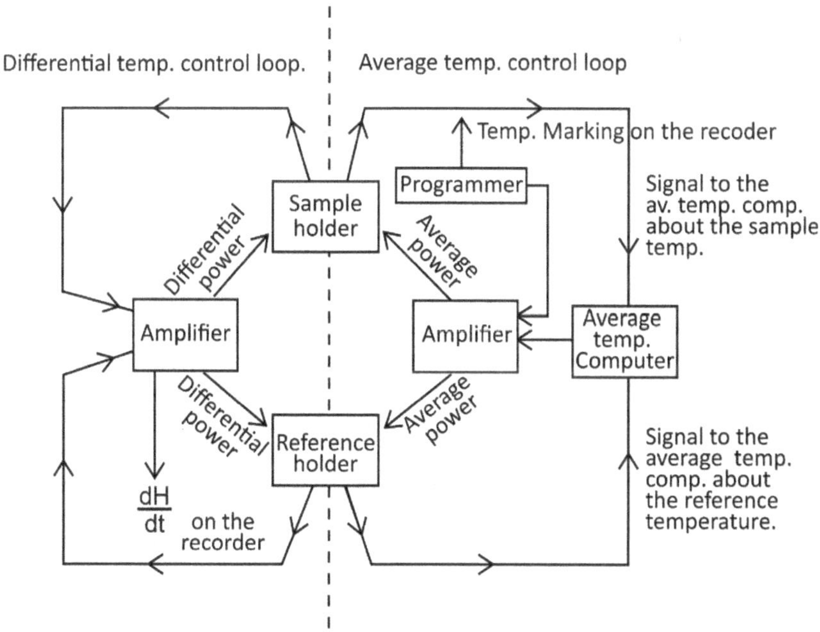

FIG. 1.2 The circuitry of a DSC (x-axis: temperature and y-axis: the difference in power supplied to the two differential heaters).

either the reference or sample chambers to keep the temperature equal. The average temperature is put on the x-axis and the difference in power supplied to the two differential heaters is displayed on the y-axis. The power difference (dH/dt) is calibrated in terms of calories per unit time [7].

When, for example, an exothermic process occurs in the sample, the average temperature circuit decreases the rate of heating of both the reference and sample equally. In the sample compartment, this decrease in the rate of heating is compensated by the heat of reaction while the differential heater in the reference compartment compensates for the decreased heating. This is recorded as an exothermic peak.

Temperatures are monitored using platinum resistance devices. Sealed sample containers eliminate atmospheric considerations in most cases. A nitrogen flow may be applied sometimes to ensure inert conditions and to achieve good electrical conductance.

The amount of heat generated by differential heaters per unit time is given by

$$P = \frac{dH}{dt} = i^2 R \tag{1.2}$$

where P is the power in watts, H is quantity of heat in Joules, i is the current in amperes, and R is the resistance in ohms. Chemical reactions, on the other hand, liberate or absorb heat according to the following equation:

$$\frac{dH}{dt} = \Delta H^\circ \frac{dn_p}{dt} \tag{1.3}$$

where n_P is the number of moles of product formed. When ΔH° is positive enthalpy change (endothermic reaction), the sample heater is energized and a negative signal is obtained. When ΔH° is negative enthalpy change (exothermic reaction), the reference heater is energized and a positive signal is obtained. The integral of the peak is equal to the heat evolved or absorbed by the reacting sample.

The heating rate, sample size, sample particle size, thermal conductivity of the cell, and sample packing, all these factors may have an effect on the rate of the reaction, particularly if large thermal gradients are allowed to develop in the sample or reference that have a severe effect on the apparent rate of reaction and any values calculated from these rates (e.g., activation parameters).

The rate at which heat is evolved in exothermic reactions must be considered in DSC. Rapid exothermic reactions may cause the rate of temperature increase of the sample to be greater than the programmed rate of heating, even when both the average and differential heaters are off. A similar problem sometimes occurs with endothermic processes, where a rapid endothermic reaction may cool the sample so severely that the average heating cannot maintain a linear heating rate and isothermal conditions. Both these situations can be rectified by adjusting the heating rate or the sample size.

Fig. 1.3A shows schematically the essential parts of a DTA apparatus, including provision for bathing the samples with a controlled atmosphere; the gas can be made to flow through the bed of particulate sample, thereby flushing away any gaseous products of decomposition. In this apparatus, as well as in some models of thermo-balance, a gas chromatograph connected to monitor

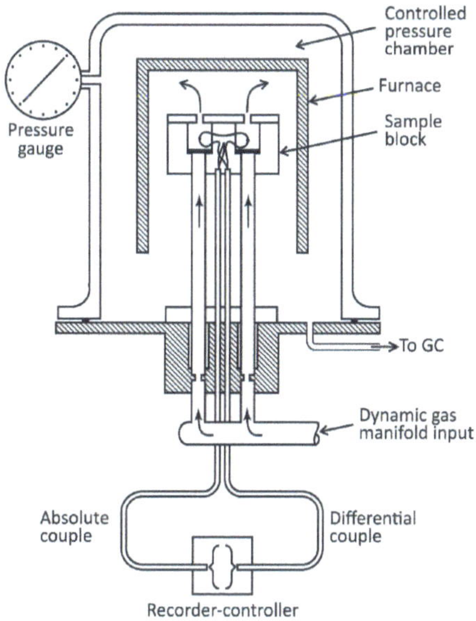

(A)

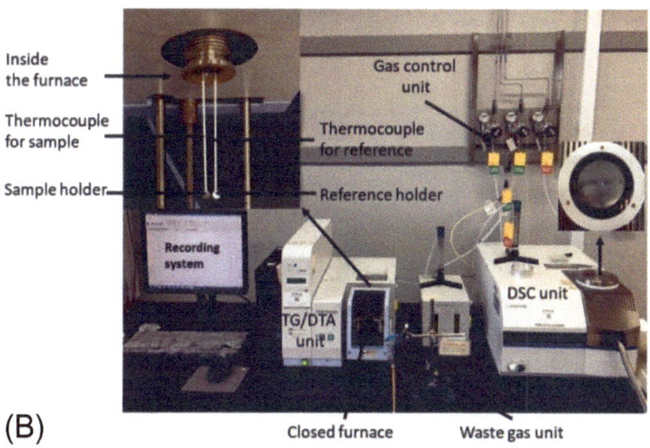

(B)

FIG. 1.3 (A) Schematic diagram of one type of DTA apparatus. (B) A photograph of TG/DTA and DSC apparatus.

evolved gases can be helpful in gathering information about a system. A photograph of TG/DTA and DSC apparatus is shown in Fig. 1.3B.

3.2 Sampling

I. Sample size is kept small to minimize temperature gradients within the sample (i.e., around 5–10 mg). In general, a sample weight resulting in a maximum heat generation in a DSC of less 8 $mJ\,s^{-1}$ is satisfactory according to the American National Standard; ANSI/ASTM E698-79.

II. Samples should be representative of the material being studied and should be fixed to achieve good thermal contact between the sample and the crucible. (Note: a thin layer of graphite can be used to cover the sample thus increasing thermal conductivity.) The escape of gaseous products through the thin graphite layer is almost free. Glass beads might also be used.

III. The sample container should be nonreactive with the sample reaction products.

IV. The reference for the sample is normally an empty container or one filled with inert material (e.g., α-Al_2O_3 powder).

V. Samples that have appreciable volatility over the temperature range of interest may require sealing (in case of DSC or DTA) in hermetic containers or a high-pressure cell, or both, to prevent vaporization interference and weight loss of unreacted material (Fig. 1.4)

VI. The sample atmosphere should closely represent the conditions of usage. Fig. 1.5 shows the relative dimensions of an empty DSC-pan and a TG/DTA-pan.

4 Differential photocalorimetry (DPC)

DPC is a new thermal analysis technique that follows the exothermic rates of irradiated samples as a function of time under isothermal conditions. It can

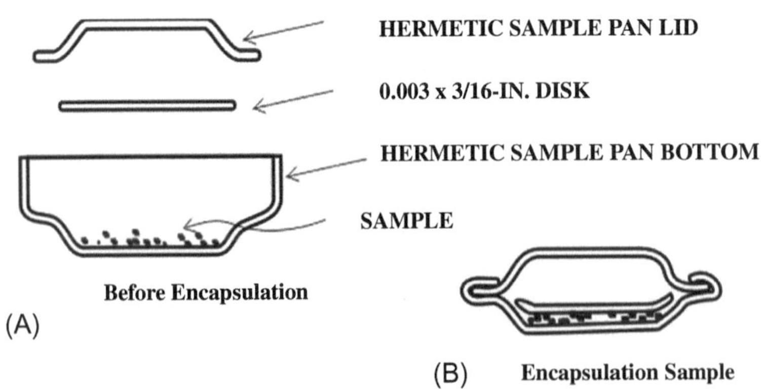

FIG. 1.4 (A and B) Arrangements for good sample contact with container.

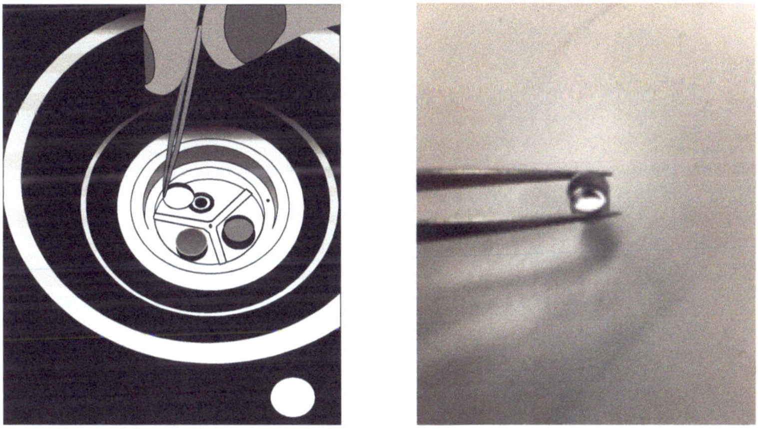

FIG. 1.5 The relative dimensions of (A) DSC-pan and (B) a TG/DTA-pan (e.g., platinum crucible).

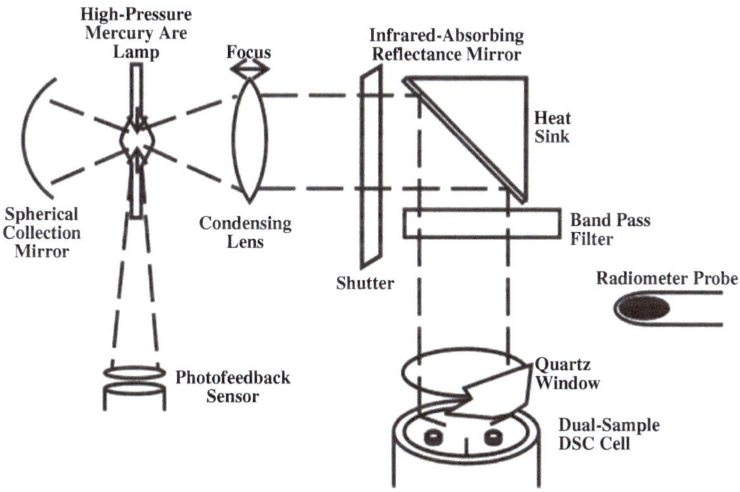

FIG. 1.6 Schematic representation of DuPont 930 differential photocalorimeter.

be applied under continuous illumination reactions or for the dark reaction following a period of illumination. Most methods of studying photopolymerization kinetics (e.g., dilatometry or monitoring of unreacted monomer concentration) are not suitable for studying bulk polymerizations and are particularly unsuitable for studying network-forming reactions. DPC has the advantage that the rate of the reaction can be measured directly, if the heat produced by the photopolymerization is proportional to the number of monomer units reacted. Fig. 1.6 shows a schematic diagram for a DPC apparatus [8].

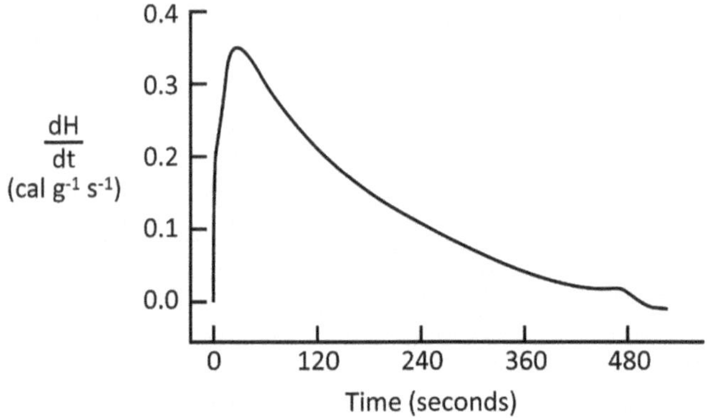

FIG. 1.7 DPC trace for photopolymerization of lauryl acrylate at 30°C.

In DPC, the sample and the reference are covered by quartz discs. A DPC curve is shown in Fig. 1.7.

As shown in Fig. 1.7, the maximum rate occurs within about 10 s of the time the shutter is opened. This delay is not the time required to reach a steady-state free radical concentration, which is much shorter, but is due to the rather low thermal conductivity of the sample and the response time of thermal analyzer [9].

DPC measures the cure rate and degree of cure of photocurable polymers. (Note: cure can be measured in terms of the percentage of the original number of functional groups which have reacted). Another concept of curing is the extent to which a cured film exhibits the desired mechanical properties. The use of radiation-curable coatings and adhesives is growing rapidly. There are several reasons for this rapid growth:

- The technique is solvent-free, thus eliminates the cost associated with solvents, reduces storage space, and eliminates the pollution caused by solvent evaporation during processing.
- Substitution of photo-cure systems for heat cure equipment saves space and lowers energy consumption.
- Depending on the thickness, photocurable polymers cure in seconds or minutes, thus allowing postpacking and immediate handling.

UV-curable polymer systems typically contain monomer or oligomer (oligomer is a lower chemical aggregate compared with polymer), a light-sensitive photoinitiator, and other additives for color and mechanical properties. Concentrations of these components determine viscosity, cure rate, adhesion, flexibility, and abrasion resistance. These properties are also affected by the degree of cure. Fig. 1.8 compares the response of two different photopolymers. Sample #1 cures very rapidly while sample #2 takes more than 50 s of exposure to reach its maximum rate.

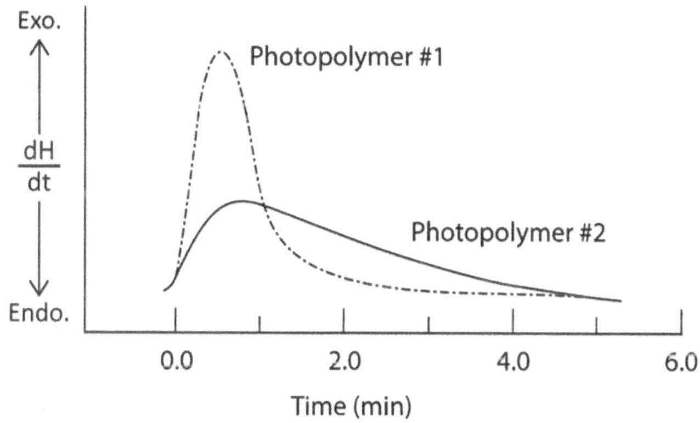

FIG. 1.8 DPC traces illustrating photo-cure of two photopolymers.

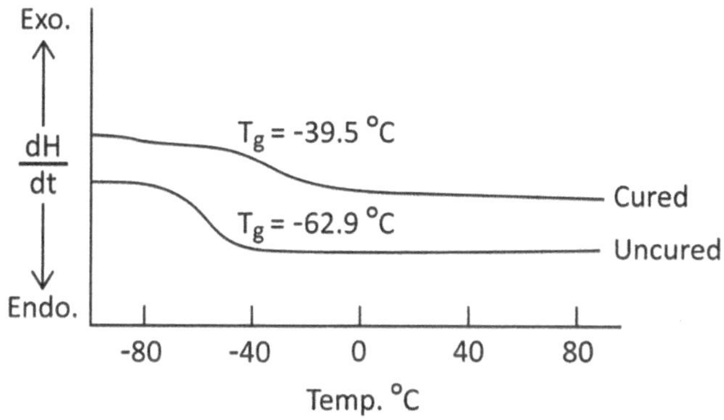

FIG. 1.9 Effect of degree of cure on the glass transition of a photopolymer.

The one component that significantly affects the cure rate of photocurable polymers is the photoinitiator. Decreasing the concentration of a photoinitiator from 0.5% to 0.3% can more than double the induction time for the reaction. Since the photoinitiation is frequently the most expensive of the photopolymer ingredients, DPC offers a fast and reliable method for optimizing and verifying the concentration of the photoinitiator.

As cure proceeds, chemical cross-linking increases and the glass transition temperature (T_g) moves toward some maximum value. Thus, T_g can be used to monitor the degree of cure (Fig. 1.9).

5 Pressure differential scanning calorimetry (PDSC)

PDSC measures heat flow and temperatures of transitions as a function of temperature, time, and pressure (elevated pressure or vacuum) [10].

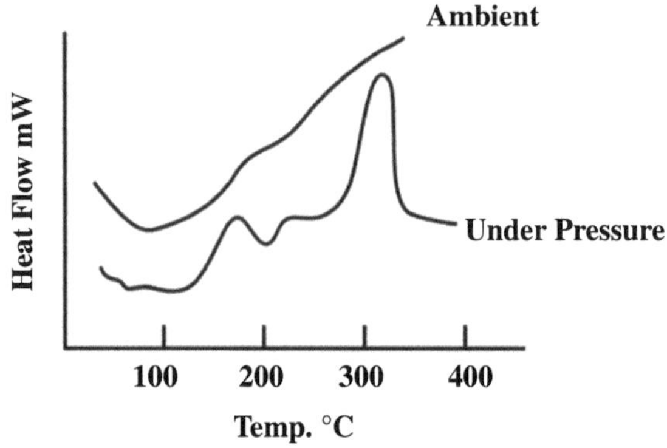

FIG. 1.10 DSC and PDSC of phenolic resin.

Antioxidants, for example, are volatile materials that might vaporize at elevated temperatures giving rise to a failure of traditional TA tests.

PDSC eliminates the volatility problem in two ways: (1) high-pressure decreases the volatility by increasing the boiling point, and (2) high pressure increases the concentration of the reacting gas. This allows lower test temperatures or provides significantly shorter test times at equivalent temperatures [3,10].

When phenolic resins are cured, they liberate water and/or ammonia simultaneously with exothermic curing reaction. Since vaporization of water and ammonia is endothermic and cure is exothermic, the two processes will be in thermal competition with one another and will result in a scan as that seen in the top curve of Fig. 1.10. There are no easily determinable transitions apparent in this scan. However, since vaporization is pressure sensitive, elevation of pressure suppresses vaporization of water and ammonia and shifts these transitions to higher temperatures, thus permitting the cure to be seen on the lower trace. With the exotherm now apparent, it is seen that cure is occurring as a two-stage process.

The measurement of oxidative stability is a very common technique that has been applied to polymeric materials, lubricating oils, greases, and vegetable oil. The oxidative stability is the measurement of time until the onset of oxidative degradation (indicated by an exothermic reaction). For particularly stable materials, the use of PDSC can shorten the analysis time since the elevation of the partial pressure of the reactive gas (e.g., oxygen) will accelerate the reaction. An example of this is the oxidative stability of vegetable oil (Fig. 1.11).

6 Low-temperature DSC

This unit permits continuous heating and cooling measurement ranging from the minus temperature region to the plus temperature region (e.g., from −175°C to +200°C). A principle diagram is shown in Fig. 1.12A and the glass

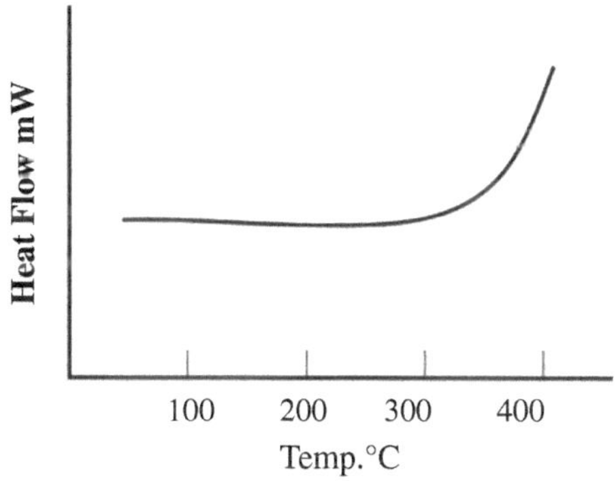

FIG. 1.11 DSC of vegetable oil in O_2 atmosphere.

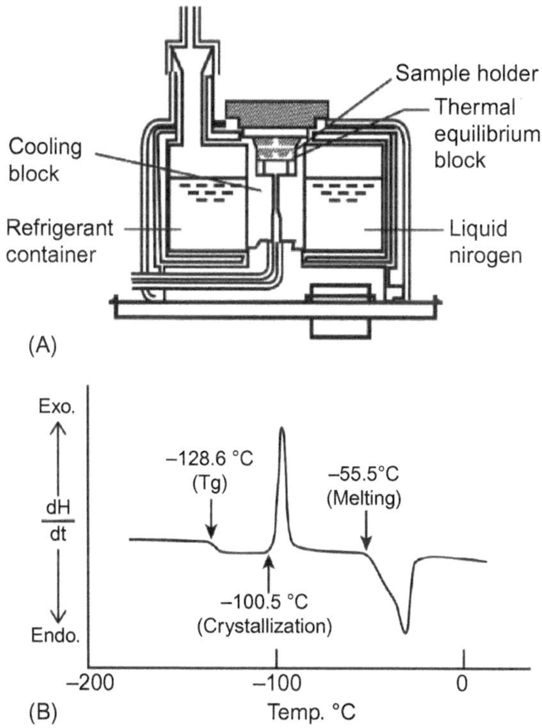

FIG. 1.12 (A) A principle diagram of a low-temperature DSC and (B) DSC calorigram showing the glass transition, crystallization, and fusion of silicon rubber.

transition, crystallization, and fusion of silicone rubber are shown in DSC calorigram (Fig. 1.12B).

7 General thermodynamic relationships

Thermal analysis is usually run under conditions of constant pressure [11,12]. The underlying thermodynamic Eq. (1.4) is the Gibbs-Helmholtz expression:

$$G = H - TS \tag{1.4}$$

where G is the Gibbs free energy (that is, the energy available to do work), T is the absolute temperature, H is the heat content (e.g., total income), S is the entropy, and the term TS represents the molecular "consumption" of energy.

Regarding chemical processes, the important quantity is not the absolute magnitude of the Gibbs free energy (G), but the change in the Gibbs free energy (ΔG) during a reaction and is given by the following equations:

$$\Delta G = G_{\text{products}} - G_{\text{reactants}} \tag{1.5}$$

$$\Delta G = \Delta H - T\Delta S \tag{1.6}$$

We have the following situations: if $\Delta G = -$ ve, the reaction proceeds spontaneously, if $\Delta G = 0$, the reaction is at equilibrium, and if $\Delta G = +$ve, the reaction will not proceed. Differentiation of equation ($\Delta G = \Delta H - T\Delta S$) with temperature is given by the following equation:

$$\frac{\partial \Delta G}{\partial T} = -\Delta S \tag{1.7}$$

In cases where ΔS is positive, $\frac{\partial \Delta G}{\partial T} = -ve$ and the increase in temperature might lead to a spontaneous reaction. In cases where ΔS is negative, then

$$\frac{\partial \Delta G}{\partial T} = +ve \tag{1.8}$$

ΔG will have a negative sign at negative temperature coefficients (i.e., a spontaneous reaction might mean that a chemical change can be forced to occur by either increasing or decreasing the temperature which is the basis of TA).

For both DTA and DSC, the change in the quantity of heat depends on the number of moles of product formed as described by the following equation:

$$\frac{dQ}{dt} = (-\Delta H)\frac{dn_p}{dt} \tag{1.9}$$

The area A under a DSC or DTA peak is proportional to the number of moles (n) changed and the enthalpy of reaction (ΔH) (1.10):

$$A\,\alpha(-n\,\Delta H), A = -c\,n\,\Delta H \tag{1.10}$$

where c is a constant depending on the instrumental conditions.

The heat capacity relation to enthalpy changes is as follows:

$$H = Q + PV \text{ or } Q = H - PV \tag{1.11}$$

Differentiating with respect to T at constant P,

$$(\partial Q/\partial T)p = (\partial H/\partial T)p - (\partial (PV)/\partial T)p \tag{1.12}$$

$$\text{and by definition}: Cp = (\partial H/\partial T)p \tag{1.13}$$

$$\text{Then}, (\partial Q/\partial T)p = Cp - (\partial(PV)/\partial T)p \tag{1.14}$$

8 Thermogravimetry (TG) or thermogravimetric analysis (TGA)

TGA measures the mass as a function of increasing temperature or time. It is suitable only for samples undergoing mass loss (Solid → Product + Gas ↑) or mass gain (Solid + Gas → Product). Plots of mass loss or gain versus temperature are called thermogravimetric curves or TG curves. The use of names is discouraged. A plot of mass versus time while the sample is being kept at a certain temperature is called "isothermal TGA curve" and a plot of the differential weight loss (that is, the weight loss per second; dW/dt) is called "differential TG" or DTG curve. Fig. 1.13 shows TGA curves accompanying the dehydration of $CuSO_4.5H_2O$ and Fig. 1.14 shows a schematic diagram of TG analyzer [13].

The components of a TGA module are (i) the balance (sensitive), (ii) the heater, and (iii) the recorder. The heater comprises three main parts namely a heating device, a temperature control, and an ambient-atmosphere control.

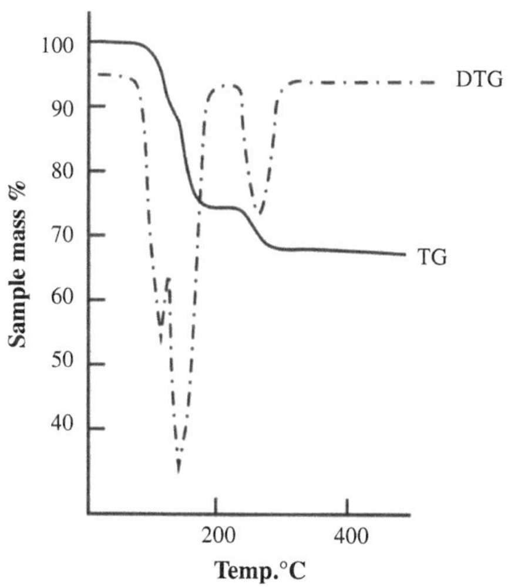

FIG. 1.13 TGA curves of $CuSO_4{\cdot}5H_2O$ in argon atmosphere with a heating rate 10°C min^{-1}. Four water molecules are lost at temperatures below 200°C and one H_2O molecule above 200°C.

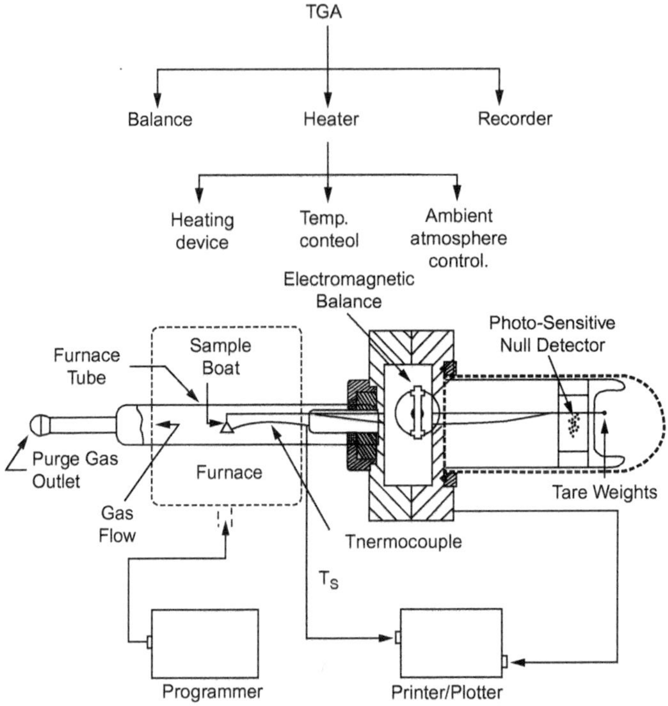

FIG. 1.14 Schematic representation of thermogravimetric analyzer.

Some experimental variables are critically important with respect to TGA instrumentation as follows.

8.1 Heating rate

Slow heating rates ensure a better resolution and enough time for the sample to attain furnace temperatures especially when the mass changes are accompanied by exothermic or endothermic processes. The slower the heating rate, the better the TG curve.

8.2 Furnace atmosphere

This is a very important factor in determining the nature of solid-state reaction in question. For instance, the presence of oxygen in the furnace atmosphere might lead to oxidation whereas pyrolysis will occur if oxygen is excluded.

8.3 Nature of the sample container

The material of the sample container may catalyze an unexpected reaction or trap some of the gases generated. The former may be evaluated by analyzing the reaction products. In the latter case, irreproducible data are obtained. Metallic sample holders reduce the chance of adsorption. Any gases generated by catalytic effect are preferably removed by flowing an inert gas past the sample during analysis. For example, an aluminum pan is used at lower temperatures (i.e., $<700°C$) while a platinum pan is used at higher temperature (i.e., until 1200°C).

8.4 Physical characteristics of the sample

This depends on the amount, particle size, and packing of the sample. Changes in these factors lead to irreproducibility of data.

8.5 TGA with heated quartz crystal microbalance (QCM)

In 1880, Jacques and Pierre Curie discovered that when a mechanical stress is applied to certain materials such as quartz, a voltage proportional to the stress is produced [14]. In 1959, Sauerbrey demonstrated the dependence of quartz oscillation frequency on the change in surface mass [15]. He coined the term quartz crystal microbalance (QCM) in late 1950s, and it was his work that led to the use of quartz plate resonators as sensitive microbalances for thin films. When voltage is applied to a quartz crystal causing it to oscillate at a specific frequency, the change in mass on the quartz surface is directly related to the change in frequency of the oscillating crystal, as shown by the Sauerbrey Eq. (1.15).

$$\Delta m = -C \cdot \Delta f \tag{1.15}$$

The change in QCM frequency (Δf) determines the mass of analyte adsorbed in ng/cm^2 [16,17].

Johnson and Mansfield demonstrated a TGA system with a silicon carbide (SC)-cut quartz crystal microbalance (QCM) at temperatures in the range from 20°C to 450°C [18]. A measurement system was built around a crystal sensor head that was mounted in a small tube furnace. Changes in third-overtone C-mode frequencies of the crystal were measured during thermally activated decomposition and combustion of a poly(methyl methacrylate) (PMMA) film with an initial mass of approximately 6 μg, and corresponding changes in mass Δm were estimated through the use of the Sauerbrey equation.

Noise and drift in frequencies and associated Δm were determined through least-squares fitting and compared with commercially available high-resolution conventional TGA systems. The microbalance-based TGA (μ-TGA) system is found to have one to two orders of magnitude lower noise than high-resolution TGA at temperatures below 200°C and to have at least an order of magnitude lower drift over the entire measured temperatures. However, increasing temperature dependence of crystal frequencies at elevated temperatures and noise in temperature measurements lead to noise in the determination of temperature-dependent mass above 400°C that is comparable in magnitude to that of high-resolution TGA. Enhancements in the performance of μ-TGA depend primarily on the implementation of reliable piezoelectric resonators with low-temperature dependence over the entire measured range. The SC-cut crystal and the crystal-sensor head were originally designed for monitoring the deposition of thin films at elevated temperatures. The performance of this system is explored through demonstration measurements on poly(methyl methacrylate) (PMMA) [18].

9 Thermomechanical analysis (TMA)

TMA is a technique incorporating a wide range of sample configuration options that is used to characterize dimension changes of materials as a function of temperature. The principle of operation is described by the schematic diagram shown in Fig. 1.15 [19].

The heart of the instrument is a moveable-core linear variable differential transformer (LVDT) whose output is proportional to the linear displacement of its core. A thermocouple in direct contact with or near the sample measures sample temperature. The sample and the probe are surrounded by a temperature-controlled cylindrical heater and Dewar assembly. The final output is a plot of probe displacement or its derivative versus either sample temperature or time.

Various probe configurations allow the TMA to be operated in differing modes such as expansion, compression, penetration, tension, dilatometric, stress relation, and fiber tension [19].

The types of application to which TMA is ideally suited include expansion coefficient, viscosity/flow, stress relaxation/viscoelasticity, composite delamination,

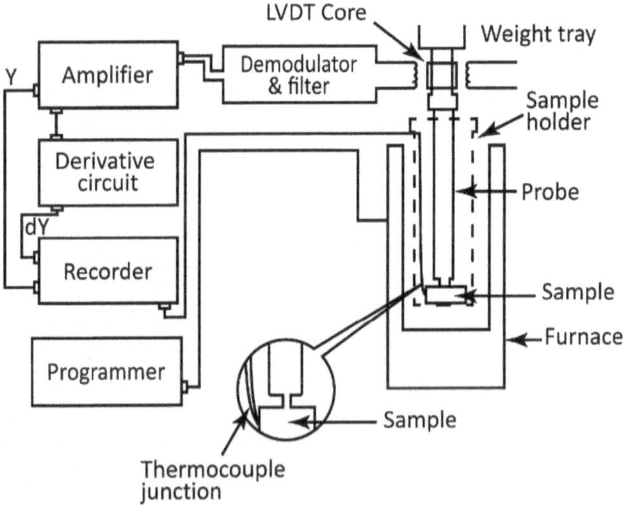

FIG. 1.15 Schematic representation of thermomechanical analyzer.

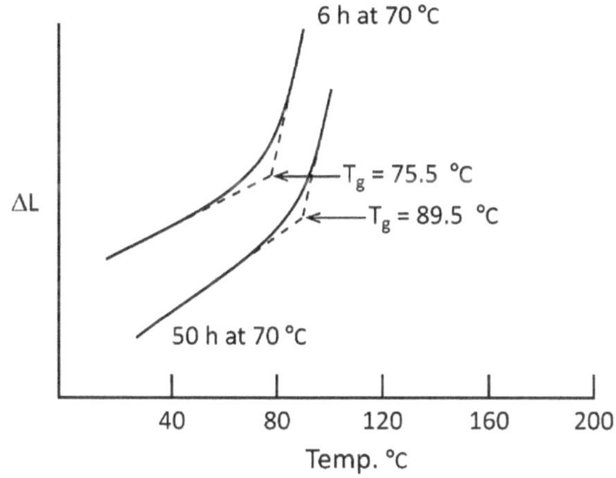

FIG. 1.16 TMA trace for epoxy samples curves at 70°C for various times. The glass transition temperature (T_g) increases as the cure time increases.

glass transition temperature (T_g), polymer creep, fiber or film shrinkage, polymer softening temperatures, tensile modulus, compressive modulus shrink temperature, and glass transition [20]. It is widely used in dental materials research since a selected filler must have thermal expansion coefficients like that of human tooth.

Thermomechanical analysis is widely used to investigate dental materials, encapsulating materials for electronic chips, etc. Fig. 1.16 shows a TMA trace for the expansion of epoxy that is used as an encapsulating material in electronic industry. The expansion coefficient (α) increases in passing through the glass transition (T_g). It is this change in (α) at T_g which permits its detection by TMA. Various probe configurations allow the TMA to be operated in differing modes.

Note: the glass transition temperature (T_g) is the temperature at which the polymeric material changes from a relatively hard, brittle to a softer more plastic material. T_g is a second-order transition representing a change from a more ordered to a less ordered arrangement of the molecules in a noncrystalline state. The higher the glass transition temperature, the more cross-linked the polymer.

9.1 Methods of TMA

9.1.1 *Compressed loading method*

This method is designed to measure the thermal expansion of solid materials. It can monitor expansion coefficients in the order of 10^{-7} m. Data by the compressed loading method for a glass sample is shown in Fig. 1.17.

9.1.2 *Tensile loading method*

This method is suitable for fibers and film samples. Both ends of the sample are caught by metallic chucks and measurement is conducted under the condition

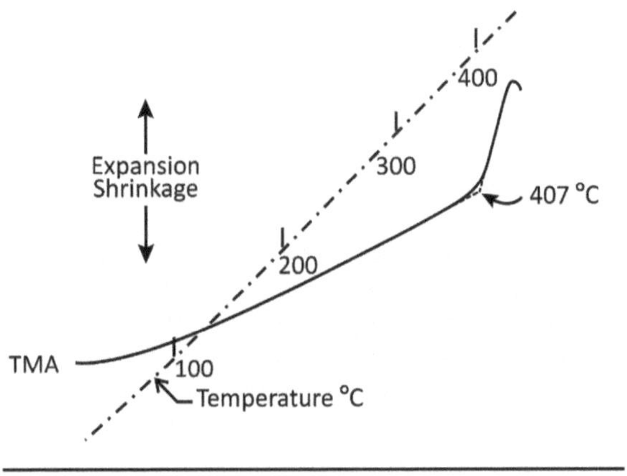

FIG. 1.17 Compressed loading thermomechanical analysis curve for a glass sample.

that a tensile load is applied. Measurement of high accuracy can be made to prevent the expansion of the chucks from being added to the expansion and shrinkage of the sample.

Loading to the sample can be set by the addition or removal to and from a balance. A complete set of jigs are usually delivered to facilitate fitting of the sample to the chucks. Accurate setting of the sample length can be made also at the time. Fig. 1.18 shows data by the tensile loading method for an amorphous metal sample (heating speed 20°C min^{-1}).

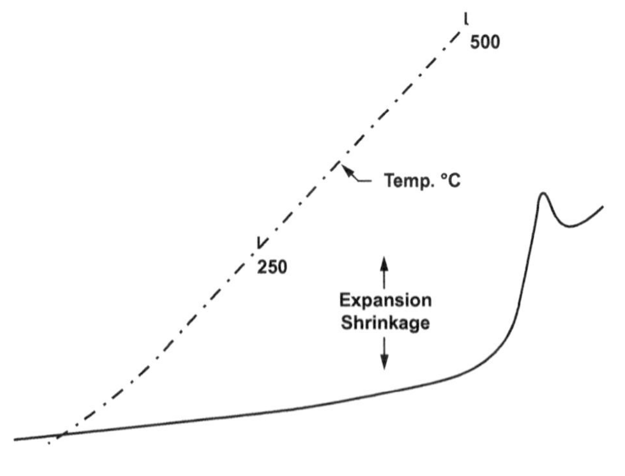

FIG. 1.18 Tensile loading thermomechanical analysis curve for an amorphous metal sample.

9.1.3 Penetration method

In this method, a pin is applied to the sample and heating is carried out under the condition that a constant load is applied in this way. The process of penetration of the pin into the sample is measured with respect to temperature. This method is classified further into two methods as described below according to the condition of heating the sample.

Dynamic method

Because of nearly discontinuous changes of the cubic volume and viscosity of material during the process of its thermal change, the speed of penetration of the pin into the sample will change. By learning temperatures at which such changes occur, it is possible to measure softening, glass transition, fusion, crystallization temperature, etc. For example, Fig. 1.19 shows a graphic plotting of the measured result of the viscosity of NBS 710 glass.

Isothermal method

This is a method in which a sample that shows Newton's flow is held at a constant temperature. The viscosity of this sample can be known by measuring the penetration speed under this condition. Calibration of equipment constants are to be made with a reference sample supplied.

Differential penetration method

This high-sensitivity method utilizes differential system. As a result, measurement of glass transition, softening, and the like can be performed of samples of several micron thickness. For example, the glass transition temperature of thermosetting resin film coated on a metal plate is important information with

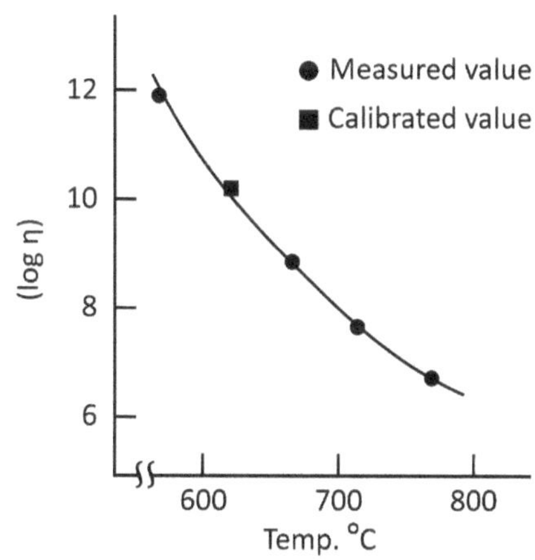

FIG. 1.19 A graphic plotting of the measured result of the viscosity of NBS 710 glass.

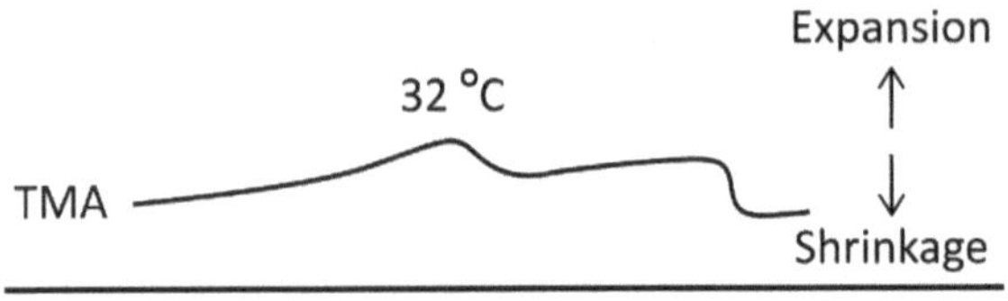

FIG. 1.20 Data by the high-sensitivity differential penetration method for a film coated on a metal plate (heating rate = 5°C min^{-1}).

which to decide the thermosetting condition of the coating film. By use of a metal plate (with no coated film) as the reference sample, the expansion of the metal plate can be canceled. Fig. 1.20 shows data obtained by the high sensitivity differential penetration method for a film coated on a metal plate (heating rate = 5°C min^{-1}).

Compression mode

Four compression probes are available (Fig. 1.21): flat and hemispherical tips for linear expansion studies, an extended tip for penetration studies, and macro-expansion tip. These probes, along with the high sensitivity, calibrated derivative, and linear heating rate, enable direct determination of expansion coefficients and examination of films and coating without removal from the substrate.

Tension mode

Tension probes are for use with films and strips of materials. Sample hooks shown in Fig. 1.22 are for larger samples. Stainless steel holders that accommodate samples of dimensions 1.3 mm × 1.0 mm × 0.1–0.5 mm are used for smaller samples.

The fiber probe is a double-hook/split-ring device designed to grasp a pair aluminum spheres that have been crimped on the ends of the fiber sample. The fiber probe and the tension probe cannot be used with either the fiber tension spectrometer or the stress relaxation accessory.

The fiber/film probe has a double hook arrangement on both movable and fixed portions of the probe designed to grasp a pair of aluminum spheres that have been crimped on the ends of the sample. The fiber/film probe can be used

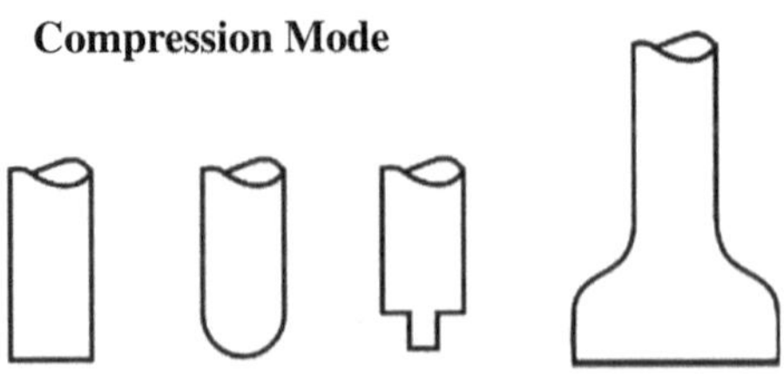

FIG. 1.21 Comparison probes used in thermomechanical analysis.

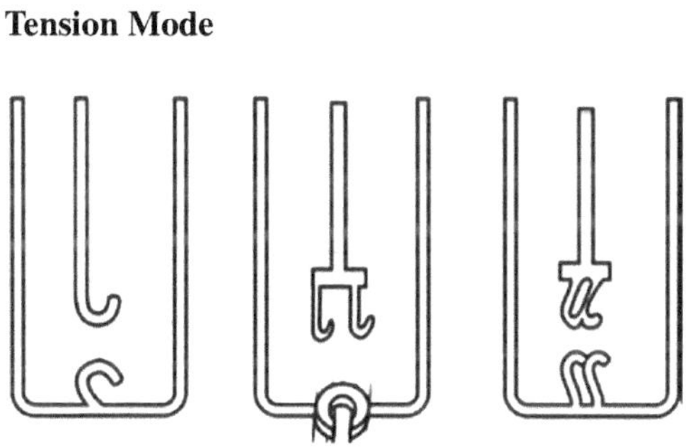

FIG. 1.22 Tension probes used in thermomechanical analysis.

alone and with both the fiber tension spectrometer and stress relaxation accessory.

Dilatometry

A dilatometer probe is used to determine volume coefficient of expansion. A representation of its design and function is shown in Fig. 1.23.

Thermodilatometry (TD)

TD is a technique in which the dimensions are measured as a function of temperature [21]. Dilatometry is known as measurement of the expansion of solids

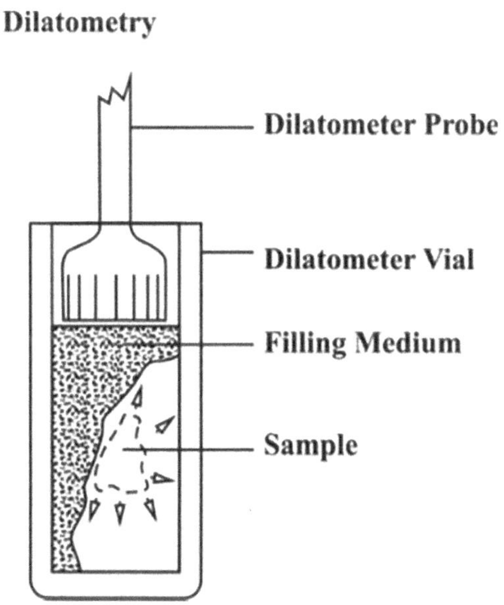

FIG. 1.23 The dilatometer probe used in thermomechanical analysis.

and liquids caused by various imposed conditions, such as absorption or chemical reaction or other time-related processes at a fixed temperature. When these expansions are measured with the change in temperature, it is called TD. Linear thermodilatometry and volume thermodilatometry are distinguished based on the dimensions measured [22].

Roose and Bergen have applied TD as a tool to evaluate the thermal behavior of laminated glass assemblies where two glass layers are bonded together by a polymer interlayer (Fig. 1.24A and B) [23]. They also investigated the effect of heating rate to estimate the onset temperature at interfacial failure from the

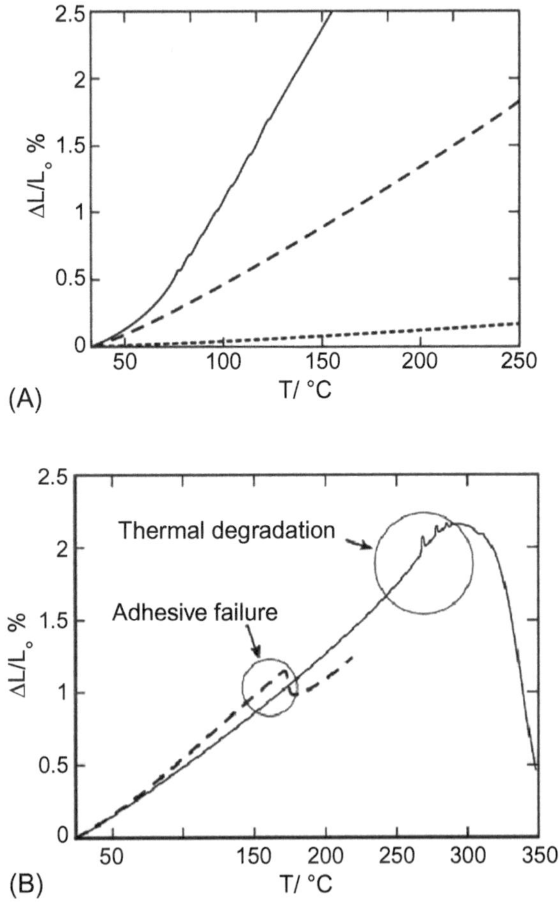

FIG. 1.24 (A) Comparison of the thermal expansion curves of a 4/1/4 glass laminate (———), a free-standing interlayer (—–), and a stacked glass sample of 2 × 4 mm (.......). The curves were obtained at a heating rate of 10°C min^{-1}. (B) Thermal expansion curves of a 4/1/4 glass laminate. The first and second heating runs are represented by the *dashed* and *solid lines*, respectively. The curves were obtained at a heating rate of 10°C min^{-1}.

TD data. They found that TD can be used successfully to evidence interfacial failure in layered systems under thermal stress. Moreover, they found that the dimensional variations of the polymer interlayer should also lie within the measurement range of the instrument otherwise delamination will hardly be detected (such as for thin interlayers). During heating, compression stress builds up due to the mismatch in thermal expansivity of the outer glass and the inner polymer layer, which enabled them to study the stress-strain history of the interlayer up to the debonding temperature.

9.2 TMA accessories

A wide selection of optional accessories broadens the capabilities of TMA including the following:

- Fiber tension spectrometer: This is used to measure shrinkage tension of fibers (held under constant elongation) as a function of temperature. In this case, the primary transducer is a load cell of low spring constant in series with the linear variable differential transformer (LVDT) sensor of the TMA. This is a very useful accessory for investigating thermal history of fibers (heat set, texturing, and draw ratio effects).
- Stress relaxation spectrometry: This is used for the measurement of relaxation modulus of viscoelastic materials versus time. It uses a transducer like that of the fiber tension spectrometer, but with a high-force load cell and invar steel (that is, nickel-iron alloy) probes to withstand up to 10N (1kg) load.
- Parallel plate rheometry: This is used for the analysis of low shear viscosity of viscous materials (10–10^7Pa). This is particularly useful in following cure advancement of thermosets to optimize processing times and temperatures.

10 Dynamic mechanical analysis (DMA)

DMA is the measurement of the mechanical properties of materials as they are deformed under periodic stress. Deformation frequencies can be projected far beyond the range possible with any practical testing technique (for example, a rocket engine mount, which can be subjected to vibration as high as 10,000Hz during maximum acceleration) [24].

DMA appeared to the practical use in the late 1970s by DuPont. It measures modulus (see notes) and damping (energy dissipation) directly, thus provides a rapid technique for evaluating reenforcement efficiency and determines transition temperatures [25].

DMA is also valuable in locating transition temperatures. Secondary transitions, such as the glass transition, are accompanied by changes in physical properties such as heat capacity, expansion coefficient, modulus, and damping. In composites, an added reinforcing agent often dilutes the magnitude of these changes, reducing the possibility their measurement by classical techniques.

Because modulus and damping properties measured by DMA change more dramatically than heat capacity or thermal expansion, DMA is more sensitive to secondary transitions.

The major damping peak in each DMA trace represents the glass transition for the material and, as expected, is easily detected. Very often, DMA observes secondary transitions too subtle for often thermal analytical techniques. Fig. 1.25 shows a schematic diagram of a dynamic mechanical analyzer and Fig. 1.26 shows a typical DMA thermogram of graphite/epoxy composite. The resonant frequency, represented by the solid line, displays relative modulus. It is converted to quantitative Young's modulus using the following equation:

$$E = \frac{4\pi^2 f_o^2 J_o - K}{2W\left[\frac{L}{2}+D\right]^2}\left[\frac{L}{T}\right]^3 \tag{1.16}$$

where E is Young's modulus (Pa), f_o is the DMA frequency (Hz), J_o is the moment of inertia of arm (kg m^2), K is the spring constant of pivot (Nm/rad), D is the clamping distance (m), W is the sample width (m), T is the sample thickness (m), and L is the sample length (m).

DMA is particularly advantageous in the detection, measurement, and interpretation of polymer transitions. As a rule of thumb, DMA offers approximately one to two orders of magnitude more sensitivity than DSC and TMA do for suitable transitions.

Fig. 1.27 compares the loss moduli (damping) of linear (high-density) and branched (low-density) polyethylene. Three transitions are seen: α at 80°C, β at −5°C, and γ at −110°C. Each of these transitions corresponds to specific

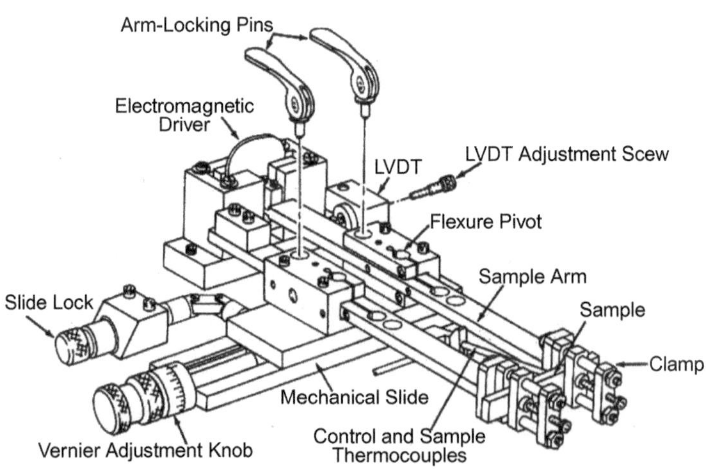

FIG. 1.25 Schematic representation of DMA electromechanical system.

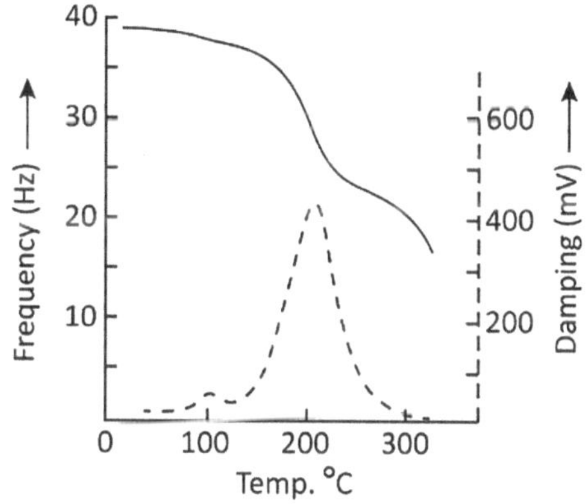

FIG. 1.26 DMA thermogram of graphite/epoxy composite.

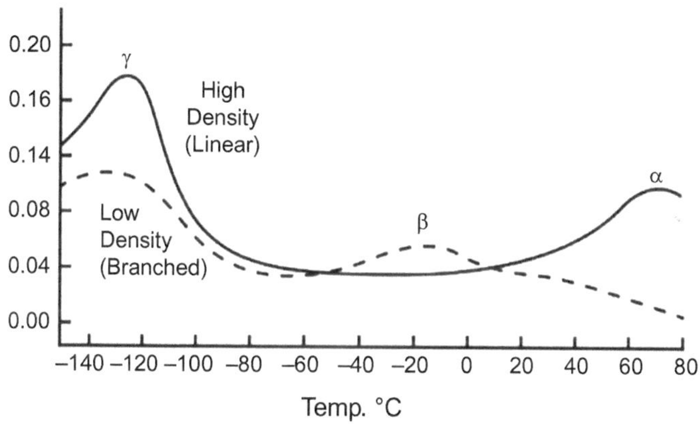

FIG. 1.27 DMA comparison of low- and high-density polyethylene.

molecular motions, which have significance in terms of structure-property relationships:

- Alpha transition (α) is associated with crystalline relaxations occurring below the melting point of polyethylene.
- Beta transition (β) is due to the motion of the amorphous region side chains or branches from the main polymer backbone. The intensity of the beta-transition varies with the degree of branching.
- Gamma transition (γ) is a result of crankshaft rotation of short methylene main chain segments and can influence the low-temperature impact stability of polyethylene.

N.B.

Stress

A force per unit area. When a stress is applied to a body (within its elastic limit), a corresponding strain is produced, and the ratio of stress to strain is a characteristic constant of the body (see elastic modulus) [26].

Strain

When a body is deformed by an applied stress, the strain is the ratio of the dimensional change to the original or unstrained dimension. The strain may be the ratios of lengths, areas, or volumes [27].

Elastic modulus

Modulus of elasticity is the ratio of stress to strain in a given material. The strain may be a change in length (see Young's modulus), a twist or shear (see rigidity modulus) or a change in volume (see bulk modulus); the stress required to produce unit strain is in each case expressed in Newtons per square meter (m^2) or dynes per square centimeter (cm^2).

Young's modulus

It is the elastic modulus applied to a stretched wire or to a rod under tension or compression; the ratio of the stress on a cross section of the wire or rod to the longitudinal strain (named after Thomas Young 1773–1829) [28].

Tensile strength (tenacity)

It is the tensile (pulling) stress that must be applied to a material to break it. It is measured as a force per unit area (e.g., Newtons per m^2 or dynes per cm^2) or pounds or tons per square inch.

In DMA, the sample is rigidly clamped between two arms, making the arms and sample part of a combined resonant system. The position of one arm and pivot is fixed, and the other arm and pivot are moveable by means of a precision mechanical slide to accommodate a wide variety of sample lengths. A pair of arm locking pins is provided to immobilize the arms and maintain them in parallel during sample mounting. After the pins are removed, the system is displaced and set into oscillation by a driver at an amplitude selected by the operator. As the system oscillates, the sample is subjected to flexural deformation. The amplitude is measured by a linear variable differential transformer (LVDT), the core of which is attached to the fixed-pivot arm. The output of the LVDT contains frequency and amplitude information needed to calculate the viscoelastic properties measured by the analyzer.

Normally, a system so displaced would oscillate at the system's resonant frequency, with constantly decreasing amplitude due to the loss of energy (damping) within the sample. The electronics of the analyzer are designed to compensate for this loss of energy in the sample. The amplitude signal from the LVDT is fed into a circuit, which in turn provides an output signal to the electromechanical driver. This supplies additional energy to the driven arm, forcing the system to oscillate at constant amplitude.

The frequency of oscillation is directly related to the modulus of the sample being investigated, while the energy needed to maintain constant amplitude oscillation is a measure of damping within the sample (like walking on sand, you get tired). Microprocessor data-reduction techniques provide graphical and tabular outputs of these properties as function of time or temperature.

The applications for DMA are many and varied including curing of thermosets, polymer blend compatibility, correlation of impact stability with damping, observation of plasticizer effects, measurement of subtle transitions, sound and vibration dissipation correlation with damping, engineering design-mechanical data, characterization of stiff composites and metals, analysis of soft elastomers, and characterization of supported systems (coating and adhesive).

11 Dielectric thermal analysis (DEA)

DEA is a technique in which the dielectric constant and dielectric loss are measured. DEA is mostly suitable for thin film, semisolid, or liquid (a cup electrode is provided as standard) materials and is also useful for powders.

A sample is placed between two parallel plate electrodes and an alternating current is passed between the two electrodes (Fig. 1.28). Materials possess a dielectric constant (ε') characterizing the extent of electrical polarization, which can be induced in the material by an electrical field. If an alternating electrical field is applied, the polarization lags the field by a phase angle δ. This results in partial dissipation of the stored energy. The dissipated energy is proportional to the dielectric loss (ε'') and the stored energy to the dielectric constant (ε'). ε'' and ε' are measured by DEA and are comparable to loss and storage modulus measured mechanically by dynamic mechanical analysis (DMA). DMA is like DEA, in which a mechanical force is used instead of an electrical field. The ratio of $\varepsilon''/\varepsilon' = \tan\delta$ in the same manner as DMA in which the elastic modulus is known as the ratio of the stress to the strain amplitudes and represents the stiffness of the material (please see Section 10).

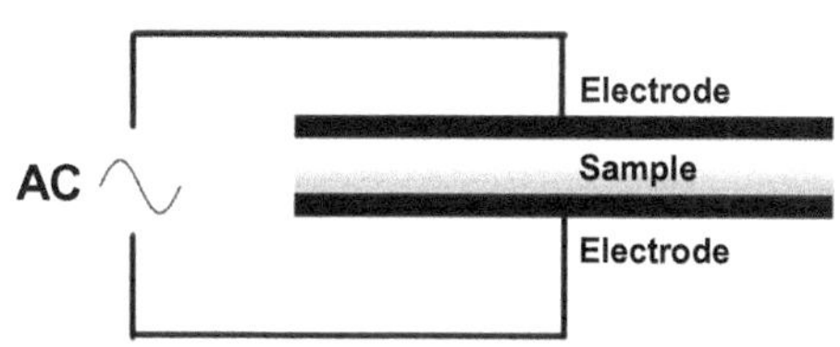

FIG. 1.28 Schematic illustration of dielectric thermal analyzer.

12 Thermally stimulated current (TSC)

TSC is a technique in which the current is measured. In the TSC experiment, the external electric field is applied to the sample during the heating and the recorded current maxima are due to charge carriers released from traps that avoid geminate recombination [29].

A different model, Holmarc's TSC spectrometer is used to study energy levels in semiconductors or insulators, especially solar cells. The sample energy levels are first filled by exposing it to a halogen lamp light source for a user specified time. The temperature of the sample at this time can be set as low as 70°C. The sample can then be heated gradually up to 150°C at a specified ramping rate, which causes emission of electrons and holes in the sample.

TSC, also known as thermally stimulated discharge currents (TSDC) or thermally stimulated dielectric depolarization (TSDD), is a less conventional relaxation technique with sensitivity comparable to DMA and a.c. dielectric. TSC is applicable to study various materials, such as semiconductors, semimetals, polymers, plastics, biaxial films, polymer-clay composites, and amorphous materials. It is also applied for conductivity, dielectric relaxation, defect states, and electrical behavior of polymers measurements [30].

TSC is a new tool that also can be used to analyze pharmaceutically important molecules. TSC studies are usually conducted to provide additional information about molecular mobility in the solid state, and as a result characterize phase transitions that are related to thermal transitions in the crystalline (polymorphic) and amorphous phases. The ability of TSC to probe molecular mobilities, previously undetected in materials, and link them to the stability of different phases has sparked immense scientific interest in this technique.

In recent years, the physical sciences have made great progress toward understanding the properties of pharmaceutically important amorphous and polymorphic materials. The major focus of this work is to utilize the advantages that they may bring to formulated products (e.g., faster solubility of amorphous drugs compared to crystalline counterparts) and at the same time, to overcome stability problems (e.g., tendency to recrystallize on storage) that they may demonstrate.

Although the main use of TSC is to study relaxation processes such glass transition, it can also be used to study first-order solid-solid transitions (polymorphic transitions). TSC has been used to distinguish between two different polymorphic forms and an amorphous phase. In addition, researchers have used TSC to analyze new compounds and have been able to assign one of the relaxation processes to the solid-solid transformation between polymorphs.

One comprehensive TSC investigation has been undertaken on polymorphs of caffeine [31]. The TSC study revealed new, previously unknown relaxation processes that are responsible for the rearrangements of the molecules in the crystalline structure prior to the main relaxation process. Caffeine exists in two enantiotropic polymorphic forms (form I and form II) implying that there is a clear transition point (temperature) at which one polymorph converts into the other. The TSC data indicate clear differences between the two forms, as shown in Fig. 1.29A and B.

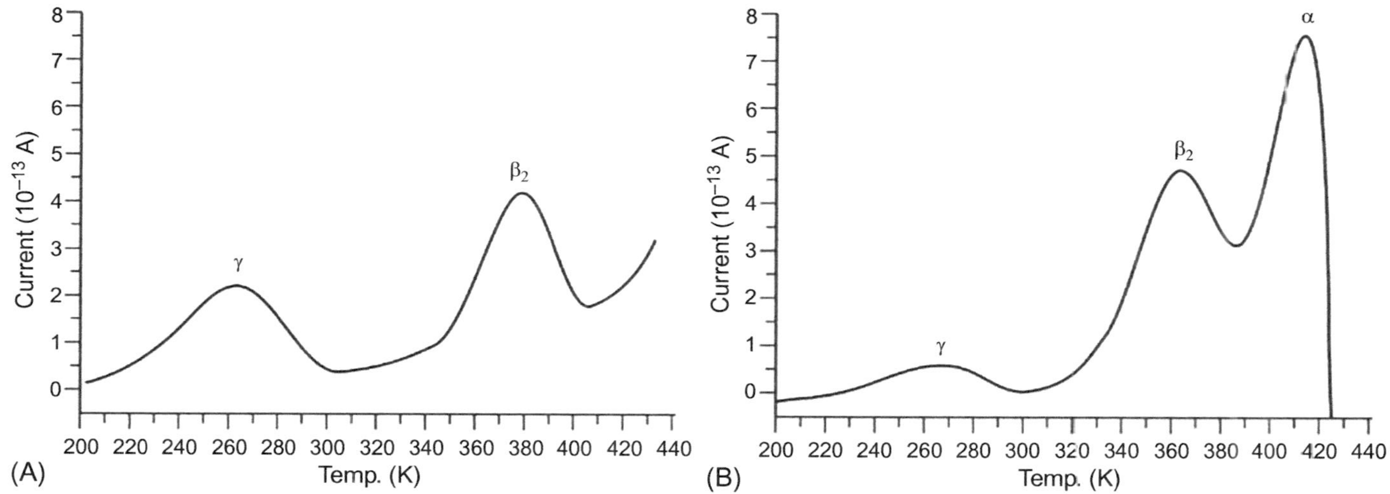

FIG. 1.29 (A) The global TSC spectrum of caffeine form I. (B) The global TSC spectrum of caffeine form II.

13 Emanation thermal analysis (ETA)

ETA is a technique in which the trapped gas within the sample is released and measured. The principle of ETA is based on the measurement of radioactive inert gas released from solids previously labeled with trace concentrations of an inert radioactive nuclide. The high detection sensitivity of radioactive nuclides used for the measurement makes it possible to use very low concentrations of the inert gases (10^{-14} atom %) so that no influence of the inert gases on the properties of the solids can be supposed. The inert gases may be incorporated into the solid samples by several techniques, such as by the diffusion of the inert gas into the solid at higher pressure and temperature, by ion implantation, or by means of nuclear reaction.

Fig. 1.30 shows the ETA curve of ultrafine SiC powder (size 20 nm), which was measured during heating in argon [32]. The SiC sample was prepared by laser-assisted synthesis from silane hydrocarbon mixture. It was found that the surface of SiC nanosized powders was covered by a layer of SiO_2 thin layer (thickness approximately 4 nm). From the ETA curve, it is supposed that thin layer of SiO_2 can melt at about 1180°C, therefore the break observed on the ETA curve in the temperature range 1180–1350°C can be ascribed to the densification of the nanosized SiC powder due to the interaction of the melted silica surface layer.

As it follows from the ETA curve measured during subsequent cooling, the sample was perfectly densified. Character of the cooling curve corresponds to very low diffusion of radon, which is characteristic for the sintered solid covered with the fused silica layer. From the ETA curve, it follows that in the temperature range 100–800°C, the diffusion of radon in the intergranular space of the sample is the controlling mechanism of the radon release from the sample.

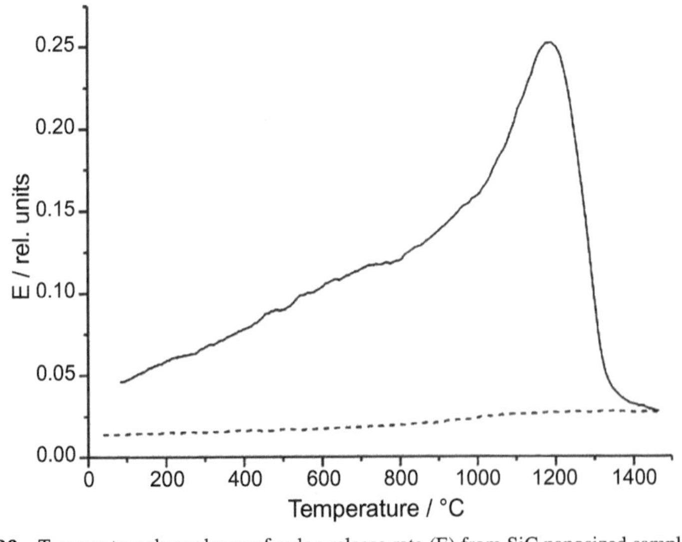

FIG. 1.30 Temperature dependence of radon release rate (E) from SiC nanosized sample during heating *(full line)* and cooling *(dashed line)* in argon at the rate 6°C min^{-1}.

14 Thermoluminescence (TL)

TL is a technique in which the emitted light is measured. TL refers to the emission of light during heating of a material previously excited by ionizing radiation. The light energy released is derived from electron displacements within the crystal lattice of such a substance caused by previous exposure to high-energy radiation. Heating the substance at temperatures about 450°C and higher enables the trapped electrons to return to their normal positions, resulting in the release of energy. The intensity of the emission can be correlated to the length of time that a given substance was exposed to radiation; the longer the time allowed for the radiation to build up an inventory of trapped electrons, the greater the energy released. Because of this feature, thermoluminescence has been exploited for dating various minerals and archeological artifacts.

TL means not temperature radiation but enhancement of the light emission of materials already excited electronically by the application of heat. The phenomenon is observed with some minerals and, above all, with crystal phosphors after they have been excited by light.

For example, Fig. 1.31 shows a glow curve of the annealed quartz after irradiation to 50 Gy. The inset (i) shows the same glow curve on a semilogarithmic scale to better show presence of other peaks. For comparison, the glow curve for the unannealed sample is shown in inset (ii). The glow curve of the annealed and

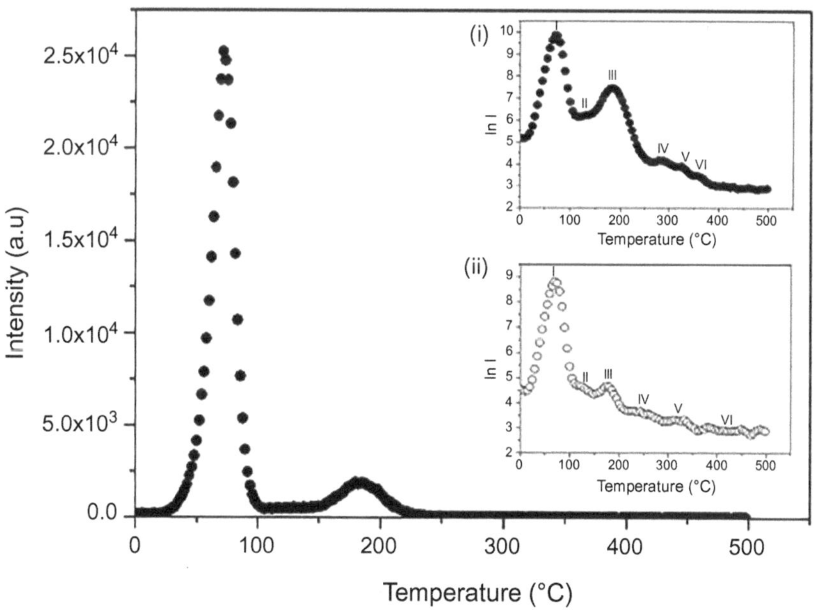

FIG. 1.31 A glow curve of the annealed natural quartz measured at 1°C s^{-1} following a dose of 50 Gy. The insets show the glow curves of the (i) annealed and (ii) unannealed samples in a semilogarithmic scale.

unannealed samples each consists of six peaks. The main peak (labeled I) in both cases is at 72°C. The secondary peaks of the annealed quartz are at 120°C, 182°C, 282°C, 334°C, and 364°C whereas in the unannealed sample, these are at 126°C, 174°C, 254°C, 330C, and 418°C. The peaks are labeled II, III, IV, V, and VI, respectively, as shown (Fig. 1.31) [33].

Using the $T_m - T_{stop}$ method (where T_m is the position of the peak temperature when an irradiated sample is linearly preheated up to a maximum temperature (T_{stop}) in the range 200–300°C), the position of peak I was determined to be independent of T_{stop} at 72.0 ± 0.1°C for the annealed sample and at 72.5 ± 0.4°C for the unannealed one. Measurements were made for samples irradiated to 50 Gy. The results show that peak I is single and follows first-order kinetics. Moreover, the order of kinetics of peak I was further assessed using the dependence of peak position T_m on dose. It was found that T_m is independent of dose in both cases confirming that peak I is of first-order kinetics. The intensity and measurement temperature are related by the equation $I(T) = C' \exp(-E/kT)$, where E is the activation energy, k is Boltzmann's constant, and C' is a constant. The method was applied on peak I giving $E = 0.91 \pm 0.01$ eV and $E = 0.93 \pm 0.03$ eV for the annealed and unannealed samples, respectively. The values are consistent suggesting that annealing has little effect on the activation energy.

15 Fast scanning calorimetry (FSC)

FSC is a technique used for the measurement of phase transformations over a wide scanning rate range. It allows for heating and cooling at rates of up to $MK\,s^{-1}$. The required sample size for FSC is very small compared to DSC, around 5–100 ng, which is a great advantage of FSC, especially in the field of protein research where samples are often only available in minute amounts [34].

The absolute heat capacity of the sample $C_p(T)$ is calculated from the difference between the heat flux on heating and cooling scans in a temperature region outside any transition. The absolute heat capacity of the sample is measured on a FSC chip sensor and specific heat capacity of the sample is measured using a conventional DSC. Then the mass is calculated from the ratio between absolute and specific heat capacities (C_p) using the following equation:

$$m_s = \frac{C_p\left[\frac{J}{K}\right]}{C_p\left[\frac{J}{gK}\right]} \tag{1.17}$$

For example, Mukhametzyanov et al. have measured the endothermic effect of lysozyme unfolding in glycerol solution (ca. 50 mg/mL) using conventional DSC (Fig. 1.32) and FSC (Fig. 1.33) in the range of scanning rate from $0.08\,K\,s^{-1}$ to $8000\,K\,s^{-1}$ [35]. The dependence of the transition temperature

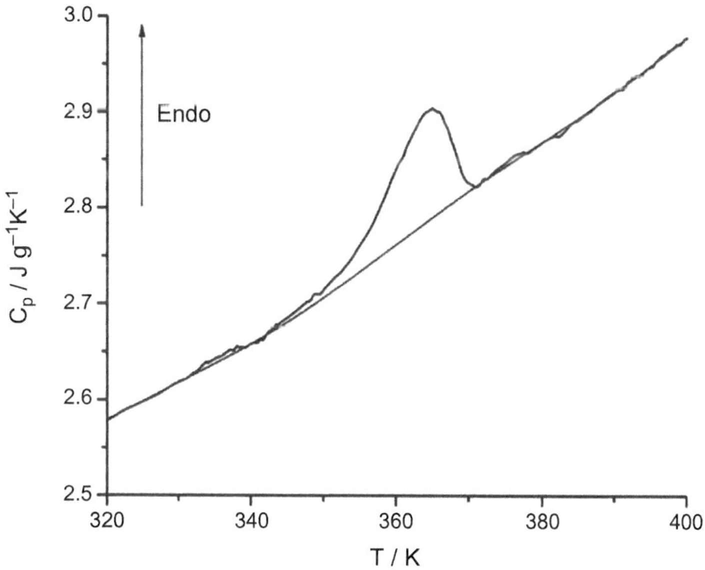

FIG. 1.32 Typical DSC thermogram of lysozyme solution in glycerol (ca. 50 mg/mL) measured by a conventional DSC with scanning rate 10 K/min. The *blue curve* represents the assumed peak baseline. The original heat flow curves were recalculated into specific heat capacity (C_p J g^{-1} K^{-1}).

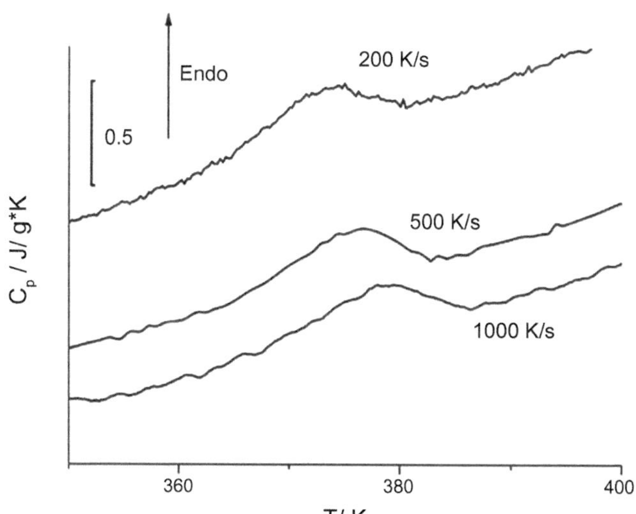

FIG. 1.33 Typical FSC thermograms of lysozyme solution in glycerol (ca. 50 mg/mL) with scanning rates of 200, 500, and 1000 K s^{-1}. The original heat flow curves were recalculated into specific heat capacity (C_p J g^{-1} K^{-1}).

expressed as the extremum of the endothermic peak on the heating rate forms two clusters in a Kissinger plot, both well approximated by nearly parallel straight lines. Apparent activation energy is close in both cases: (261 ± 13) and (251 ± 9) kJ mol^{-1} for FSC and conventional DSC, respectively. The difference between the two clusters demonstrates that unfolding proceeds at a lower temperature on the chip sensor than what could be expected from an extrapolation of the results of conventional DSC toward higher temperatures. The experiments with droplets of different sizes placed on the chip sensor suggest that relative surface area of the sample contributes to this effect. The values of enthalpy of unfolding measured by FSC and conventional DSC are similar and greater than the apparent activation energy. Therefore, the unfolding process appears not to be a single-step two-state process.

16 Microcalorimetry

Microcalorimetry is a technique that uses a suite of techniques to directly measure enthalpy and heat capacity changes that arise when chemical reactions occur. In aqueous solutions, heat flux into or out of the sample almost happens on reaction. These reactions may involve a wide variety of situations, e.g., the interaction of two molecules (such as a cyclodextrin and its guest), changes in the conformation of complex macromolecules (such as proteins or DNA), or even in the structure of very complex multimolecular colloidal drug delivery systems (such as liposomes) [36].

Calorimetry relies on the fact that all chemical reactions involve a change in energy, usually accompanied by heat release (exothermic) or absorption (endothermic). Microcalorimetry is an ultrasensitive development of the technique that measures very small heat changes in small sample volumes, making it suitable for biomaterials.

Applications range from confirming intended binding targets in small molecule drug discovery to the development of stable biotherapeutics. These processes are often studied using two calorimetric techniques: isothermal titration calorimetry (ITC) and differential scanning calorimetry (DSC).

ITC directly measures the heat released or absorbed during a biomolecular binding event. This allows accurate determination of binding constants (K_D), reaction stoichiometry (n), enthalpy change (ΔH), and entropy change (ΔS). DSC allows understanding the stability of proteins and other biomolecules. It has wide application in protein engineering, rational drug design, and biopharmaceutical production, where developing stable proteins is a critical goal. DSC is a key thermal analysis technique that measures heat changes, which occur in the biomolecule during a controlled increase or decrease in temperature, making it possible to study materials in their native state.

Isothermal microcalorimetry as a tool has been applied to predict seed longevity by directly measuring the heat flow produced as seeds age. Differences in heat flow reflected differences in longevity in response to a priming treatment, and because of the storage environment (water content) [36]. Heat flow was recorded in primed and nonprimed (control) seeds of *Ranunculus sceleratus* L., aged in a microcalorimeter at 35°C at three different seed water contents [ca. 0.12, 0.075, and 0.045 g H_2O (g dw)$^{-1}$]. The rate of heat flow and total heat generated (an indicator of extent of reaction) were generally greater in control seeds, which aged at a fast rate than in primed seeds. Total heat generated over a given period also increased with increasing water content. The power-time curves did not indicate first- or second-order rate kinetics, consistent with the probability that seed aging is complex and involves several reactions (Fig. 1.34). Even after the capacity to germinate had ceased, there was a residual power signal.

As a method, isothermal microcalorimetry gave consistent results using independent samples at different times. Therefore, short-term experiments at relatively high-water contents and/or temperatures may have the potential to predict the relative longevity of seed lots, at least within a species.

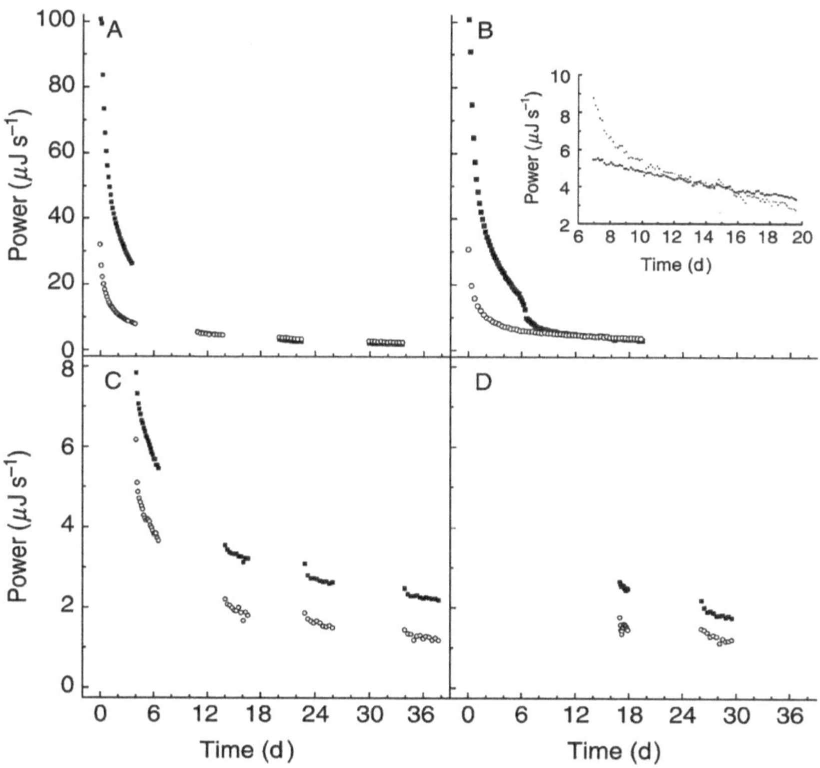

FIG. 1.34 Heat flow curves measured on samples of approximately 4100 primed (○) or control (■) seeds of *R. sceleratus* aged at 35°C. Seeds preequilibrated at 5°C over LiCl: (A and B) 230 g L^{-1}, (C) 410 g L^{-1}, or (D) 580 g L^{-1}. The inset in panel B expands the data between 7 and 20 d. For clarity, only a small proportion of the recorded data is shown [36].

17 Laser flash method. Thermal constant measurement (LF/TCM) system

This system is commercially available as a unit from the Rigaku TAS 100 Thermal Analyzer system.

It is designed to measure thermal constants (thermal diffusivity, specific heat, thermal conductivity) of solids by laser pulse method. The unit can perform high precision measurement quickly over a wide temperature range from ambient to 1500°C. In this method, the sample is subjected to a high intensity short duration radiant energy pulse. The power source can be a laser or a flash lamp. The energy will then be absorbed by the specimen and emitted again on the top of the sample. This radiation results in a rise in temperature on the surface of the sample. This increase in temperature is recorded from an infrared (IR) detector against time.

18 Moisture evolution analyzers (MEAs)

In moisture evolution analysis, the sample is heated to the operator-selected temperature in an oven, to drive off water. Temperature is included on a direct-reading digital pyrometer; moisture from the heated samples is carried by a continuously flowing stream of externally dried nitrogen and passed into the electrolytic cell for measurement. Water content of a sample is indicated accurately on a six-digit light-emitting diode (LED) integrator that gives a direct indication of moisture levels up to 99.999.9 μg in 0.1 μg increments. Integration time is controlled by a timer that can be set for up to 1 h. Most analysis can be conducted in 10–25 min.

Polymer moisture present during processing of polymeric materials affects the quality of the final product. In the field of pharmaceuticals, high moisture levels can affect the stability and appearance of many pharmaceuticals. By recording the output of the electrolytic cell, the instrument can differentiate between adsorbed and crystalline water if temperature difference is enough. Many inorganic chemicals contain adsorbed water and crystalline water, which can affect physical properties. The moisture evolution analyzer can determine moisture content and hydration levels.

19 Thermal analysis data-processing system

This system is designed to be combined with different types of thermal analyzers to automatically perform collection and operation of necessary data and to graphically record the result. It offers a useful tool for manpower saving, prompt handling, and enhancement of the reproducibility and measurement accuracy.

DTA/DSC	- Calculations of peak areas and reaction calories. - Detection and calculations of temperatures at which the reaction starts and ends and the peaks of transitions.
TGA	- Calculations of the amount and weight change rate. - Detection and calculations of temperatures at which the weight change starts and ends
TMA	- Calculations of the mean expansion coefficient in an optional temperature range - Detection and calculations of transition temperatures and softening temperatures

20 Summary

This chapter describes the thermal analysis techniques since the first appearance of DTA by Austen in 1899 using the equipment developed by Kurnakov and Saladin to the most advanced TA techniques to date. It provides introductory to the essential principles, definitions, theories, IUPAC abbreviations, and examples supporting each technique of the 15 mentioned thermal analysis methods. In the meantime, instrumentation, sampling, typical operation, and the main parts of typical TA machines are described. The general thermodynamic relations of parameters related to annealing are derived and reported as well.

References

[1] T. Lever, P. Haines, J. Rouquerol, E.L. Charsley, P.V. Eckeren, D.J. Burlett, ICTAC nomenclature of thermal analysis (IUPAC recommendations 2014), Pure Appl. Chem. 86 (2014) 545–553.

[2] W.C. Roberts-Austen, Metallographist 2 (1899) 186. N.S. Kurnakov, Z. Anorg, Chemie 42 (1904) 184. E. Saladin, Iron and Steel Metallurgy, Metallography 7 (1904) 237.

[3] L.C. Thomas, International Lab, 1987. B. Wunderlich, Thermal Analysis: Encyclopedia of Materials: Science and Technology, 2001, pp. 9134–9141. M. Wagner, Thermal Analysis in Practice, Carl Hanser Verlag, Munich, 2018, pp. 158–161.

[4] D. Law, D. Zhou, in: Y. Qiu, Y. Chen, G. Zhang, L. Yu, V. Rao (Eds.), Solid-State Characterization and Techniques in Developing Solid Oral Dosage Forms: Pharmaceutical Theory and Practice, Elsevier Inc., 2017, pp. 59–84. V.A. Zhorin, M.R. Kiselev, A DSC study of the endothermic process, associated with hydrogen bonding, in polysaccharides after high-pressure plastic deformation, High Energy Chem. 54 (2020) 263–269.

[5] M. Wagner, Thermal Analysis in Practice, Carl Hanser Verlag, Munich, 2018, pp. 66–143. E. Morintale, A. Harabor, C. Constantinescu, P. Rotaru, Use of heat flows from DSC curve for calculation of specific heat of the solid materials, Phys. AUC 23 (2013) 89–94. S. Vyazovkin, N. Koga, C. Schick (Eds.), Handbook of Thermal Analysis and Calorimetry, second Ed., Vol. 6, Elsevier Science, 2018.

[6] T. Ozawa, Temperature control modes in thermal analysis, Pure Appl. Chem. 72 (2000) 2083–2099.

[7] C. Schick, Differential Scanning Calorimeter (DSC) with Temperature Controlled Furnace, PerkinElmer LAS Inc, 2005. U.S. Application No. 11/054,755 filed on February 10. EP20060734839.
[8] M.J.M. Abadie, O.O. Novikova, V.Y. Voytekunas, V.G. Syromyatnikov, A.Y. Kolendo, Differential scanning photocalorimetry studies of 1,6-hexanedioldiacrylate photopolymerization initiated by some organic azides, J. Appl. Polym. Sci. 90 (2003) 1096–1101.
[9] G.R. Tryson, A.R. Shultz, A calorimetric study of acrylate photopolymerization, J. Polym. Sci. Part B: Polym. Phys. 17 (1979) 2059–2075.
[10] S. Zhu, A. Le Bail, H.S. Ramaswamy, High-pressure differential scanning calorimetry: comparison of pressure-dependent phase transition in food materials, J. Food Eng. 75 (2006) 215–222.
[11] D.R. Olander, General Thermodynamics, CRC Press, Talyor & Francis Group, 2007, pp. 1–408.
[12] S.K. Lower, All about enthalpy calorimetry and the first law of thermodynamics: chemical energistics, in: A Chem 1 Reference Text, 2003, pp. 1–32.
[13] M. Wagner, Thermal Analysis in Practice, Carl Hanser Verlag, Munich, 2018, pp. 162–186. S. Vyazovkin, Thermogravimetric Analysis, Characterization of Materials, 2012, pp. 1–12. M. Zhang, F. Zhao, Y. Yang, H. Li, H. Gao, E. Yao, J. Zhang, T. An, Z. Jiang, Synthesis, characterization and catalytic behavior of MFe_2O_4 (M = Ni, Zn and Co) nanoparticles on the thermal decomposition of TKX-50, J. Therm. Anal. Calorim. 141 (2020) 1413–1423.
[14] J. Curie, P. Curie, Development via compression of electric polarization in hemihedral crystals with inclined faces, Bull. Soc. Minerol. France 3 (1880) 90–93.
[15] G.Z. Sauerbrey, Use of quartz vibration for weighing thin films on a microbalance, Phys. J. 155 (1959) 206–212.
[16] S.K. Vashist, P. Vashist, Recent advances in quartz crystal microbalance-based sensors, J. Sens. 2011 (2011) 571405.
[17] E. Mansfield, A. Kar, T.P. Quinn, S.A. Hooker, Quartz crystal microbalances for microscale thermogravimetric analysis, Anal. Chem. 82 (2010) 9977–9982.
[18] W.L. Johnson, E. Mansfield, Thermogravimetric analysis with a heated quartz crystal microbalance, in: 2012 IEEE International Frequency Control Symposium Proceedings, 2012, https://doi.org/10.1109/fcs.2012.6243694. Y. Wang, B. Wu, Y. Zhao, Q. Liu, J. Lei, Study on a novel solid-solid phase change materials: solvent-free preparation, thermal properties and phase separation behaviors, J. Therm. Anal. Calorim. 141 (2020) 1305–1315.
[19] J. James, in: S. Thomas, R. Thomas, A.K. Zachariah, R.K. Mishra (Eds.), Thermomechanical analysis and its applications in thermal and rheological measurement techniques for nanomaterials characterization, Elsevier Inc., 2017, pp. 159–171, https://doi.org/10.1016/B978-0-323-46139-9.00007-4. M. Wagner, Thermal Analysis in Practice, Carl Hanser Verlag, Munich, 2018, pp. 187–208.
[20] M. Reading, P.J. Haines, Thermomechanical, dynamic mechanical and associated methods, in: P.J. Haines (Ed.), Thermal Methods of Analysis, Principles, Applications and Problems, 1995, pp. 123–160.
[21] M.E. Brown, Thermodilatometry (TD): Introduction to Thermal Analysis, Springer, Dordrecht, 1988, pp. 56–62.
[22] IUPAC, Compendium of chemical terminology, in: A.D. McNaught, A. Wilkinson (Eds.), The Gold Book, second ed., Blackwell Scientific Publications, Oxford, 1997. XML on-line corrected version: http://goldbook.iupac.org created by M. Nic, J. Jirat, B. Kosata; updates compiled by A. Jenkins, 2006.
[23] P. Roose, H.V. Bergen, Int. J. Appl. Glass Sci. 2 (2011) 290–300.

[24] K.P. Menard, Dynamic Mechanical Analysis: A Practical Introduction, CRC Press, Boca Raton London New York Washington, DC, 1999.

[25] R.J. Crawford, Plastic Enginerring, third ed., St Edmundsbury Press Ltd, Bury St Edmunds, Suffolk, 1998.

[26] N. Sasaki, S. Odajima, Stress-strain curve and Young's Modulus of a collagen molecule as determined by the X-ray diffraction technique, J. Biomech. 29 (1996) 655–658.

[27] T. Saito, K. Soga, T. Hoson, I. Terashima, The bulk elastic Modulus and the reversible properties of cell walls in developing Quercus leaves, Plant Cell Physiol. 47 (2006) 715–725.

[28] A. Farsi, A.D. Pullen, J.P. Latham, J. Bowen, M. Carlsson, E.H. Stitt, M. Marigo, Full deflection profile calculation and Young's modulus optimisation for engineered high performance materials, Sci. Rep. 7 (2017), 46190.

[29] V.M. Gerrish, Characterization and quantification of detector performance, Semiconduct. Semimet. 43 (1995) 493–530.

[30] I. Glowacki, J. Jung, J. Ulanski, A. Rybak, Conductivity measurements, in: Polymer Science: A Comprehensive Reference, 2, 2012, pp. 847–877.

[31] M.D. Antonijevic, Applications of thermally stimulated current spectroscopy in pharmaceutical research, Eur. Pharm. Rev. 17 (2012). https://www.europeanpharmaceuticalreview.com/article/13693/ap.

[32] V. Balek, V. Zeleòák, T. Mitsuhashi, S. Bakardjieva, J. Šubrt, H. Haneda, Emanation Thermal Analysis of SiC Based Materials, 67, 2002, pp. 83–89.

[33] D.E. Folley, M.L. Chithambo, Influence of annealing on thermoluminescence of natural quartz: kinetic analysis and experimental study of apparent inverse thermal quenching, Radiat. Meas. 120 (2018) 53–58.

[34] J.E.K. Schawe, S. Pogatscher, Material characterization by fast scanning calorimetry: practice and applications, in: C. Schick, V. Mathot (Eds.), Fast Scanning Calorimetry, Springer, Cham, 2018, pp. 3–80.

[35] T.A. Mukhametzyanov, I.A. Sedov, B.N. Solomonov, C. Schick, Fast scanning calorimetry of lysozyme unfolding at scanning rates from 5 K/min to 500,000 K/min, Biochim. Biophys. Acta Gen. Subj. 1862 (2018) 2024–2030.

[36] H. Liu, X.-B. Zhan, Y.-M. Lu, Z.-L. Xu, T. Wu, Thermal hazard evaluation of AIBME by micro-calorimetric technique coupled with kinetic investigation, J. Therm. Anal. Calorim. 141 (2020) 1443–1452. F.R. Hay, M.A.A. O'Neill, A.E. Beezer, S. Gaisford, Isothermal micro-calorimetry: a tool to predict seed longevity? Seed Sci. Res. 16 (2006) 89–96.

Chapter 2

Temperature-modulated DSC (TMDSC)

Heat capacity (C_p) is the amount of heat required to raise the temperature of a material by 1°C. True heat capacity (no transition occurring) is completely reversible; the material releases the same amount of heat as temperature is lowered by 1°C. Specific heat capacity refers to a specific mass and temperature change for a material (J g^{-1} °C^{-1}) [1]. The subscript "p" represents "constant pressure."

Practically, conventional DSC is used for measuring the heat capacity directly. Rather, it measures the heat flow rates (Φ) in W. Assuming that the heat flow is represented by dQ (J) and the heat flow rate by dQ/dt (J s^{-1} or Watts), then the heat flow rate (dQ/dt in Js^{-1}) can be used to calculate heat capacity C_p of the sample by dividing the heat flow signal by the heating rate (dT/dt in °C s^{-1}). The relation between heat capacity C_p and heat flow rate is given by the relation:

$$\phi = \frac{dQ}{dt} = C_p \cdot \frac{dT}{dt} = c_p \cdot m \cdot \beta \tag{2.1}$$

where dQ is the heat flow (J), ϕ and $\frac{dQ}{dt}$ are the rates of heat flow (J s^{-1}, or W), C_p is the heat capacity (J °C^{-1}), c_p is the specific heat capacity (J °C^{-1} g^{-1}), $\frac{dT}{dt}$ and β are the scan rates (°C s^{-1}), and m is the sample mass (g). For simplicity, the "specific heat capacity" term will be written as "heat capacity" unless otherwise stated.

Although DSC instruments measure the rate of heat flow, yet for the purpose of simplicity, some authors use the term "heat flow" in place of "heat flow rate."

The heat capacity C_p thus obtained is more appropriately called "apparent heat capacity" because the original heat flow signal may include other interfering thermal transitions. The true value of heat capacity C_p can only be obtained in temperature regions where there are no interfering transitions [1].

The specific heat capacity (J °C^{-1} g^{-1}) (usually written as c_p) is obtained by dividing the heat capacity C_p by sample mass m (g) as shown by the following equation:

$$c_p = C_p / m \tag{2.2}$$

Thermal Analysis. https://doi.org/10.1016/B978-0-323-90191-8.00005-1

The enthalpy H (J) is given by

$$H = \int_0^T C_p dT \tag{2.3}$$

and the entropy S (J Kelvin^{-1}) is given by:

$$S = \int_0^T \frac{C_p}{T} dT \tag{2.4}$$

Conventional DSC technique has some limitations including the following:

- It measures only the sum or average value of the heat flow rate from overlapping processes. This makes quantitative analysis of the individual processes impossible. Most modern commercial materials are made from blends of different components. The temperature range for transitions in the different materials often overlap and experience is required to create methods that can assist with interpretation of results.
- It is less sensitive for detecting low-energy transitions. The only way to improve sensitivity for detecting low-energy transitions is to increase sample size and/or heating rate. This however decreases resolution (the ability to resolve transitions that occur close in temperature). DSC thus cannot optimize both sensitivity and resolution in a single experiment.

Temperature-modulated DSC (TMDSC) was introduced to resolve overlapping transitions. It differs from the standard DSC in that it applies two simultaneous heating rates to the sample. The linear or average heating rate provides the same information (total heat flow rate) as standard DSC, while the modulated heating rate (MHR) is used to determine the fraction of the total heat flow rate that responds to a changing heating rate. This fraction of the total heat flow that is frequency dependent is called reversing heat flow or heat capacity component of the total heat flow. When a process is reversible, the measured curve should be reproducible after cooling the sample down and reheating it. In contrast, heat flow that does not respond to the changing heating rate is called nonreversing heat flow or latent heat flow component. It is associated with irreversible processes and is kinetic or time dependent in nature. The heat input does not change the temperature because the heat is used for causing changes in the sample. Such changes take all the energy of the heat and therefore no heat is left for a temperature increase. As in conventional DSC, TMDSC can also be run in a cooling or heating mode [2].

The above classification and the relevant terms used in literature can be summarized as follows [3–5]:

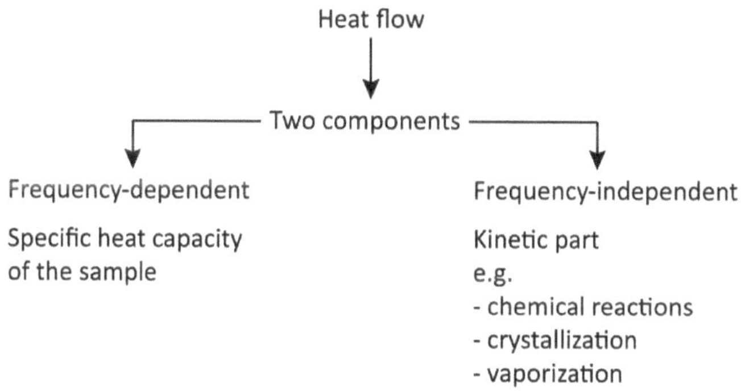
Heat flow
Two components
Frequency-dependent
Specific heat capacity of the sample
Frequency-independent
Kinetic part
e.g.
- chemical reactions
- crystallization
- vaporization

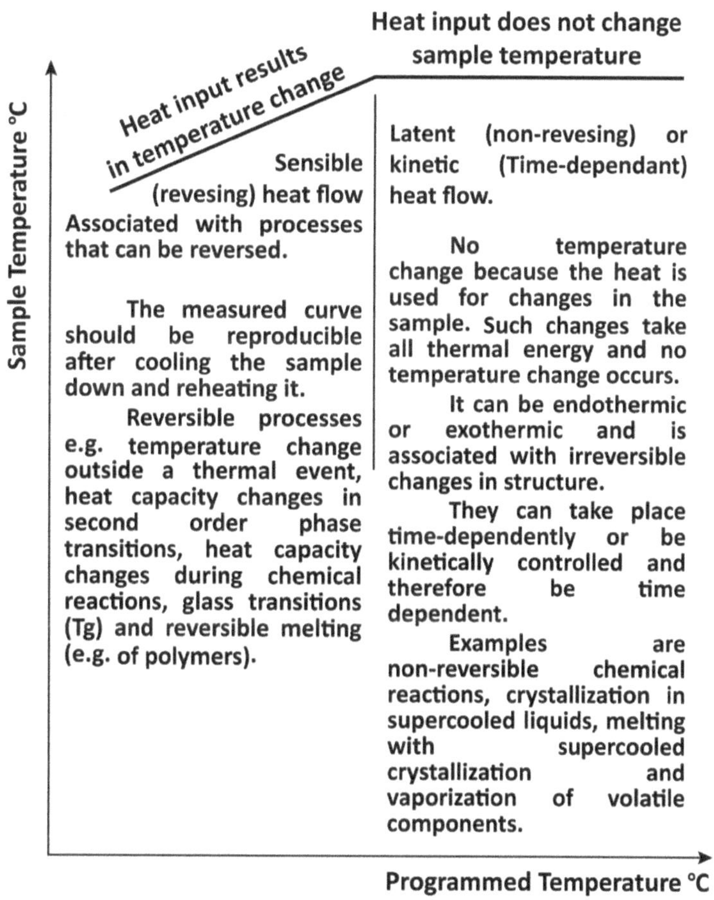
Heat input does not change sample temperature
Heat input results in temperature change
Sample Temperature °C
Sensible (revesing) heat flow
Associated with processes that can be reversed.
The measured curve should be reproducible after cooling the sample down and reheating it.
Reversible processes e.g. temperature change outside a thermal event, heat capacity changes in second order phase transitions, heat capacity changes during chemical reactions, glass transitions (Tg) and reversible melting (e.g. of polymers).
Latent (non-revesing) or kinetic (Time-dependant) heat flow.
No temperature change because the heat is used for changes in the sample. Such changes take all thermal energy and no temperature change occurs.
It can be endothermic or exothermic and is associated with irreversible changes in structure.
They can take place time-dependently or be kinetically controlled and therefore be time dependent.
Examples are non-reversible chemical reactions, crystallization in supercooled liquids, melting with supercooled crystallization and vaporization of volatile components.
Programmed Temperature °C

Mathematically, the heat flow Φ input into the sample is thus described as the sum of two components, sensible and latent heat flows:

$$\Phi(T,t) = mc_p\beta + m\Delta h_r(d\alpha/dt) \tag{2.5}$$

where m is the sample mass, c_p is the specific heat capacity, β is the heating rate of the sample, Δh_r is the specific enthalpy of a thermal event (e.g., enthalpy of reaction), and α is the extent of reaction (or more in general extent of structural change).

The first term in Eq. (2.5) depends on the heating rate. This is called sensible (reversible) heat flow, Φ_{sens}. The latent heat flow, Φ_{lat}, is the second term in Eq. (2.5). It is determined by physicochemical processes occurring in the sample, which depend on the change in internal variables far from an equilibrium state. Thus, Φ_{lat} is not primarily driven by the heating rate.

TMDSC, also known as modulated temperature DSC (MTDSC), was first introduced by Gobrecht et al. [6]. In their original paper, Gobrecht, Hamann, and Willers described a method which enables the direct measurement of relaxation times and relaxed and nonrelaxed heat capacities as a function of temperature. The apparatus was designed to be used as additional equipment to the Perkin-Elmer differential scanning calorimeter DSC-1B. In their method, the specimen is heated periodically with time instead of proportionally. The amount and phase of differential heat flows are measured. The parameters of an equivalent circuit of a heat capacity model are fitted to the measured complex heat capacity (c_p^*) values obtained at various frequencies and appropriate amplitudes of temperature. Furthermore, the applications of what they called "oscillating thermal analysis" were also proposed.

The technique was further developed in the 1990s by Reading and coworkers [7, 8]. It was introduced commercially and underwent modifications in the 1990s [7, 9–18].

TMDSC allows examining a variety of phenomena and processes in polymeric materials and their thermodynamic properties, including, but not limited to, the glass transition, cold crystallization, crystallization, reorganization, melting, and aging. TMDSC can be used to extract important characteristics such as glass transition temperature (T_g); changes in heat capacity at T_g; enthalpy relaxation; melting temperature; heat of fusion; crystallization; total, reversing, and nonreversing heat capacities; and many other parameters. All these quantities can be measured as a function of temperature, time, frequency, and other variables [19].

Different kinds of temperature modulation functions and evaluation procedures have been proposed so far:

(i) Stepwise temperature changes with isothermal segments [20–23]: These types of temperature programs are especially useful for the separation of the heat flow components. However, the time required for such experiments is relatively long because the sample should always relax to the equilibrium state during the isothermal segments.

(ii) Single-frequency modulations: The modulation function can be sinusoidal [7, 24–28]. Usually the first harmonic is used for evaluation procedure. The single-frequency modulation is used for the measurement of the frequency dependence of the heat capacity [6, 27, 29–31].

(iii) Multifrequency periodic modulations generated by superposition of multiple sinusoidal functions [32]: The simultaneous determination of the frequency dependence of the heat capacity over a broad frequency range is fulfilled in one single measurement by using stochastic temperature modulation using the TOPEM technique [33–35].

TMDSC has many significant advantages compared to the conventional DSC:

1. Separation of complex transitions into fine components more easily.
2. More accurate measurement of the crystallinity of polymers.
3. Improved resolution without compromising sensitivity.
4. Improved sensitivity to detect weak transitions.
5. Thermal conductivity measurement.
6. Determination of heat flow and heat capacity in a single experiment.

TMDSC thus provides a powerful enhancement to the standard DSC experiment and can simultaneously improve the sensitivity and resolution of the DSC experiment. To obtain the maximum benefit from the TMDSC experiment, however, optimal MDSC conditions must be chosen, such as sample size, modulation period, measuring transitions, and modulation amplitude.

1 TMDSC optimum conditions

TA Instruments in its TA Publication Number TN045 recommended the following protocol for choosing the proper TMDSC conditions with respect to sample size, modulation period, underlying heating rate, and modulation amplitude.

1.1 Sample size

To minimize thermal gradients, we usually use a thin, low mass sample. Keeping the sample thin minimizes the distance the heat needs to travel to keep the entire thickness of the sample in thermal equilibrium. Lower mass decreases the total amount of heat energy a sample must absorb due to its heat capacity. The surface area of the sample should be maximized relative to the mass. This can be accomplished by flattening solid samples in a press or cutting multiple thin pieces if hermal or mechanical history is a concern. Flattening of the sample in a press could change the sample's structure. The sample should be cut, rather than pressed. If a solid sample can be powdered, this should be done to maximize surface area. All these suggestions will improve the thermal contact of the sample and minimize gradients. The recommended sample size for polymers films and bulks are 5–10 mg and for organics, biologicals, inorganics, and metals 10–20 mg.

1.2 Modulation period

The ability of the sample to follow the imposed modulation depends greatly on the period which is chosen. In TMDSC, program periods as fast as 10 s are available. However, very few samples can follow this rapid modulation rate. The fastest modulation rate which a sample will follow depends on the heat capacity and thermal conductivity of the sample. Most samples will easily follow modulation periods of 50 s or greater. For lighter or lower heat capacity materials, the modulation period can be as low as 40 s. For heavier, higher heat capacity materials or those with low thermal conductivities, the period may be longer. Generally, a 60-s period is recommended when analyzing transitions and 100 s when the most accurate values of heat capacity are desired. Next, we select the underlying heating rate. This is determined using the width of the narrowest transition in the standard DSC scan.

1.3 Measuring transitions

Before beginning a TMDSC experiment on an uncharacterized material, a standard DSC scouting run should be performed at 5°C min^{-1} to determine if MDSC will enhance the results. After analyzing the standard DSC data, the following questions should be addressed:

- Do we need heat capacity information?
- Is the transition a heat capacity-related event?
- Is the transition masked by another event?
- Is the transition too weak or too broad to be measured by standard DSC?
- Is higher sensitivity or higher resolution needed?
- Will heat capacity change as a function of time at a constant temperature (i.e., isothermal curing)?

If the answer is "Yes" to any of these questions, then MDSC will help to enhance the results, and may provide more information than the standard DSC experiment alone. Once we have decided that MDSC will improve our understanding of the data, it is important to choose the optimal MDSC modulation conditions. The first parameter to consider is modulation period as discussed earlier and the underlying heating rate. The underlying heating rate can be determined according to the following equation:

$$\beta = \frac{T_{1/2}}{np} 60\mathrm{s}/\mathrm{min} \tag{2.6}$$

where $T_{1/2}$ is the full-width at half-height, n is the number of cycles, and p is the period.

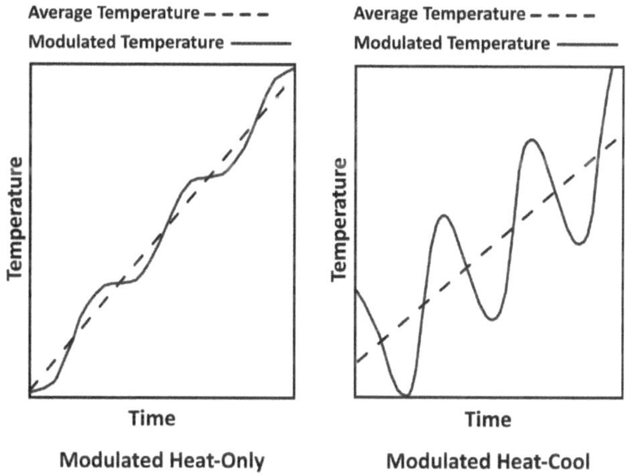

FIG. 2.1 The two major types of experiments available: modulated heat-cool (right) and modulated heat-only (left).

1.4 Modulation amplitude

After determining modulation period and underlying heating rate, the amplitude should be chosen. As a rule, larger modulation amplitudes provide for more sensitivity. When a dynamic underlying heating rate is chosen, there are two main types of experiments available: modulated heat-cool and modulated heat-only (Fig. 2.1).

In the modulated heat-only profile, the modulated temperature (solid line) never decreases, i.e., the sample is never cooled. But in the modulated heat-cool profile, some cooling of the sample is encountered on each modulation cycle. For general MDSC experiments, and for the measurement of heat capacity or the glass transition, the modulated heat-cool profile is recommended. This profile provides for the best sensitivity for weak transitions and improves the precision of heat capacity measurements. The modulated heat-only profile is a special case condition and should be used when studying melting or simultaneous melting/crystallization phenomenon. In these cases, cooling of the sample may induce crystallization. To prevent crystallization from occurring (and potentially complicating the measurement), the modulated heat-only profile is used.

2 TMDCS techniques

Currently, three TMDSC techniques are in use, namely:

1. the isothermal step (IsoStep) method
2. the alternating DSC (ADSC) method
3. the TOPEM most advanced technique

2.1 The isothermal step (IsoStep) method

The isothermal step IsoStep technique was introduced by Perkin-Elmer Instruments for power-compensated DSCs. It is the simplest type of TMDSC in which the sample is heated in a series of steps of just a few degrees separated by isothermal periods. In this technique, the temperature program consists of many dynamic segments; each of them begins and ends with an isothermal segment (Fig. 2.2). The length of the isothermal or dynamic segment is typically 30 s to 2 min and the heating rate during the dynamic segments is about 0.5–3°C min^{-1}. The isothermal and dynamic segments are usually of equal length and the same heating rate is used in all dynamic segments (Fig. 2.2).

Specific heat capacities of some common substances at 25°C are as follows [36]:

Substance	Alumina	Aluminum	Lead	Polystyrene	Water
c_p J g^{-1} °C^{-1}	0.895	0.776	0.129	1.3	4.18

IsoStep technique is used to accurately determine heat capacity using sapphire as a reference material. The heat flow within the isothermal segments is associated with kinetic phenomena such as crystallization, chemical reactions, vaporization, etc. Heat flow in dynamic segments is associated with thermal mass and heat capacity c_p. Alternative inspection of isothermal and dynamic segments allows c_p information to be separated from kinetic information.

In a typical measurement of heat capacity, c_p, the measured heat flow rate Φ (mW) is divided by the scan rate β and sample mass m, Eq. (2.7).

$$c_p = \frac{\phi}{m \cdot \beta} \tag{2.7}$$

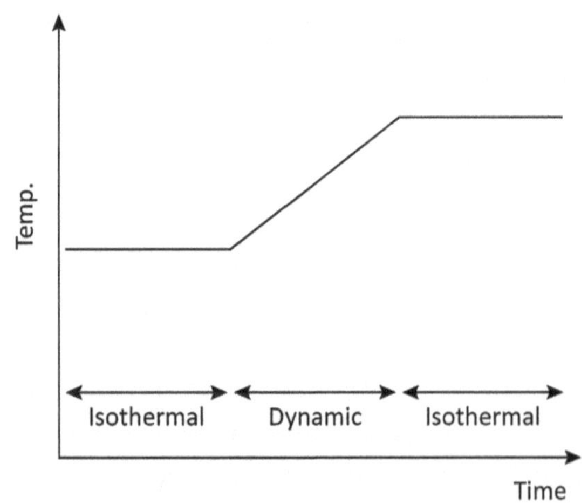

FIG. 2.2 IsoStep temperature program consisting of different isothermal and dynamic segments.

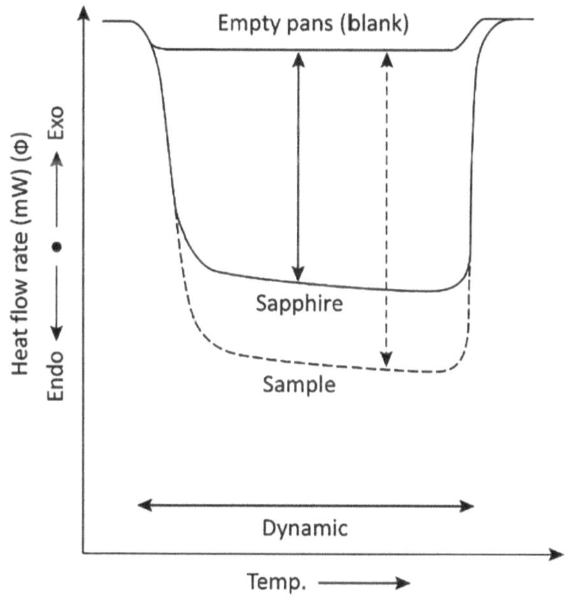

FIG. 2.3 Heat flow rate drifts for empty crucibles, standard sapphire, and unknown sample within the dynamic segment of IsoStep scans.

Three values of heat flow rates Φ are obtained as shown in Fig. 2.3: (i) heat flow rate upon using empty crucibles Φ_{blank}, (ii) heat flow rate upon using sapphire sample of mass m (g) Φ_{sapphire}, and (iii) heat flow rate upon using mass m (g) of the sample under investigation Φ_{sample}.

In real measurements, the heat capacity is not constant at different temperatures. Instead, it depends on the temperature [37] and the specific heat of most substances increases with increasing temperature [36]. By utilizing the sapphire specific heat capacity literature value at a certain temperature, $c_{\text{p(sapphire literature)}}$, a correction factor of machine response is introduced. Heat capacity values for sapphire at different temperatures are published in literature [36]

$$\text{Correction factor} = \frac{c_{\text{p (sapphire literature)}}}{\left(\phi_{\text{sapphire}} - \phi_{\text{blank}}\right)/m \cdot \beta} \tag{2.8}$$

The sample specific heat capacity at the stated temperature (T) $c_{\text{p (sample, T)}}$ can then be calculated as follows:

$$c_{\text{p sample},T} = \frac{\phi_{\text{sample}} - \phi_{\text{blank}}}{m_{\text{sample}} \cdot \beta} \times \text{correction factor} \tag{2.9}$$

In contrast, heat flow in the isothermal segments is associated with dynamic or irreversible processes. A drift in heat flow values within these segments (either in the endothermic or exothermic directions) gives the heat flow rates associated with irreversible kinetic process in question (Fig. 2.4).

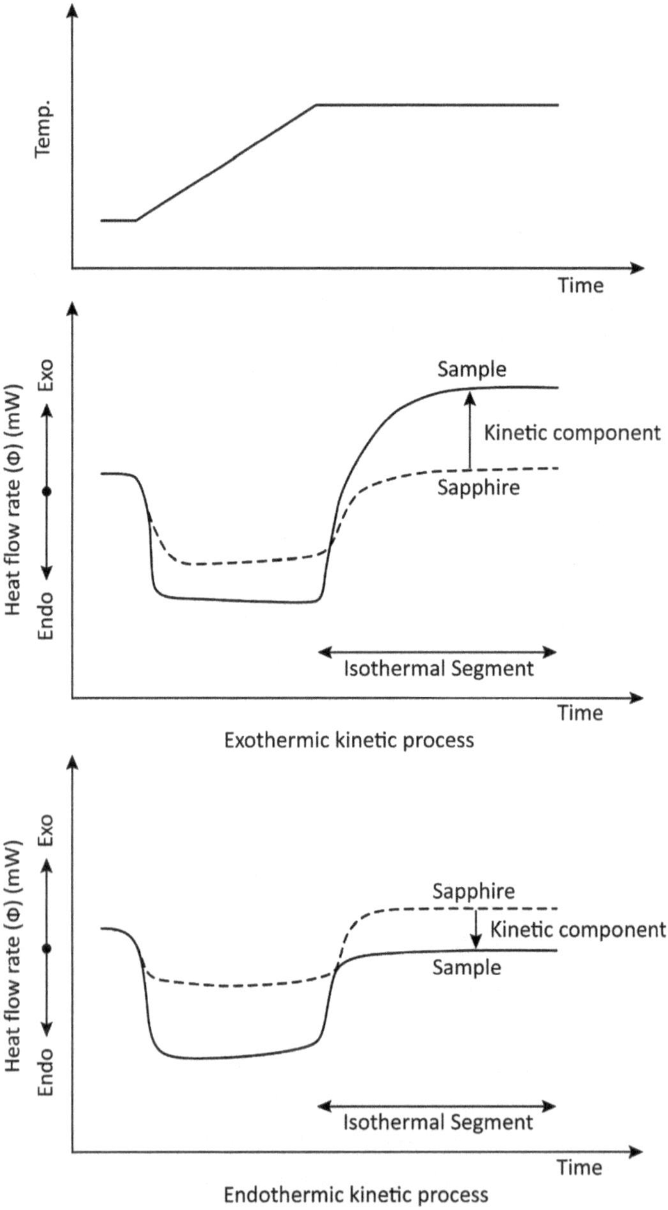

FIG. 2.4 Heat flow rate drifts for sapphire and unknown sample within the isothermal segment of IsoStep scans showing irreversible (kinetic) exothermic (middle) and endothermic (bottom) processes.

Because the StepScan DSC approach requires rapid response, the technique is only feasible with the power compensated DSC, which allows for fast heating and thermal equilibration. The application of the StepScan approach to a large mass furnace, heat flux DSC instrument would be difficult or technically unfeasible due to the inability to rapidly respond and equilibrate.

The other advantage of the StepScan DSC approach is that it provides a direct heat capacity measurement using the traditional and time-proven means without the need for deconvolution or the extraction of sine wave amplitudes. Other techniques, such as TMDSC (vide infra), as applied to a large mass, heat flux furnace, frequently have problems due to distorted sine waves and phase lag. Because of the direct nature of the StepScan DSC experimental approach for assessing the thermodynamic heat capacities, the results are not plagued with these experimental difficulties [38].

IsoStep technique for separation of heat capacity from overlapping kinetic effects

By performing this heat-hold sequence multiple times, over a temperature interval, the thermodynamic or reversible heat capacity data can be obtained. Fig. 2.5 shows the StepScan DSC approach with the application of a repetitive sequence of short heat-hold segments.

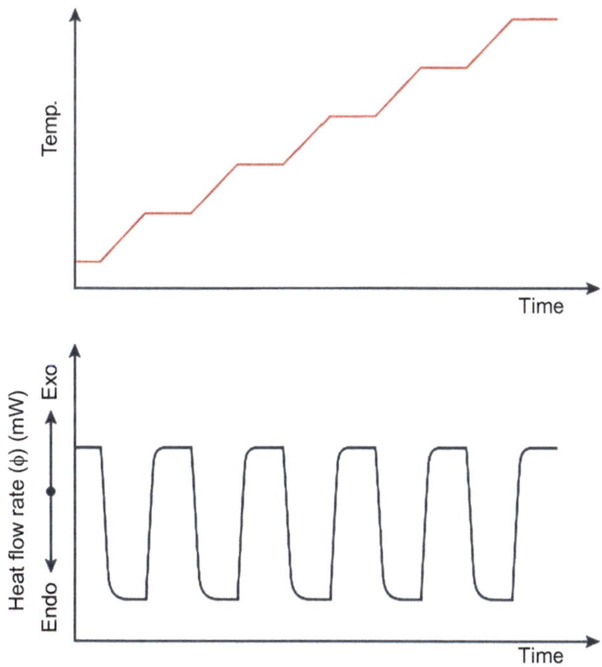

FIG. 2.5 A repetitive sequence of short heat-hold segments in an IsoStep scan.

With the application of small temperature increments in the heat segments, and by holding for a short time interval (e.g., 60 s) in the hold segments, the heat capacity that is yielded reflects the reversible aspects of the sample. Kinetic or irreversible effects (on the time scale of the experiment) are eliminated in the thermodynamic heat capacity data set. For example, if a sample has a glass transition, T_g, which has an overlapping enthalpic relaxation such as moisture loss or crystallization event, the thermodynamic C_p signal will show the classic stepwise change in the heat capacity, which makes it simple and straightforward to analyze and interpret. Therefore, the use of StepScan DSC helps in the clearer interpretation of transitions (e.g., T_g) that may be masked or obscured with interfering irreversible events, such as enthalpic relaxation, stress relief, moisture evolution, crystallization, curing, etc. StepScan DSC thus provides a useful means for the enhanced characterization of the thermal properties of samples. The technique uses a series of heat-hold steps, which encompass the temperature range of interest. The StepScan DSC results provide thermodynamic or reversible heat capacity signal along with the average or standard DSC data set.

The StepScan DSC approach offers the following advantages over other methods, such as TMDSC [39]:

- direct measurement of thermodynamic C_p (no deconvolution required)
- straight forward
- free of experimental problems such as sine wave distortions or phase lag
- accurate
- hassle-free

Fig. 2.6 shows a typical experiment to illustrate the application of IsoStep technique for separation of heat capacity c_p from overlapping kinetic effects using PET as an example. Using apparatus software, one can follow repetitive dynamic segments whereby c_p curves can be separately monitored. Reversing heat flow curves can also be monitored by programmed multiplication of the corresponding c_p by heating rates (β) and sample mass (m),

$$\Phi = c_p \cdot \beta \cdot m \quad (2.10)$$

Both c_p and reversing heat flow curves contain the same information. A blank curve is first obtained using empty crucibles. This curve will be subtracted from both c_p and reversing heat flow curves to compensate for any asymmetry in measuring system giving blank-corrected c_p and reversing heat flow curves. We then move to the isothermal segments whereby nonreversing or dynamic curve for a PET sample is obtained.

Fig. 2.6 shows the blank-corrected curves of the sapphire sample (the form of c_p and reversing heat flow curves) and PET curves are also measured (the two uppermost curves). The heat capacity and reversing heat flow curves of the PET sample show a glass transition at about 75°C. In the nonreversing heat flow curve, a small endothermic peak can be seen at roughly the same temperature. This is a result of enthalpy relaxation during storage of the sample.

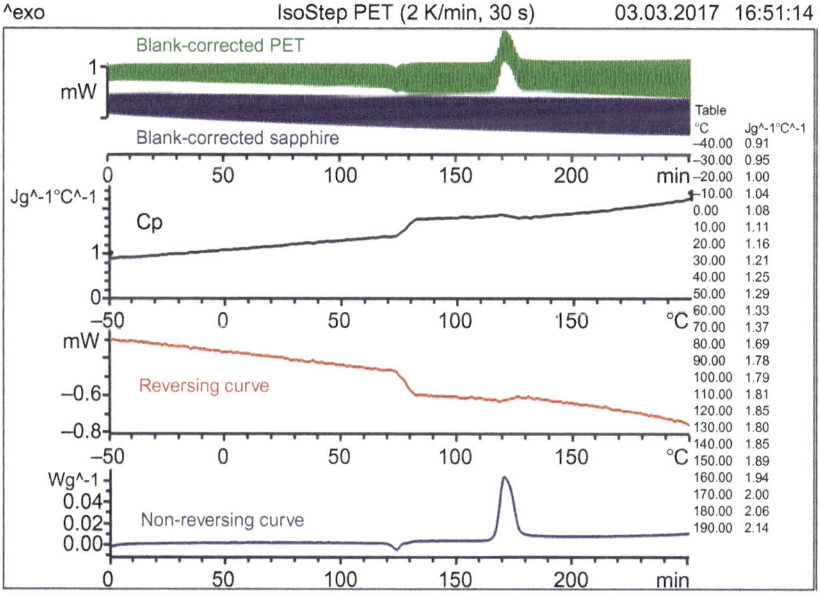

FIG. 2.6 IsoStep measurements of PET.

At about 125°C, an exothermic peak occurs due to cold crystallization. The crystallization of the sample affects the specific heat capacity, which is slightly lower for the semicrystalline material than for the liquid (step in the c_p curve at about 130°C).

2.2 The alternating DSC (ADSC) method

The alternating DSC (ADSC) method is a type of modulated temperature DSC in which the same heat flux DSC cell is used but a periodically changing (usually sinusoidal of just one frequency) temperature modulation of small amplitude is overlaid on the conventional linear temperature ramp (Fig. 2.7). Other periodic modulations such as a saw tooth (ISM DSC) are possible. This results in the heating rate at times being faster or slower than the underlying linear heating rate. It should not be understood that sample cooling takes place as the temperature does not decrease. Only, there is a periodic change in heating rate. This variation in instantaneous heating is illustrated in Fig. 2.7. The actual heating rate depends on three experimental variables: the underlying heating rate, the amplitude of modulation, and the period (frequency) of modulation. Typical values for these parameters are a heating rate of 1–5°C min^{-1}, an amplitude of ±0.5°C to ±1°C, and an oscillation period of 40–60 s. Higher resolution can be achieved by decreasing the heating rate, increasing the amplitude, and decreasing the oscillation period. This achieves 4–6 cycles during the thermal event of interest.

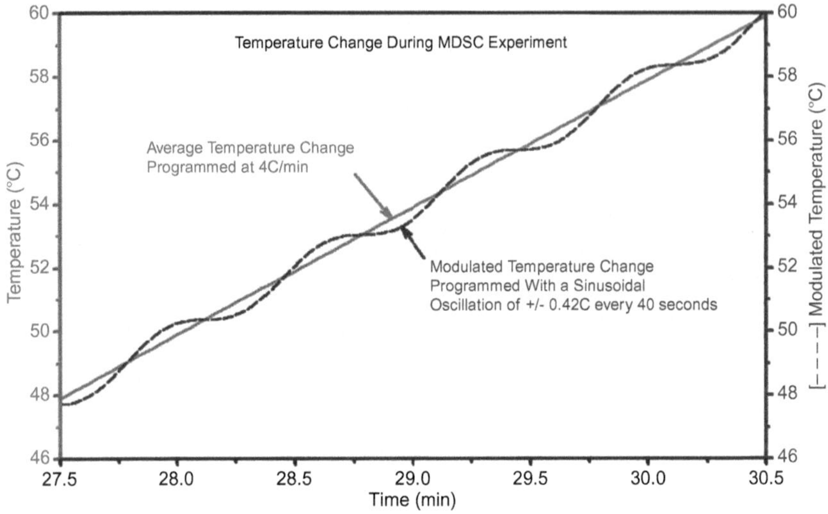

FIG. 2.7 Temperature profile from an MDSC experiment. A sinusoidal temperature variation is overlaid on a linear heating.

Modulated DSC thus differs from standard DSC in that it applies two simultaneous heating rates to the sample. The linear or average heating rate provides the same information (total heat flow rate) as standard DSC, while the sinusoidal (modulated) heating rate is used to determine the fraction of the total heat flow rate that responds to a changing heating rate (Fig. 2.8). In general, this heat flow rate is caused by heat capacity, changes in heat capacity, and by most melting. This fraction of the total heat flow is called reversing heat flow or heat capacity component of the total heat flow. Heat flow that does not respond to the changing heating rate is determined by subtracting the reversing signal from the total signal. This difference signal is the nonreversing heat flow or the kinetic (time-dependent) component.

This technique is commercialized in several versions including modulated DSC (MDSC, from TA Instruments) and alternating DSC (ADSC, from Mettler-Toledo), and they are usually referred to as TMDSC techniques. Despite their differences, all are fundamentally based on the same principle, which involves a Fourier transformation of the modulated heating rate (MHR) and heat flow signals, from which one can obtain certain information in addition to that available from conventional DSC. TMDSC permits the evaluation of two different kinds of heat capacity:

(i) The "thermal" heat capacity from the underlying heating or cooling rate, which is identical to that available from conventional DSC at the same rate.

(ii) The "dynamic" heat capacity from the amplitudes of the heating rate and heat flow modulations, which depends on the frequency of the modulations.

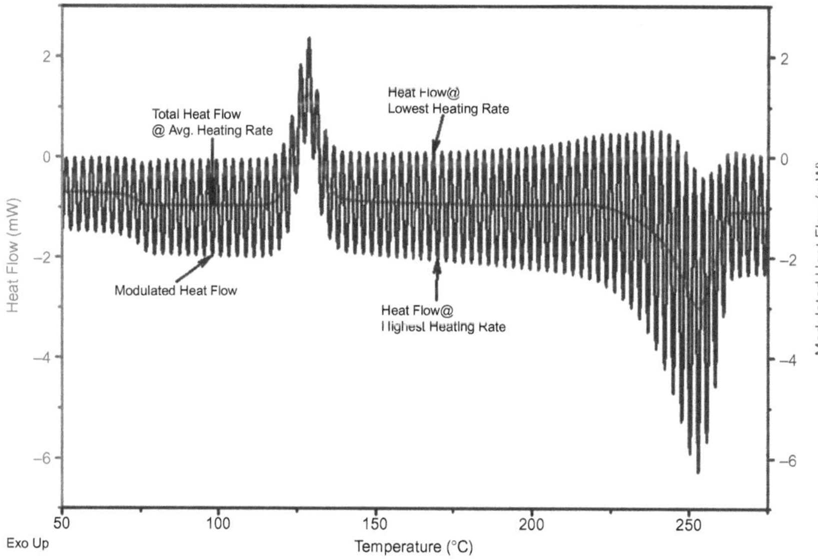

FIG. 2.8 The total heat flow signal calculated from the average value of the measured modulated heat flow signal.

As a result of the sinusoidal change in both heat flow and temperature change, two signals are obtained in a TMDSC experiment: (i) modulated heat flow (MHF) and (ii) MHR. Both signals are shown in Fig. 2.9. All other MDSC signals are calculated from these two measured signals.

Now, what makes difference between MHF rates is the change in their *amplitudes* according to different reversible thermal events such as heat capacity (c_p), glass transition (T_g), and some melting processes. At the same time, the corresponding heating rate amplitudes are not (or slightly) affected. Dividing the heat flow amplitudes (mW using Fourier transform) by the corresponding heating rate amplitudes (°C s^{-1}, also using Fourier transform) gives the heat capacity (or reversible) curve.

$$C_p(\text{reversible}) = \frac{\text{Heat Flow Amplitude}}{\text{Heating Rate Amplitude}} \times \text{Calibration Constant} \qquad (2.11)$$

The calibration constant is determined by comparing the measured c_p values for the total and reversing signals at a temperature in the middle of the range of interest with those of the theoretical values for sapphire at that temperature and a calibration constant is determined as: Calibration Constant = Theoretical ÷ Measured Values. To obtain the most accurate heat capacity measurements, it is recommended that the reversing c_p signal be calibrated with sapphire prior to running actual samples as shown above.

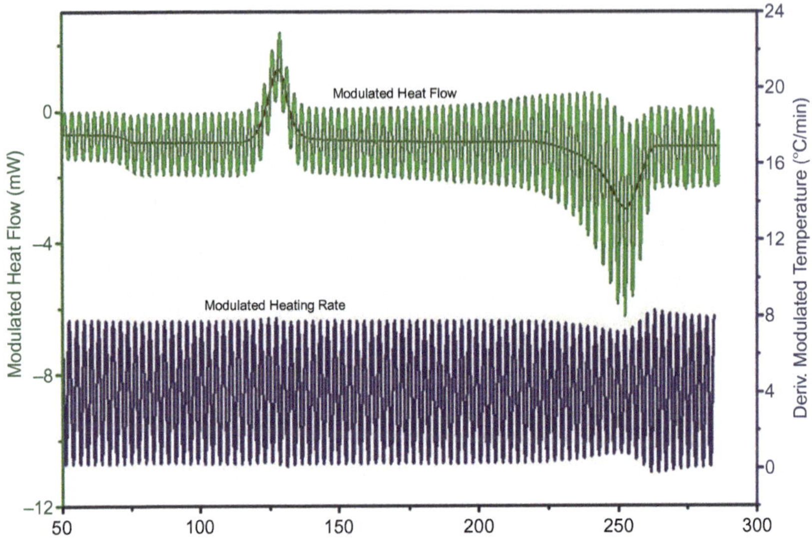

FIG. 2.9 Modulated heat flow (MHF; top), and modulated heating rate (MHR; bottom) signals obtained in a TMDSC experiment.

Averaging the heat flow amplitudes gives the total heat flow curve as shown in Fig. 2.9. The process of determining the average involves the use of Fourier transform analysis on the sine wave. The average is continuously calculated (every 0.1 s) rather than using the simple average that would significantly limit resolution. The total heat flow (like ordinary DSC) is the summation of both c_p (reversible) and kinetic (irreversible) curves:

$$\frac{dQ}{dt} = C_p \cdot \frac{dT}{dt} + f(T, t) \tag{2.12}$$

where $\frac{dQ}{dt}$ is the total heat flow rate (mW, which is mJ/s), C_p is the sample heat capacity (that is, specific heat capacity c_p × sample mass) (J °C^{-1}), dT/dt is the heating rate β (°C min^{-1}), and $f(T, t)$ is the heat flow rate that is a function of temperature and time (mW).

The left-hand side of Eq. (2.12) gives the total heat flow $= dQ/dt$ (J s^{-1} or W). The right-hand side is the sum of two types of heat flow, namely: (i) reversing heat flow $= C_p \times dT/dt$ (also called heat capacity component) in (J °C^{-1}) × (°C s^{-1}) = J s^{-1} or W and (ii) nonreversing heat flow $= f(T, t)$ (also called kinetic component) also in J s^{-1} or W.

Accordingly, the kinetic (irreversible) heating curve is obtained by subtracting the c_p (reversible) curve from the total heat flow curve as given by the following scheme:

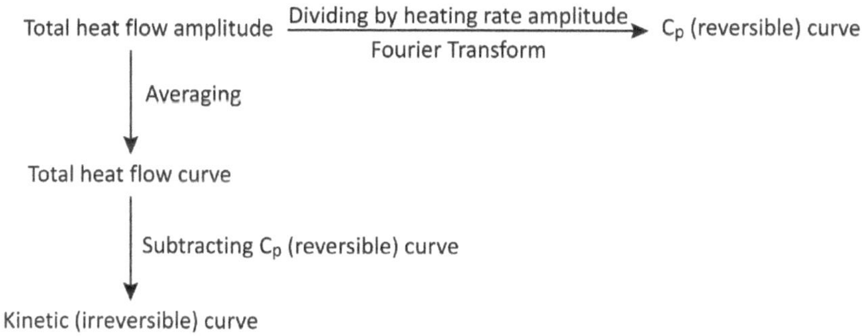

The nonreversing heat flow signal is calculated by subtracting the reversing heat flow signal from the total heat flow signal. All three signals are shown in Fig. 2.10 on a sample of quench-cooled polyethylene terephthalate (PET) as a classical model in this area.

In Fig. 2.10, the thermal events occurring in region (A) are associated with glass transition (T_g) process. In the c_p reversible curve (bottom curve), an endothermic heat flow rate is observed accompanying transition from solid to soft PET sample change with increased molecular mobility. In the kinetic nonreversing curve (upper curve), this event is manifested as a tiny endothermic peak that accounts for enthalpic recovery. In region (B), the kinetic nonreversing curve (upper curve) shows an exothermic peak that accounts for cold crystallization of amorphous PET. A slight change in c_p reversible curve is observed in this region since amorphous materials have a slightly higher c_p values compared

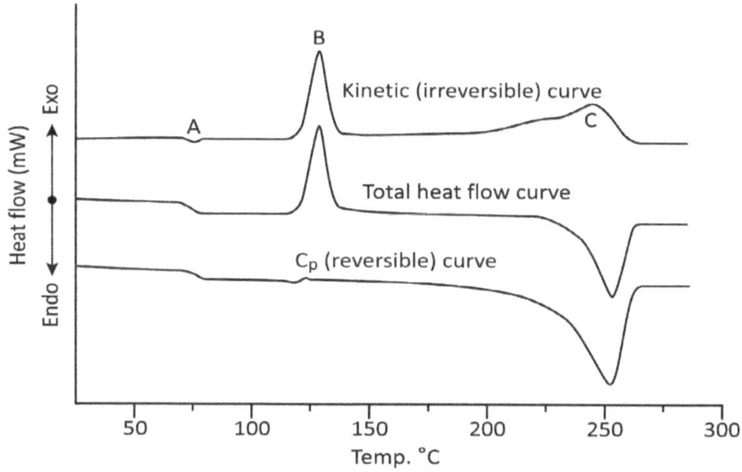

FIG. 2.10 The total heat flow, the kinetic (nonreversing) heat flow, and the c_p (reversing) heat flow curves for a quench-cooled polyethylene terephthalate (PET) sample.

with crystalline ones. In region (C), the kinetic reversing curve (lower curve) shows an endothermic peak that accounts for melting of crystalline PET sample. A corresponding exothermic peak is observed in the upper kinetic irreversible curve that is attributed to crystal perfection accompanying the melting process [39].

TMDSC thus provides all the benefits of DSC but is not limited by its natural limitations. In general, TMDSC provides the following advantages:

- Separation of overlapping transitions.
- Improved sensitivity for detecting weak transitions.
- Improved sensitivity and resolution in a single experiment.
- More accurate measurement of polymer initial crystallinity.
- Direct measurement of heat capacity.
- Quasiisothermal measurement of heat capacity changes during reactions or kinetic processes.

Some applications on TMDSC

It has been shown that traditional DSC provides a single signal (dH/dt) that is the sum of all thermal events occurring at a specific temperature or time during the experiment. This often causes difficulty in the detection of small transitions or in data interpretation. MDSC improves upon traditional DSC since it measures the total heat flow plus its heat capacity component and obtains the kinetic component from their difference. Its ability to resolve complex transitions into specific components improves data interpretation.

The previous examples shown in Fig. 2.10 used a single, unblended, polymer (PET). Another sample with slightly different physical properties was analyzed by DSC and the data (Fig. 2.11) might incorrectly lead to the conclusion that the material was only PET. However, this sample was a known mixture of PET and polycarbonate (PC). PC is amorphous with a T_g near 140°C. In this sample, however, T_g is undetected by traditional DSC since the heat flow due to cold crystallization in the PET occurs at the same temperature. This interpretation error would also cause an incorrect calculation of PET crystallinity (12.2 J g^{-1} from Fig. 2.11). In fact, it should be 0 J g^{-1} for a quench-cooled sample.

TMDSC is superior to DSC since it can separate kinetic events (e.g., a cold crystallization) from changes in heat capacity (e.g., a T_g). Fig. 2.12 shows TMDSC data for the above sample. It clearly reveals T_g for PC in the reversing heat flow signal and allows the correct calculation of crystallinity for quench-cooled PET from the sum of crystallization and melting (sum of nonreversing and reversing signals) [39].

Similar or more complex analyses can occur with pharmaceutical materials [39]. Drug delivery systems are usually mixtures of the drug and other excipient materials [39]. The drug can be amorphous or exist in multiple polymorphs. It can also be anhydrous or hydrate, solvate or salt. As with polymers, the ability

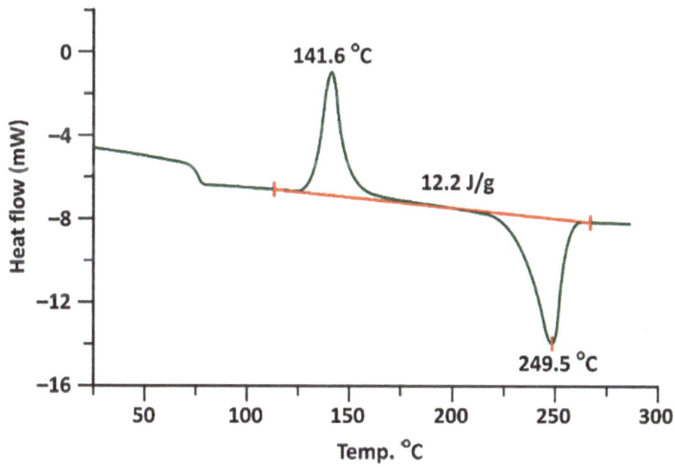

FIG. 2.11 Conventional DSC thermogram of a sample of PET and PC mixture.

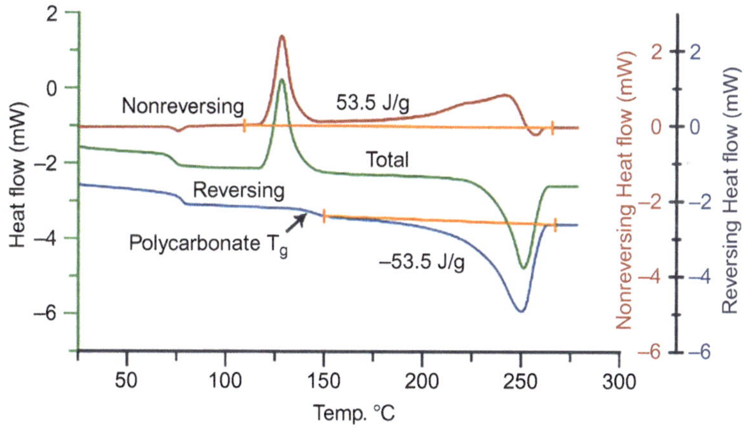

FIG. 2.12 MDSC data for mixed PET and PC sample.

of TMDSC to enhance resolution of overlapping events while maintaining high sensitivity can greatly simplify interpretation or eliminate errors in analyses. Fig. 2.13 shows the analysis of a drug delivery system based on amorphous, biodegradable polymer microspheres. Here, the total signal, which is the analog of standard DSC, makes it very hard to find the polymer T_g. The TMDSC reversing signal has the kinetic processes (enthalpic recovery and water evaporation) removed and permits an easy analysis of the polymer's T_g.

MDSC was used in the characterization of the thermal properties of active pharmaceutical ingredients to select appropriate physical forms for developing and defining proper manufacturing, handling, and storage conditions of those

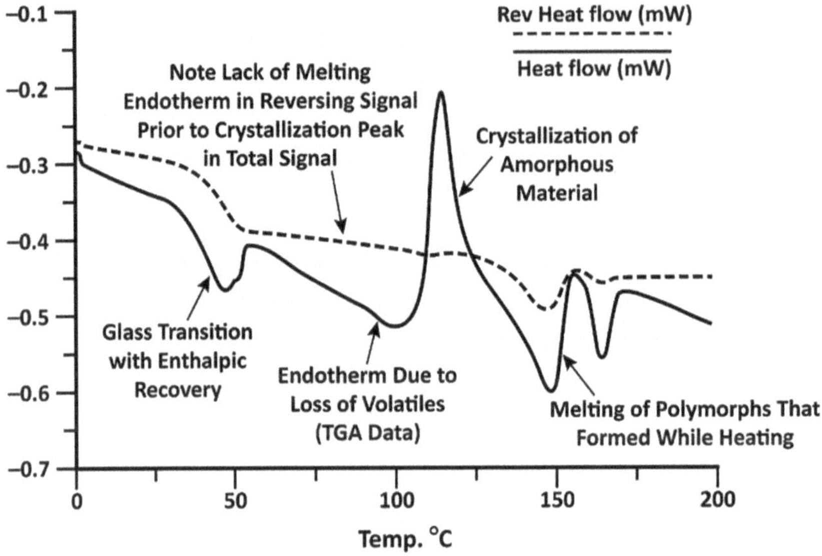

FIG. 2.13 MDSC data for the analysis of a drug delivery system based on amorphous, biodegradable polymer microspheres.

chemical entities. It has proven to be an effective tool in the thorough characterization of thermal behavior of compounds in preformulation studies. This analytical method is utilized in (a) the determination of glass transitions, (b) characterization of desolvation and degradation processes, and (c) the study of polymorphic transformations and crystallizations [40].

(a) Determination of glass transition temperature (T_g)

The determination of glass transition temperature (T_g) of pharmaceutical products, excipients, and active drug substance is critical to the consideration of proper handling, manufacture, and storage conditions of these materials. T_g represents the temperature below which the molecular mobility of a glassy amorphous solid is dramatically reduced and above which the amorphous material takes on a "rubbery" character with increases in the number and magnitude of molecular motions. Given that increased molecular mobility can lead to increased chemical reactivity and the propensity for a metastable amorphous material to crystallize over time, amorphous solids should be handled and stored well below the glass transition temperature, which makes accurate determinations of T_g imperative in preformulation and formulation development. Often the presence of extraneous thermal events such as enthalpic relaxation or moisture loss can confound the measurement of T_g. The utility of MDSC in these situations has been shown for polymer blends and nylon and in formulation applications. Enthalpic relaxation originates from thermally induced atomic or molecular rearrangement at the glass transition, which results in structural

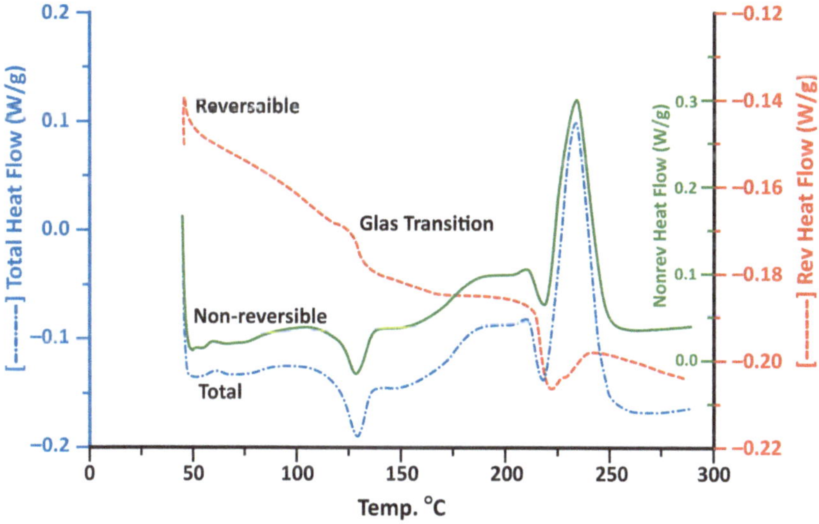

FIG. 2.14 MDSC thermogram separating the enthalpic relaxation at 107.2°C (shown in the nonreversible signal) from the glass transition (shown in the reversible signal) [40].

relaxation toward equilibrium. An example of this phenomenon is shown in Fig. 2.14 in which the enthalpic relaxation is observed at 107.2°C in the total and the nonreversing heat flow signals and the underlying glass transition can be seen in the reversing signal [40].

Above the glass transition, the molecular mobility increases and crystallization of the drug substance occurs, as witnessed by the presence of a major exothermic event above 190°C. Moisture in amorphous material can act as a plasticizer and therefore may have profound lowering effects on values for the glass transition. Furthermore, the loss of moisture upon heating typically results in broad endotherms that may not allow facile determination of the glass transition by DSC since a preconditioning heating step is required before the glass transition can be determined. Fig. 2.15 shows a MDSC thermogram for a polymeric drug substance that contains residual moisture. The volatilization of water upon heating results in an extremely broad endotherm from 50°C to 125°C, which overlaps the glass transition at 87.6°C. Furthermore, the plasticizing effect of water renders a broad glass transition spanning from 65.6°C to 98.2°C. The use of conventional DSC to measure T_g of this compound requires the sample to be subjected to a heat:cool:reheat cycle that results in the acquisition of a glass transition on material that has a significantly altered water content. Given the plasticizing effect of water and the fact that T_g is lowered in the presence of water, determination of T_g by conventional DSC may misrepresent T_g giving artificially high values in the absence of water and therefore MDSC may give a more meaningful value of T_g [40].

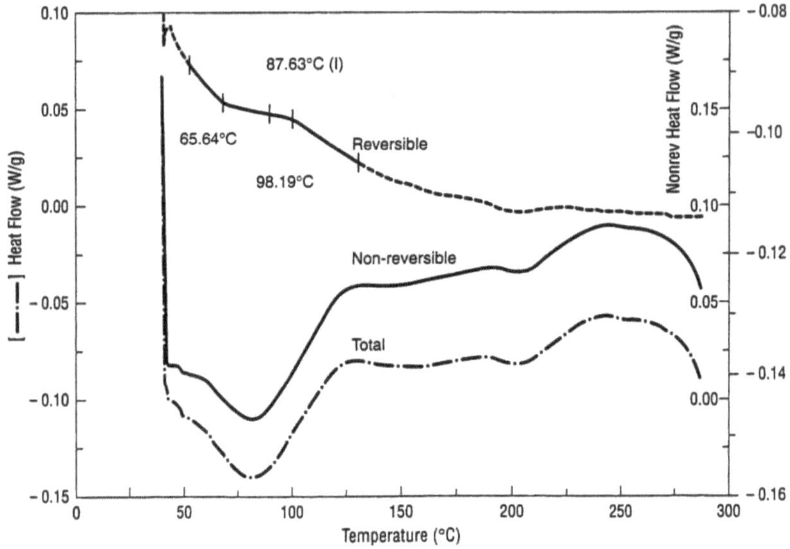

FIG. 2.15 MDSC thermogram of a polymeric drug substance illustrating the loss of moisture in the broad nonreversible endotherm (bottom signal) and the underlying glass transition in the reversible signal (top signal). The total signal is shown in the middle [40].

(b) Desolvation and degradation

Examples of nonreversible processes include, but are not limited to desolvation, crystallization phenomena and decomposition. An example of a complex thermogram showing multiple nonreversible processes is shown in Fig. 2.16. The compound was initially a stable dihydrate, and the release of the water of hydration from the crystal lattice is observed in the total and nonreversing signals at 105.5°C. The sensitivity of MDSC allows the detection of broad nonreversible exotherm at c.150°C, which corresponds to the decarboxylation of the compound. Finally, a major nonreversible exotherm attributed to further decomposition was observed at 222.9°C [40].

Fig. 2.17 illustrates another example of decomposition in which the high-energy exotherm at 216°C represents a nonreversible thermally induced cleavage of a hydroxamic acid functional group. Commonly, melting of drug substances is accompanied by decomposition. TMDSC is capable of deconvoluting overlapping melting and decomposition processes as shown in Fig. 2.18 where the total heat flow signal shows inadequate resolution of the melt endotherm and the decomposition exotherm. The separation of the events by TMDSC clearly shows the reversible endothermic melt component at 198.4°C and the nonreversible decomposition exotherm at 200.1°C [40].

(c) Polymorphic transformations

Although X-ray powder diffraction is generally required for the confirmation of different crystalline forms, TMDSC has been a useful tool along with hot stage

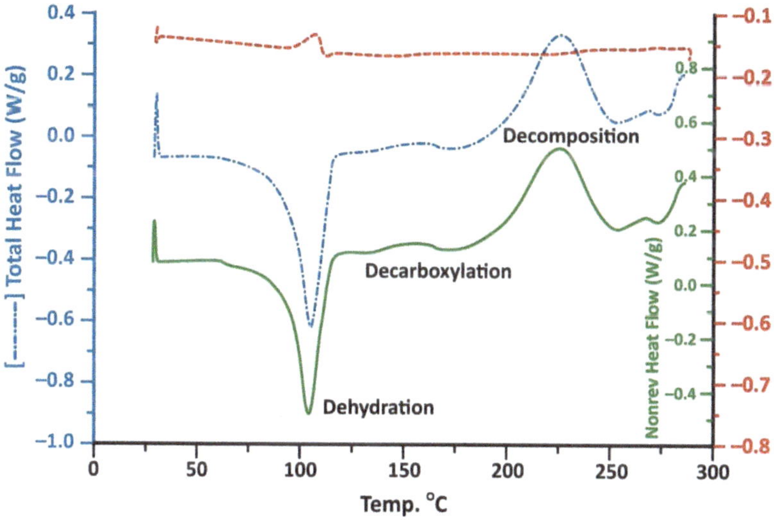

FIG. 2.16 Modulated DSC thermogram of a drug substance undergoing multiple nonreversible events including dehydration at 105.5°C (bottom signal), decarboxylation at 150°C, and further decomposition above 200°C. Both reversible (top) and total (middle) signals are shown [40].

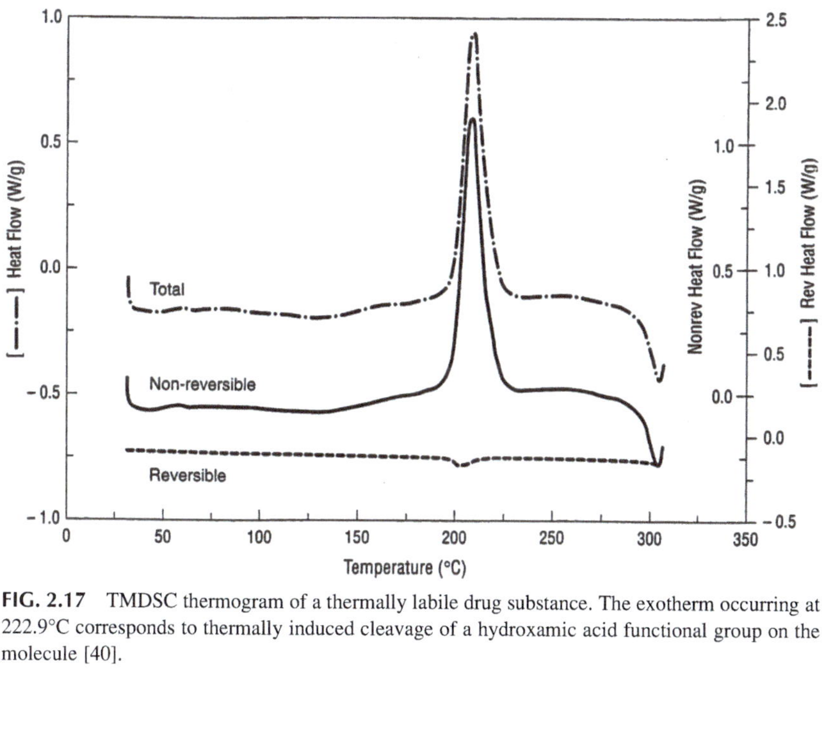

FIG. 2.17 TMDSC thermogram of a thermally labile drug substance. The exotherm occurring at 222.9°C corresponds to thermally induced cleavage of a hydroxamic acid functional group on the molecule [40].

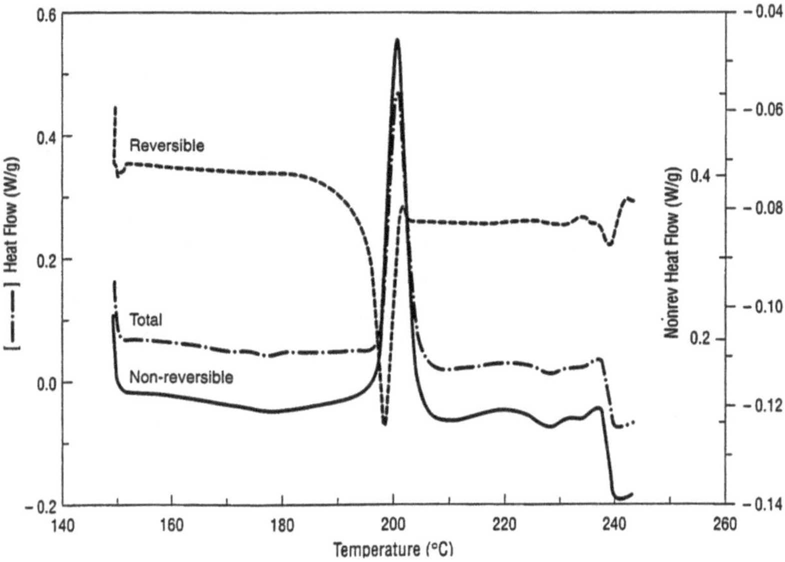

FIG. 2.18 TMDSC thermogram of a drug substance exhibiting overlapping melting with decomposition. The melt endotherm is seen in the reversible signal and can clearly be resolved from the decomposition exotherm in the nonreversible heat flow [40].

microscopy to provide insight into various types of polymorphic transformations. The solid-solid polymorphic interconversion of losartan occurs in the absence of melting and has been previously reported [40]. The TMDSC thermogram for this compound is shown in Fig. 2.19, where the kinetically controlled molecular reorganization is responsible for the broad endotherm at 237.4°C seen in the nonreversible heat flow and virtually no reversible transition is observed. In addition to the polymorphic conversion, decomposition after melting of the compound is also observed as an exotherm at 270.5°C in the nonreversing signal [40].

Another type of a polymorphic conversion is illustrated in Fig. 2.20 in which the transformation occurs via a melt followed by recrystallization. The endotherm present at 175.8°C in the reversing signal represents the initial melt prior to the nonreversible recrystallization exotherm at 177.3°C. The presence of an endotherm coinciding with the initial melt was also observed in the nonreversing heat flow. While this occurrence is quite common in MDSC, the cause for the signal is not widely understood by those using the technique, but maybe due, in part, to initiation of the polymorph recrystallization upon initial liquefaction (melting) of the material [40].

2.3 The most advanced technique, TOPEM

Despite the advantages afforded by TMDSC in various areas, ADSC, for example, requires a minimum of two separate scans, a "blank" and a "sample," for

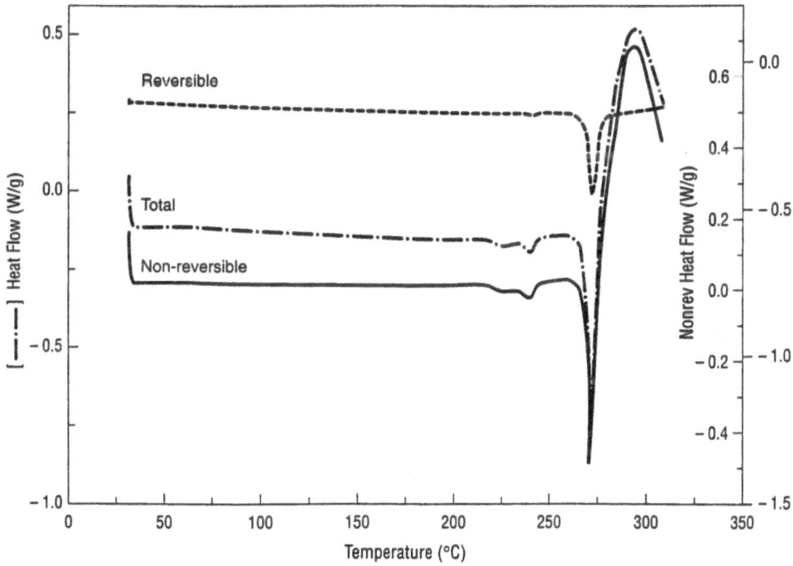

FIG. 2.19 TMDSC thermogram of losartan, which initially undergoes a nonreversible solid-solid polymorphic conversion, shows an endothermic melt transition in the reversible signal, and finally decomposes as seen in the nonreversible exotherm [40].

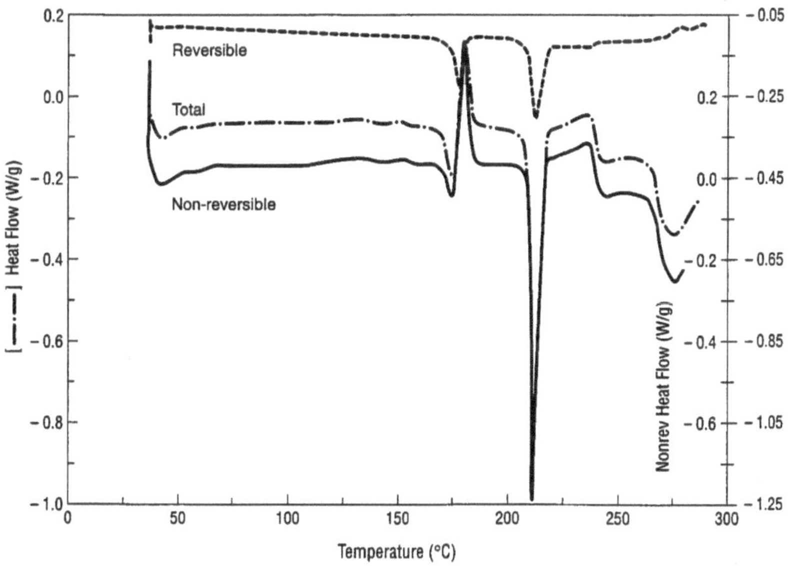

FIG. 2.20 TMDSC thermogram of a polymorphic drug substance that melts, as seen by the reversible endotherm at 175.8°C, with recrystallization following in the nonreversible signal at 177.3°C [40].

any given set of experimental parameters, which is a time-consuming procedure. Moreover, it applies only one (usually sinusoidal) frequency. A more advanced new temperature-modulated DSC technique (TOPEM) was introduced by Mettler-Toledo in late 2005. The term "TOPEM" seems to derive from the word "TOP" referring to the technique being an advanced multifrequency TMDSC technique and the initials "PEM" refer to the "parameter estimation method," that is, the state-of-the-art mathematical method used in its data analysis. The extended PEM technique eliminates instrumental influences and extends the measurable frequency range [5, 33, 41–47].

The basic idea of TOPEM is to overlay the isothermal or ramped temperature with a time series of stochastic (random) temperature pulses of different durations as shown in Fig. 2.21. TMDSC methods used before have overlaid the isothermal temperature or heating ramp with a (usually) sinusoidal temperature modulation of just one frequency. In contrast, TOPEM, the new advanced multifrequency temperature modulation technique, uses many different frequencies (broadband). As a result, the underlying heating rate is modulated with a broad band of frequencies. The modulations consist of temperature pulses, of fixed magnitude and alternating sign, with random durations within limits specified by the user. Sample properties can be determined as a function of time and temperature over a wide frequency range using PEM method.

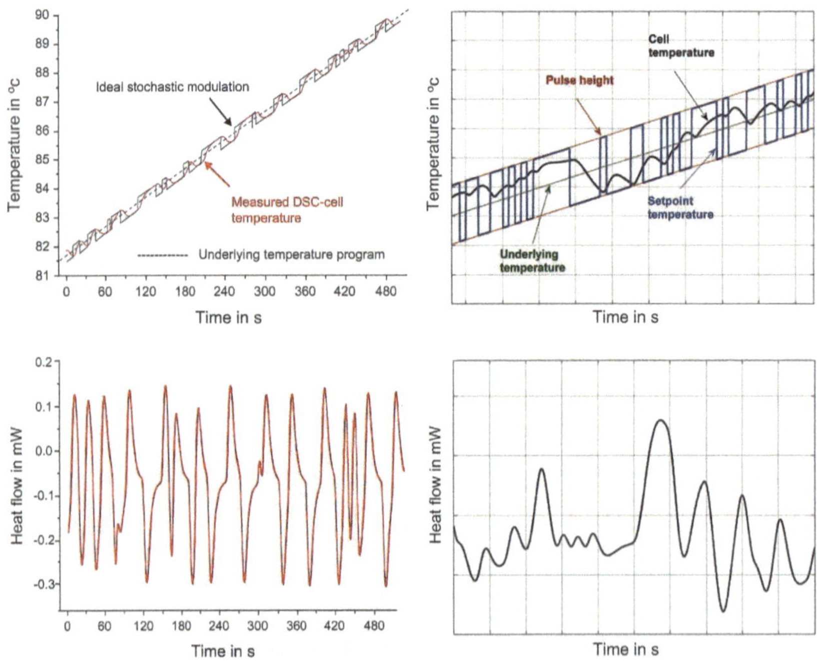

FIG. 2.21 Different presentations of input temperature profile (top) and corresponding output heat flow (bottom) in a TOPEM measurement.

In a single scan, it is possible to distinguish frequency-dependent phenomena (such as glass transition) from frequency-independent phenomena (such as cold crystallization). The small temperature modulation hardly influences nonreversing effects, but with reversing effects gives rise to a corresponding modulation of the heat flow. The broad frequency response implies that TOPEM can determine the complex heat capacity over a range of frequencies in a single scan. Sample properties that shift with frequency can easily be distinguished from frequency-independent effects. This greatly simplifies the interpretation of samples that exhibit overlapping effects. At the same time, TOPEM allows the quasistatic frequency-independent heat capacity to be measured.

A magnified sector of the stochastic input temperature profile is shown in Fig. 2.22 in which the pulse widths change at random between a chosen minimum and maximum. Also, the pulse height must be chosen.

The chosen temperature program defines the input signal. The output signal is the measured heat flow. A mathematical procedure calculates the correlation between the input (heating rate) and output signal (heat flow). After this procedure, a component is obtained related to the input signal and one that does not. The correlated one is the reversing heat flow, and the noncorrelated one is the nonreversing heat flow [44, 47].

The heating rate cannot be too high so that enough data points are obtained. The switching time of the pulse width is set between 15 and 30 s. For low-frequency measurements, a high switching time is needed and vice versa. Too low or too high switching times result in noisy results. The pulse height is determined by the behavior of the sample. The sample must be able to follow the oven. A sample can follow a little pulse height better than a high pulse height because it needs less time to equilibrate. In contrast, more information is gained with higher pulse heights because the sample is heated for a longer period. This means that a compromise must be found between higher and lower pulse heights.

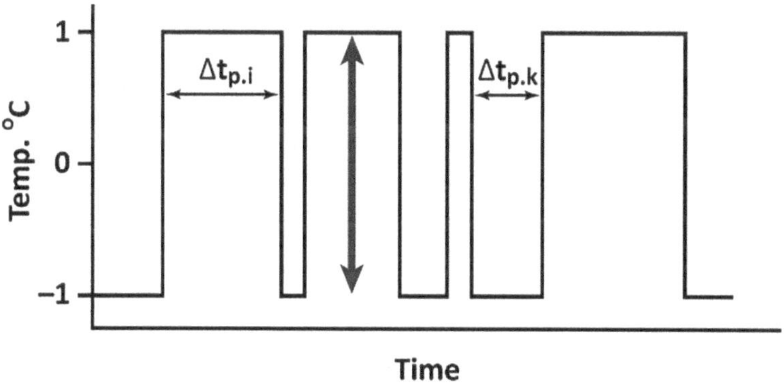

FIG. 2.22 Example of possible pulses in TOPEM. The vertical arrow is the pulse height. The horizontal arrows are two possible pulse widths (Mettler-Toledo, 2009).

If the reversing heat flow is measured dynamically, it is in general less than the sensible heat flow because all time-dependent processes that take longer than the characteristic measuring time (e.g., 60 s) are not measured. In quasi-static measurements, the difference between the sensible and the reversing heat becomes minimal. For this reason, using TOPEM, it is possible to obtain an excellent separation of sensible and latent heat flows.

The TOPEM technique includes four steps:

Step 1: The input signal is the measured temperature of the DSC resulting from a superimposition of a conventional temperature program with a small stochastic temperature perturbation. The measured heat flow is output signal.

Step 2: Analysis of the correlation between the input signal and the output signal is done by using a mathematical procedure. As a result, the heat flow is separated into two components, a component which is correlated to the input signal and another component which is not correlated to the input signal. The correlated component characterizes the linear behavior of the system sample-instrument. The heat flow component which is noncorrelated to the heating rate is determined by the excess heat capacity. This is the nonreversing heat flow.

Step 3: From the characterized system, the cross-correlation function between heating rate and heat flow is determined. This function is equivalent to the heat flow signal measured as the response to a small temperature step normalized by the height of the temperature step. The integration of this function delivers the quasistatic heat capacity $c_{p,0}$ and the reversing heat flow $\Phi_{rev} = mc_{p,0}\,\beta_u$. The total heat flow is the sum of reversing and nonreversing heat flow.

Step 4: Using the multifrequency evaluation of the characterized system, the frequency-dependent complex heat capacity can be determined for any frequency.

Using TOPEM, the dynamic behavior of the sample can be analyzed in a wide frequency range by only one single measurement. The reversing heat flow signal based on the quasistatic heat capacity and the nonreversing heat flow signal is a direct result of the correlation analysis instrumentation.

3 Heat capacity terms

Before going to the theory of TOPEM, different heat capacity terms are as follows:

- quasistatic heat capacity, $c_{p,0}$
- complex heat capacity, c_p^*
- in-phase heat capacity, c_p'
- out-of-phase heat capacity, c_p''

In thermodynamics, a quasistatic process is a thermodynamic process that happens slowly enough for the system to remain in internal equilibrium, thus a quasistatic process is also a reversible process. Infinite slowness is characteristic feature of quasistatic process. In quasistatic and reversible processes, the system remains in internal equilibrium yet in the former it is possible for system to increase its entropy during a process. So, the difference between the two is referred to as entropy generation. Heat capacity values are temperature dependent [5, 33, 43, 44, 47]. In TOPEM, small temperature modulation hardly influences nonreversing effects, so they instantaneously remain in equilibrium. Due to modulation time intervals, heat capacity values are assumed constant and the sample is assumed to be under instantaneous equilibrium. The heat capacity value under such conditions is termed as the quasistatic heat capacity, $c_{\mathrm{p},0}$. The quasistatic heat capacity can thus be written as

$$c_{\mathrm{p},0} = c_p^*(\omega \to 0) = c_p(t \to \infty) \tag{2.13}$$

where ω is the angular frequency ($\omega = 2\pi f$, f is the frequency) and c_p^* is the complex heat capacity [5].

A further aspect of TMDSC, in general, and TOPEM, in particular, is the analysis of relaxation processes by measuring the frequency dependence of thermal events. For these measurements, the thermal response of the sample is analyzed at different frequencies. The simplest kind of a thermal relaxation is the thermal lag due to heat transfer in the instrument and the sample. The result is that the heat capacity apparently becomes frequency dependent and a phase shift between heating rate and heat flow occurs. The information gained from the data can be used for the calibration of the TMDSC signals [48].

In addition to simple relaxations due to heat transport, the sample can show thermal relaxation processes due to the lag in heat transfer from the fast external to slow internal degrees of freedom. An example of such thermal relaxation processes is the entropy fluctuation due to the cooperative rearrangements in glass formers. This results in the heat capacity becoming time or frequency dependent. An example of such a relaxation process is glass transition [49–52]. Analogous to the dielectric permittivity or the mechanical compliance, the frequency-dependent heat capacity can be described by a complex function

$$c_\mathrm{p}^*(T, \omega) = c_\mathrm{f}(T) + c_\mathrm{s}^*(T, \omega) = c_\mathrm{f}(T) + c_\mathrm{s}'(T, \omega) - ic_\mathrm{s}''(T, \omega) \tag{2.14}$$

where c_f is the part of the heat capacity due to the fast-internal degrees of freedom, which are primarily vibrational modes. c_s is due to the slow internal degrees of freedom, which cause the frequency dependence of the heat capacity. This contribution of the heat capacity is related to all types of structural relaxations that occur close to (metastable) equilibrium. Because of the phase shift between heating rate and heat flow, the frequency-dependent heat capacity is a complex function containing a real part, $c_s{}'$, and an imaginary part, $c_s{}''$; i is the imaginary unit. The frequency-dependent heat capacity provides information about the dynamics of the underlying molecular processes [52–54].

Depending on the focus of interest, two different evaluation approaches can be applied to TMDSC measurements. These are the determination of the frequency dependence of the complex heat capacity and the separation of the heat flow into sensible and latent heat flow components. These two components as determined by the single-frequency conventional TMDSC cannot be directly discussed in terms of thermodynamic properties. The reason is the frequency dependence of the nonreversing heat flow. This problem is discussed in Ref. [55]. The demands for advanced TMDSC technologies can be summarized as follows:

(i) The measurement of the quasistatic heat capacity (Eq. 2.13).
(ii) The frequency-independent determination of the nonreversing heat flow.
(iii) The ability to separate latent and sensible heat flow contributions.
(iv) The simultaneous determination of the frequency dependence of the heat capacity over a broad frequency range. These requirements can be fulfilled in one single measurement by using stochastic temperature modulation using TOPEM [28].

This imperative can be measured by applying frequency mode to the instrument. For a periodic changer as in ADSC and TOPEM, the introduction of complex heat capacity is convenient. The complex specific heat capacity measured at frequency ω, c_p^*, accounts for the real (reactive or in-phase heat capacity) part c_p' and the imaginary (out-of-phase or dissipative heat capacity) part ic_p'' with $i=\sqrt{-1}$. One must use the following equation to make $c_p''(\omega)$, the dissipative part, positive

$$c_p(\omega)=c_p'(\omega)-ic_p''(\omega) \qquad (2.15)$$

A flow diagram of the TOPEM evaluation was given by Mettler-Toledo as follows [44]:

$T(t)$ and $\Phi_{measured}(t)$	→ Pulse response $g(t)$ (system characterization)	→ Result curve $c_{p,0}$, φ_{tot}, $\varphi_{reversing}$ and $\varphi_{nonreversing}$	Frequency selection →	Frequency-dependent curves including: c_p^*, c_p', c_p'' and phase angle φ

Mettler-Toledo reported several interesting application fields foreseen for TOPEM technique [44]:

Industry	Effects that can be analyzed with TOPEM
Automobile and aerospace	Curing reactions, influence of moisture, glass transition, vitrification
Chemical	Exothermic reactions (safety investigations), glass transition, kinetics, crystallization behavior, polymorphism, drying, heat capacity
Electronics	Curing reactions, glass transition, vitrification
Paints	Curing reactions, influence of moisture, glass transition, drying, vitrification
Rubber (elastomers)	Glass transition, phase separation, melting, vulcanization
Plastics (thermoplastics, thermosets, fibers, films, textiles, adhesives, packaging, and cables)	Curing reactions, influence of moisture, enthalpy relaxation, glass transition, cold crystallization, phase separation, melting, melting and crystallization, vitrification, heat capacity
Foodstuffs	Influence of moisture, gelation, glass transition, stickiness, polymorphism, drying
Pharmaceuticals	Influence of moisture, glass transition, melting (isothermal step melting point), crystallization behavior, polymorphism, drying, heat capacity, decomposition behavior, stability
Research and development	Everything given above plus the physical nature of transitions

4 Theory of TOPEM

In TOPEM, the technique is based on a thorough analysis of the response of a DSC (including the apparatus and the sample) toward a stochastically modulated underlying temperature program. Fig. 2.23 is a block diagram showing how the TOPEM technique processes the inputs from the DSC into the more understandable outputs of reversing and nonreversing heat flow [56].

With TOPEM, the input signal is a superposition of an underlying temperature ramp ($T_0 + \beta_u t$) (where T_0 denotes the start temperature, β_u is the underlying heating rate, and t is time) and a stochastic temperature modulation δT(t). The measured heat flow consists of a reversing and a nonreversing component $\varphi_{rev}(t)$ and $\varphi_{non}(t)$, respectively. The system (i.e., the instrument including the sample and the pans) is described by the transfer function $g(t)$.

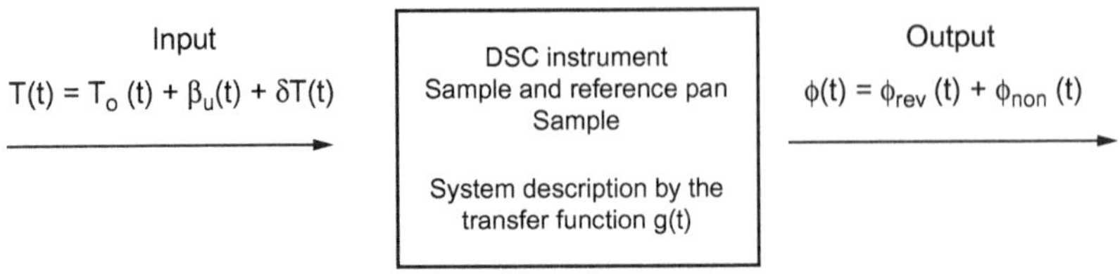

FIG. 2.23 Schematic representation of the signal flow in a TOPEM showing the stochastic modulated input and the heat flow output where T_0 is the start temperature, β_u is the constant underlying heating rate, δT is the temperature perturbation, $\varphi(t)$ is the measured heat flow, $\varphi_{rev}(t)$ is the reversing heat flow component, and $\varphi_{non}(t)$ is the nonreversing heat flow component.

The analysis of the DSC system is based on the following assumptions:

1. The DSC is considered as a linear, time-invariant system during a sufficiently long-time interval and for a reasonably small temperature modulation.
2. The sample response to the temperature program can be separated into a linear response (reversing heat flow) and a nonlinear response (nonreversing heat flow). It is assumed that the nonreversing processes are slow. This assumption corresponds to an assumption regarding the sample: it is only allowed to exhibit slow nonreversing effects. An effect will be considered slow if its time scale is larger than what has been called a sufficiently long-time interval.

These assumptions are the same for all known TMDSC techniques.

To probe the response of the DSC system, TOPEM makes use of a stochastic temperature modulation. Ideally, the modulation consists of step-like temperature increments and decrements with random interval lengths along the underlying heating rate. The actual DSC cell temperature as well as the resulting heat flow are recorded.

5 TOPEM signal analysis

In the absence of nonreversing phenomena and assuming time-invariant linear behavior, a DSC system is completely characterized by its transfer function $g(t)$. Thus, the output heat flow $\varphi(t)$ is related to the input temperature program $T(t)$ according to the following equation:

$$\phi(t) = g(t)*\beta(t) = \int_0^\infty g(\tau)\beta(t-\tau)d\tau \tag{2.16}$$

where the sign (*) refers to convolution of $g(t)$ and $\beta(t)$ functions producing a third function expressing how the shape of one is modified by the other.

Here, $\beta(t)$ denotes the heating rate, i.e.,

$$\beta(t) = \frac{dT(t)}{dt} = \beta_u + \frac{d\delta T(t)}{dt} \tag{2.17}$$

β_u being the underlying heating rate.

Physically, the transfer function corresponds to the response of the system toward a step-like temperature change at its entry. Due to the assumed time invariance of the system, the integral of the transfer function is equivalent to the specific heat capacity of the sample. The temperature and heat flow signals are sampled at a certain rate. Furthermore, time invariance of the system is only valid within a limited time interval and the assumption of linear response is violated if nonreversing thermal effects occur. Neglecting, first, nonreversing processes, the technical problem requires the determination of $g(t)$ $(0 \leq t < \infty)$ from discrete data in a sufficiently long-time frame in which the assumption of time invariance is at least a good approximation. A standard technique to solve this problem is the use of a discrete Laplace transformation referred to as z-transformation (that is related to the z-plane). Thus, the z-transformed Eq. (2.18) expresses the functions in simple multiplication rather the convolution format that reads as

$$\widetilde{\phi}(z) = H(z)\widetilde{\beta}(z) \tag{2.18}$$

where $\widetilde{\phi}(z)$ is the z-transform of $\phi(t)$, $H(z)$ is the z-transform of $g(t)$, and, $\widetilde{\beta}(z)$ denotes the z-transform of $\beta(t)$. The sign ($\tilde{}$) indicates Laplace of a certain function.

A convenient way to describe $H(z)$ is by setting

$$H(z) = \frac{B(z)}{A(z)} \tag{2.19}$$

where $B(z)$ and $A(z)$ are polynomials of degree q and p, respectively.

Thus,

$$A(z) = a_0 z^p + a_1 z^{p-1} + a_2 z^{p-2} + \cdots + a_n z^{p-n}, \quad a_0 \neq 0 \tag{2.20}$$

$$B(z) = a_0 z^q + a_1 z^{q-1} + a_2 z^{q-2} + \cdots + a_n z^{q-n}, \quad a_0 \neq 0 \tag{2.21}$$

Substituting Eq. (2.19) into Eq. (2.18), applying the inverse z-transformation to get original function, and considering the discrete nature of the recorded data one obtains

$$\sum_{\ell=0}^{p} a_\ell \phi(t - \ell \Delta t) = \sum_{k=0}^{q} b_k \beta(t - k\Delta t) \tag{2.22}$$

where Δt corresponds to the sampling time interval and a_ℓ and b_k denote the unknown coefficients of the polynomials $A(z)$ and $B(z)$. Making use of the time invariance of the system in a sufficiently long time interval, this equation can be numerically solved by a least-squares fitting procedure in the considered time interval.

In the presence of nonreversing phenomena, an extra term is added, and Eq. (2.16) is modified as follows:

$$\phi(t) = \int_0^\infty g(\tau)\beta(t - \tau)d\tau + \phi_{\text{non}}(t) \tag{2.23}$$

Here φ_{non} (t) is the nonreversing heat flow, which describes the nontime invariant heat flow contribution. Applying again the mathematical procedure outlined above, one finds

$$\sum_{\ell=0}^{p} a_\ell \phi(t-\ell\Delta t) = \sum_{k=0}^{q} b_k \beta(t-k\Delta t) + \phi_{non}(t, c_n) \tag{2.24}$$

where φ_{non} (t, c_n) is the nonreversing heat flow that is described by an appropriate expression containing parameters c_0, $c_1, \ldots, c_n$. Again, this equation is solved numerically in a suitable time interval. As a result, the nonreversing heat flow, as well as a parametric description of the transfer function $g(t)$ (coefficients a_ℓ and b_k), are available.

As already noted, integration of the transfer function delivers the quasistatic heat capacity, $c_{p,0}$, of the sample, i.e.,

$$c_{p,0} = \frac{1}{m} \int_0^\infty g(t) dt \tag{2.25}$$

Here, m denotes the sample mass. The quasistatic heat capacity corresponds to the frequency-independent heat capacity that could also be determined using conventional DSC techniques (e.g., the sapphire method) if no excess heat-producing thermal effects occur.

Multiplying the quasistatic heat capacity by the sample mass and the underlying heating rate delivers the quasistatic reversing heat flow, $\varphi_{rev}(t)$, i.e.,

$$\phi_{rev} = m \cdot \beta_u \cdot c_{p,0} \tag{2.26}$$

Adding the reversing and nonreversing heat flow yields the total heat flow:

$$\phi(t) = \phi_{rev}(t) + \phi_{non}(t) \tag{2.27}$$

which corresponds to the result of a conventional DSC measurement run at the underlying heating rate. In contrast to other TMDSC techniques, TOPEM offers a simple consistency check. In fact, the moving average of the TOPEM DSC signal in the time interval used for the TOPEM signal analysis agrees with the total heat flow as calculated from the reversing and nonreversing heat flow.

Frequency dependence of heat capacity, introducing the complex heat capacity term, a complex number Z, can be presented as $Z = a + ib$ with $i = \sqrt{-1}$. For the description of periodic changer, as in MTDSC, the introduction of complex quantities is convenient. The complex heat capacity has proven useful for the interpretation of systems slowly responding to internal degrees of freedom and may be used for representing the slow response in the glass transition. For the specific complex heat capacity measured at frequency ω, $c_p(\omega)$ with its real (reactive) part $c_p'(\omega)$ and the imaginary part ic_p'', one must use equation Eq. (2.15) to make $c_p''(\omega)$, the dissipative part, positive.

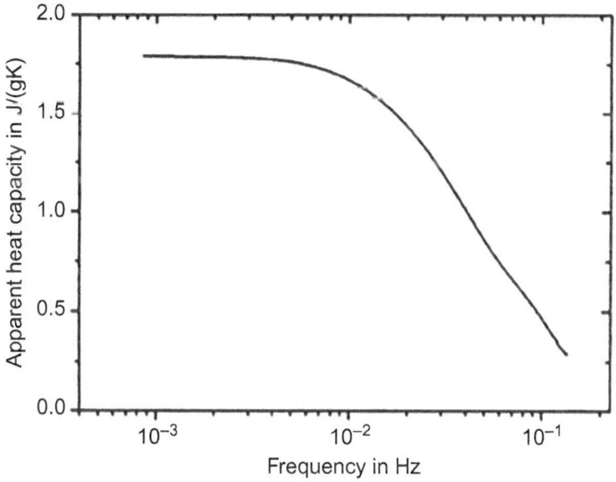

FIG. 2.24 Frequency dependency of the apparent heat capacity of a sample in thermal equilibrium.

Schawe et al. investigated nonequilibrium systems under temperature modulation in light of the stochastic thermodynamics [47]. They showed that, for small amplitudes of the temperature oscillations, the heat flux behaves sinusoidally with time, a result that allows the definition of the complex heat capacity. The real part of the complex heat capacity is the dynamic heat capacity, and its imaginary part was shown to be proportional to the rate of entropy production.

If the transfer function $g(t)$ is known, the response of the system can be calculated for any temperature program for a sinusoidal modulation on an underlying temperature ramp. Thus, the well-known frequency-dependent complex heat capacity (including c_p^*, c_p', c_p'' and the phase angle) can be determined from one single TOPEM experiment in a wide frequency range. This is illustrated in Fig. 2.24, which shows the apparent heat capacity as a function of the applied frequency for a sample in thermal equilibrium.

In thermal equilibrium, the specific heat capacity is frequency independent and corresponds to the quasistatic heat capacity, $c_{p,0}$. The decrease in the apparent heat capacity with higher frequencies (see Fig. 2.24) is therefore due to the limited heat transfer capability of the system. During relaxation phenomena such as glass transitions, the heat capacity becomes a frequency-dependent property of the sample. In such a case, the measured frequency dependency of the heat capacity must be corrected for the frequency dependency due to instrumental effects. This can be done using the time-dependent quasistatic heat capacity.

In comparison to other TMDSC techniques, TOPEM thus offers the following advantages:

- A quasistatic heat capacity and quasistatic reversing and nonreversing heat flows are available.

- The frequency dependency of the complex heat capacity (c_p^*, c_p', c_p'' and phase angle) is determined from one single experiment.
- The frequency-dependent complex heat capacity can be calibrated using the quasistatic quantities.

The quasistatic heat capacity $c_{p,0}$ is sometimes expressed as Eq. (2.13).

TOPEM therefore is an effective tool for heat capacity measurements (quasistatic and frequency dependent) as well as for the separation of the heat flow into a reversing and nonreversing component.

6 Thermodynamic properties of polymers by TMDSC

The thermodynamic properties of polymers cover a wide range of physical properties of polymeric materials. Polymers as other materials are well described by two types of information based on structure/composition and energetics. Each of these two information sets can be separated into macroscopic and microscopic parts. The macroscopic description deals with a phase on a large scale and, on the other hand, the microscopic one deals with matter on the atomic scale. The thermodynamic properties cover the macroscopic description of energetics. However, for a full understanding, it needs the link to the microscopic description, which is achieved through statistical thermodynamics and calculation. Another connection to the microscopic description is offered by the kinetics that can be examined macroscopically, but for its understanding and explanation, a microscopic-level model must be valid [57].

For many years, polymer researchers/scientists have recognized the equilibrium states and utilized nonequilibrium states to explore the history and details of their polymeric samples. For a nonequilibrium sample, it is possible, for example, to thermally establish how it is transferred into the solid state (e.g., determination of the thermal and mechanical decomposition history). Recently, with the use of TMDSC analytical method, it was discovered that within the global, nonequilibrium structure of semicrystalline polymers, locally reversible melting and crystallization processes are possible on a nanoscale phase level [58]. TMDSC method allows examining several phenomena in polymeric materials such as glass transition, cold crystallization, crystallization, reorganization and melting, aging, and other process (Fig. 2.25). Using TMDSC enables the measurement of characteristics as glass transition temperature (T_g); changes in the heat capacity at T_g; enthalpy relaxation; melting temperature; heat of fusion; crystallization; total, reversing, and nonreversing heat capacities; and many other parameters. All these quantities can be measured as a function of temperature, time, frequency, and other factors and variables [59].

At the same time, the integral thermodynamic functions include enthalpy, H, entropy, S, and free enthalpy, G (Gibbs function) and information on pressure, volume, and temperature is of importance. The transition parameters of pure, one-component systems are first order such as glass transitions. Mesophase transitions, in general, were reviewed as well as liquid crystals [60].

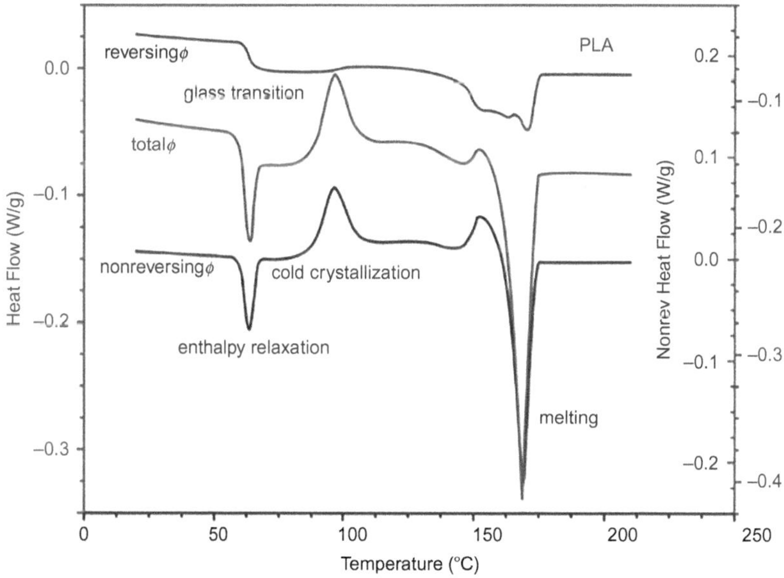

FIG. 2.25 Total, reversing, and nonreversing heat flow rates as a function of temperature from standard TMDSC for poly(lactic acid) (PLA) [63].

7 Some applications of TMDSC in polymer science

7.1 Amorphous polyethylene terephthalate

Fig. 2.26 shows a measurement of amorphous PET by TMDSC as an example. The total heat flow signal shows a shift of glass transition at 74.7–82.1°C, an exothermic peak due to cold crystallization at 132.1°C, and an endothermic peak due to melting at 255.8°C. Glass transition and melting are observed in the heat capacity component and cold crystallization and melting are noticed in the kinetic component. Accordingly, phenomena measured by DSC are divided into a heat capacity component and kinetic component. Melting phenomena are shown to contribute to both components [61].

7.2 Glass transition with enthalpy relaxation

Fig. 2.27 shows the result of an epoxy resin measurement (cured sample) as an example of glass transition with enthalpy relaxation using TMDSC method. Polymer glass transition may be included with the endothermic peak because of enthalpy relaxation due to the thermal history of the sample. The endothermic peak in Fig. 2.27 is large, resulting in the possibility of it being mistaken for a melting peak. The existence of the endothermic peak makes finding an accurate glass transition temperature difficult. Nonetheless, shifting of the DSC curve by glass transition and endothermic peak by enthalpy relaxation,

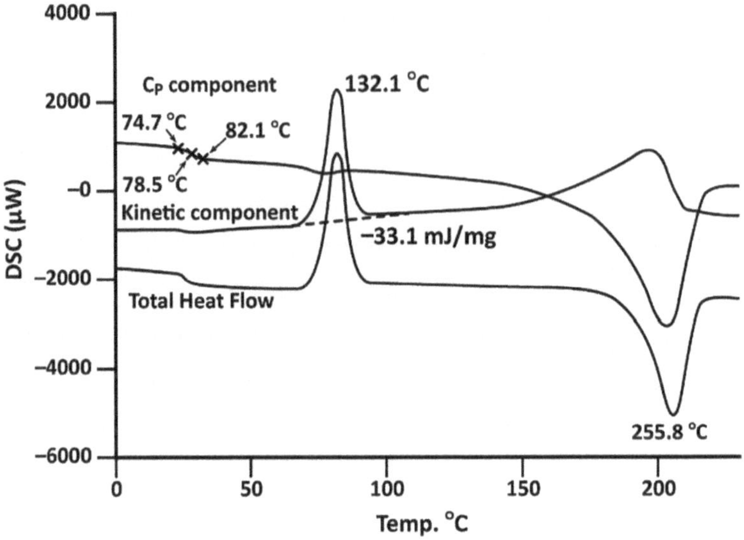

FIG. 2.26 TMDSC measurement of amorphous PET.

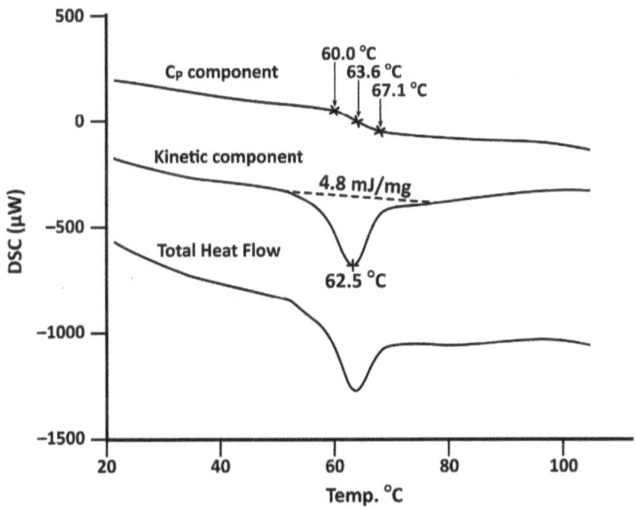

FIG. 2.27 TMDSC measurement of glass transition with enthalpy relaxation.

as shown in Fig. 2.27, are separate in TMDSC measurements; therefore, shifting by glass transition can be obtained clearly. Furthermore, because the quantity of heat can be found from the area of the enthalpy relaxation peak separated as the kinetic component, the size of the relaxation phenomenon can be estimated [61].

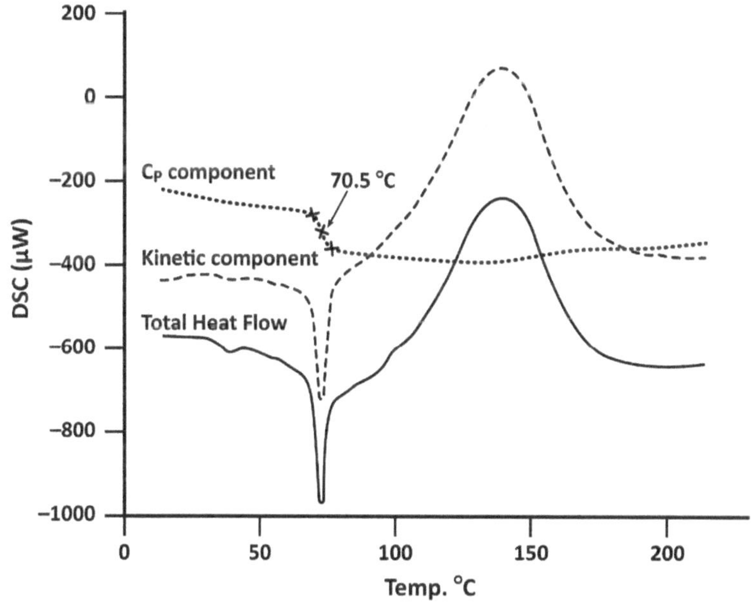

FIG. 2.28 TMDSC measurement of epoxy resin (uncured sample).

7.3 Curing reaction of thermoset polymer

Fig. 2.28 shows the result of an epoxy resin measurement (uncured sample), as an example of curing reaction of a thermoset resin. With total heat flow, glass transition with enthalpy relaxation appears around 70°C after which simultaneously the exothermic peak from thermal curing reaction across a wide temperature range near 180°C is noticed. The curing peak appears only in the kinetic component when separated into each component, because the thermal curing reaction of the thermoset polymer is an enthalpy change. This makes possible accurate reading of the end temperature of the glass transition and the start temperature of curing reaction [61].

7.4 Immiscible polymer blends

Fig. 2.29 shows measurement of a PET and PC (polycarbonate) polymer blend as an example of measuring an immiscible polymer blend. Around 110–130°C of total heat flow, cold crystallization of PET and PC glass transition overlap. In such cases, accurate analysis of both PET cold crystallization and PC glass transition is difficult. However, separating data into each component shows the glass transition of PC into the heat capacity component and PET cold crystallization into the kinetic component. Thus, it is possible to read the glass transition of PC at 114–123°C [61].

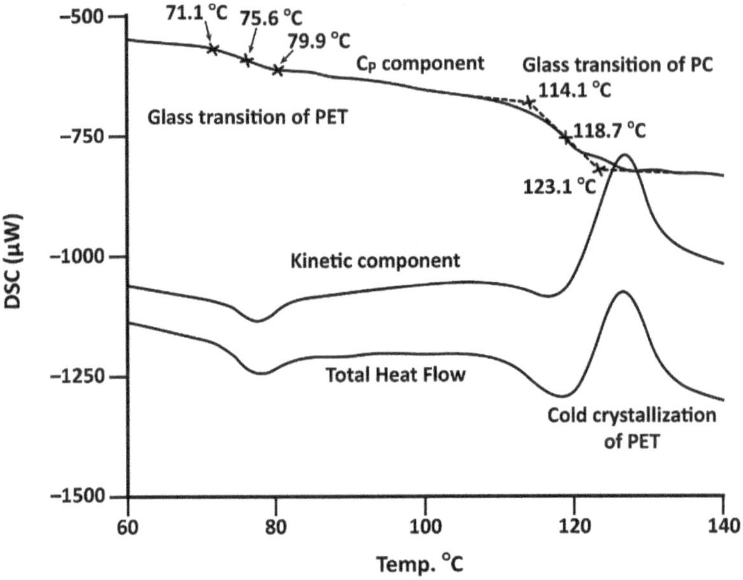

FIG. 2.29 TMDSC measurement of PET and PC polymer blend.

7.5 Vaporization with volatilization system

Fig. 2.30 shows performed measurement of nylon fiber (nylon 6) as an example of measuring phenomenon with vaporization. Nylon easily absorbs moisture because the amide group, which is a hydrophilic function group, is included in the molecule. During heating up nylon, vaporization occurs giving an endothermic peak from vaporization over a wide range of temperature as noticed by the total heat flow as shown in Fig. 2.30. Since the glass transition of nylon overlaps this endothermic peak, the glass transition from this data cannot be observed. However, the glass transition can be clearly observed in the separated heat capacity component from around 0°C to 40°C. Vaporization is a kinetic phenomenon; therefore, the endothermic peak from this does not appear in the heat capacity component. Thus, measurement of the glass transition of sample that includes water or solvent can be measured by TMDSC [61].

7.6 Crystal polymorphism

Because of the presence of metastable phases of crystals in pharmaceuticals, there are many substances in which crystal polymorphism is examined by DSC. Fig. 2.31 shows an example of a pharmaceutical having crystal polymorphism measured by TMDSC. A broad endothermic peak from 40°C to 80°C is noted in total heat flow. This peak cannot be observed in the heat capacity component; therefore, from this, vaporization of trace moisture or volatilization of residual solvent is estimated. Furthermore, the overlapping endothermic and

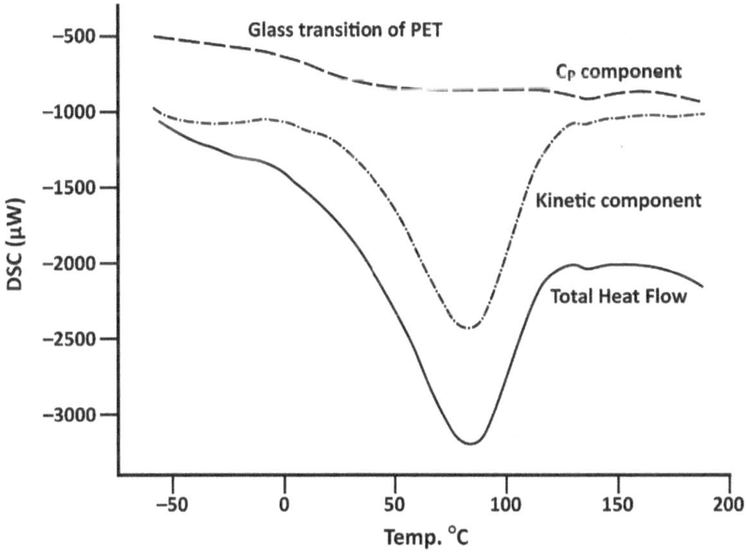

FIG. 2.30 TMDSC measurements of nylon fiber.

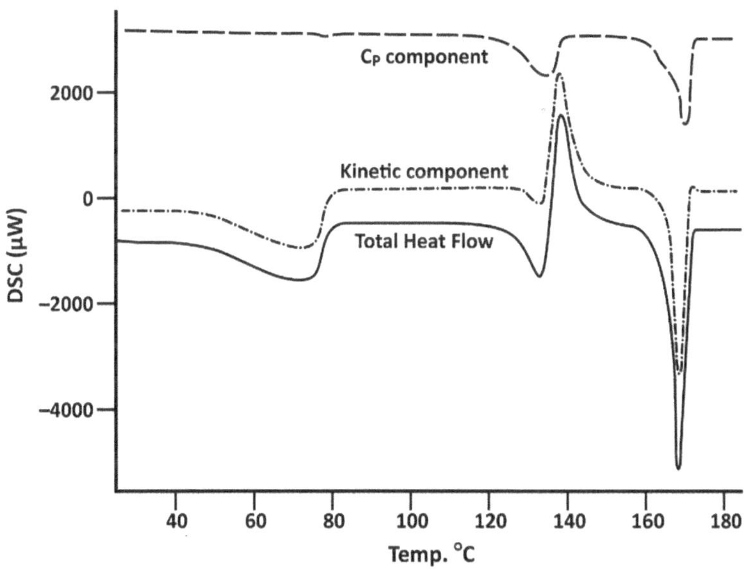

FIG. 2.31 TMDSC measurement of crystal polymorphism.

exothermic peaks at around 130°C together with an endothermic peak at 167°C are noted. Endothermic and exothermic actions near 130°C express the fusion of the metastable phase continuing transitions to stable crystals and the endothermic peak at 167°C is considered a fusion of stable crystals. The endothermic peak is noticed and exothermic peak is not observed by the heat capacity component; therefore, the endothermic phenomenon that can be seen in the total heat flow is melting, and the exothermic phenomenon is thought to capture crystallization, which corroborates investigations mentioned earlier [61].

7.7 Heat capacity for solid-state polymers

Heat capacity c_p is a necessary quantity not only for the evaluation of the integral thermodynamic properties, such as enthalpy (*H*), entropy (*S*), and free enthalpy (Gibbs function) (*G*) (Chapter 1), but also for the precise separation of latent heats of transition from motional energies, for the determination of glass transitions, description contents of phases of polymers, such as determination degree of crystallinity, mobile and rigid amorphous phases, etc. Experimentally, in the low-temperature region, heat capacity measurements are dominated by a low-temperature AC calorimeter and a quantum design product for physical properties measurement system (PPMS) [62]. The PPMS is an automated low-temperature and magnet system used for the measurement of material properties such as specific heat capacity. For higher region of temperature, DSC, TMDSC, and FSC have gained increasing prominence. Typical precision of measurements is now 3% in the temperature range above 10 K for both solid and liquid heat capacities, although modern methods of a TMDSC may increase the precision to levels as high as 0.1% by eliminating heat losses through measuring the change in the calorimeter response as a function of frequency response of the calorimeter (TMDSC) [63].

Below 10 K, the heat capacity c_p, although of interest for the theory of motion, does not contribute much to the integral *H* at higher temperature. It depends on the physical state (crystalline, semicrystalline, or amorphous), whereas c_p of solids, from about 50 K to the glass transition temperature T_g, is largely independent of the physical structure. The theory of the heat capacity of solids is well established [62]. In case of solids, we consider vibration of atoms, ordering of atoms (such as defects and crystal lattice), and conduction of electrons. However, the vibrations of atoms are the most significant point. The heat capacity is largely of vibrational origin (specially below glass transition temperature, T_g) and to a great extent described by the harmonic oscillator approximation. For example, Pyda et al. preformed a study of the glass transition of an amorphous and a semicrystalline poly(lactic acid) (PLA-H) using DSC and TMDSC [64]. They evaluated the reversing, total, and nonreversing apparent heat capacities of samples with different contents of L- and D-lactic acid and with various thermal histories (Figs. 2.32 and 2.33).

7.8 Heat capacity for liquid-state polymers

The liquid heat capacity of polymeric materials can be measured above the glass transition or the melting transition temperature. Quantitatively, knowledge of liquid heat capacity of polymers is rather poor [65] and needs still more research

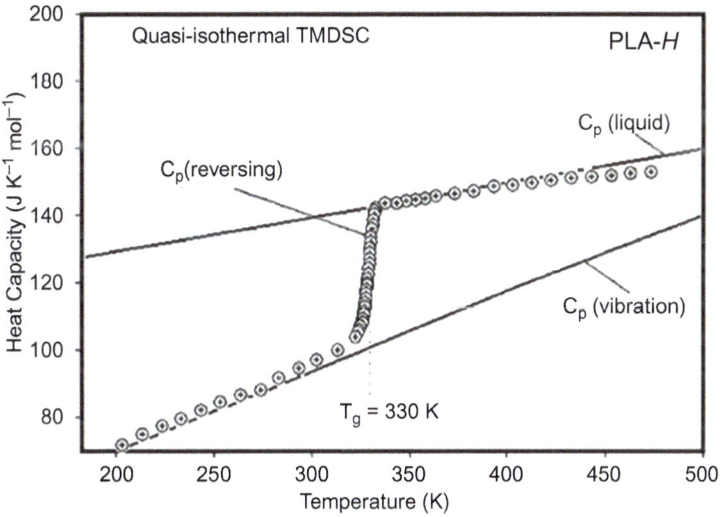

FIG. 2.32 Reversing molar heat capacity C_p of amorphous PLA-H by quasiisothermal TMDSC in the glass transition region (amplitude of modulation 0.5 K and period 100 s) [64].

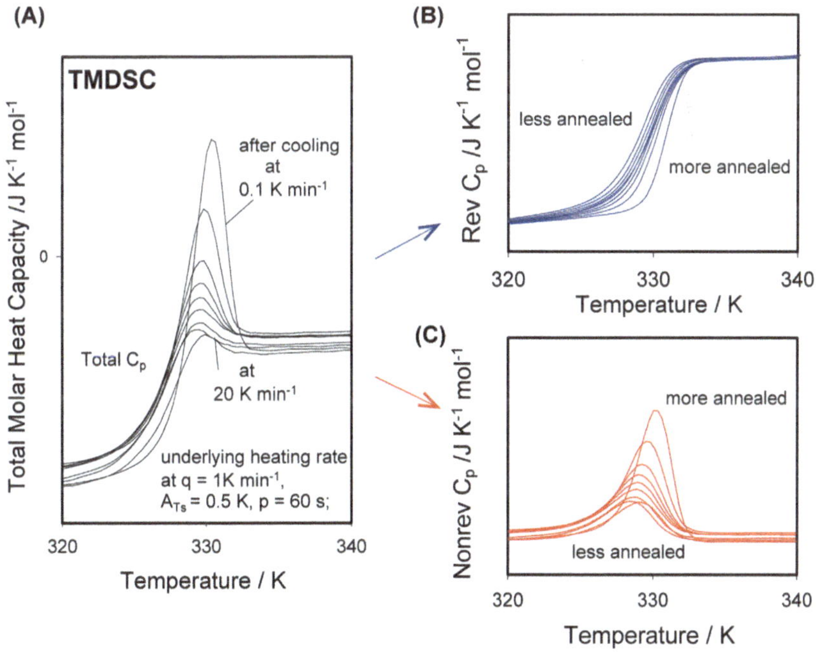

FIG. 2.33 (A) Total, (B) reversing, and (C) nonreversing C_p of PLA-H with different thermal histories by TMDSC in the glass transition region on heating [64].

and description at the microscopic molecular level. Empirically, it was often observed that the liquid heat capacity is a linear function of temperature. For example, for poly(N-isopropylacrylamide) (PNIPA), the liquid heat capacity can be calculated from the following relation [66]:

$$C_p^{liquid}(calc) = N_{CHCH_2}(10.4 + 0.1314T) + N_{CONH}(113.13 - 0.0284T) + N_{CH(CH_3)_2}(18.79 + 0.2013T) \quad (2.28)$$

where N_{CHCH_2}, N_{CONH}, and $N_{CH(CH_3)2}$ are the numbers of the respective groups in the repeating unit of PNIPA (Table 2.1). Based on Eq. (2.28) ($N_{CHCH_2}=1$, $N_{CONH}=1$, and $N_{CH(CH_3)2}=1$), the calculated liquid heat capacity for PNIPA was found to be (Fig. 2.34)

$$C_p^{liquid}(calc) = 0.3043T + 142.32 \quad (2.29)$$

Fig. 2.34 shows a comparison between this calculated liquid molar heat capacity, C_p^{liquid}(calc), and experimental data. The experimental liquid heat capacity of fully amorphous PNIPA was approximated by a linear regression and expressed as [66]

$$C_p^{liquid}(exp) = 0.3272T + 132.84 \text{in} JK^{-1} mol^{-1} \quad (2.30)$$

TABLE 2.1 Liquid heat capacities of different structure group of polymers.

Structure group of polymers	C_p (liquid) of group (J $K^{-1}mol^{-1}$)
Methylene (—CH_2—)	17.91 + 0.0411 *T* or 13.91 + 0.0487 *T*
Dimethylmethylene [—$C(CH_3)$—]	18.79 + 0.2013 *T*
Carbonyl (—CO—)	32.73 + 0.07119 *T*
Carboxyl [—(CO)O—]	64.32 + 0.002441 *T*
Carbonate [—O(CO)O—]	84.54 + 0.06446 *T*
Phenylene (—C_6H_4—)	73.13 + 0.1460 *T*
Naphthylene (—C_6H_{10}—)	114.49 + 0.2527 *T*
Dimethylphenylene [—$C_6H_2(CH_3)_2$—]	111.41 + 0.2378 *T*
Vinyl (—$CHCH_2$—)	10.4 + 0.1314 *T*
Amide (—CONH—)	18.79 + 0.2013 *T*
Oxygen (—O—)	28.13 − 0.00711 *T*
Selenium (—Se—)	52.408 − 0.049766 *T* + 0.000032608 T^2
Sulfur (—S—)	46.59 − 0.02028 *T*

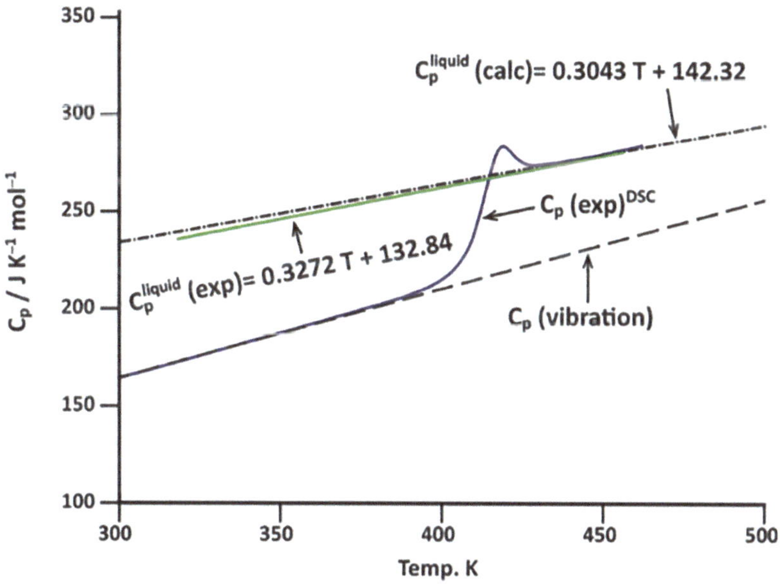

FIG. 2.34 Comparison of the calculated heat capacity of solid and liquid PNIPA and experimental data. The experimental data of apparent heat capacity, molar heat capacity C_p (exp) DSC, was performed by standard DSC ($\beta = 10\,K/min$) [66].

This was then compared to the estimated linear contributions of polymers that have the same constituent groups using the addition scheme developed for the Advanced Thermal Analysis System (ATHAS) Polymer Heat (Table 2.1). Additionally, the evaluation of liquid heat capacity of polyethylene based on vibration, C_v (vibration), conformation C_v (conf), and anharmonic, C_v (ext) contributions is shown in Fig. 2.35.

7.9 Enthalpy (*H*), entropy (*S*), and Gibbs function (*G*)

All integral thermodynamic functions such as enthalpy (H), entropy (S), and Gibbs function (G) can be obtained through the thermodynamic relationships mentioned in Chapter 1 through knowing equilibrium solid and liquid heat capacities and transition parameters of linear polymers. For perfect crystalline solids, the entropy at 0 K is zero. Thus, absolute entropies can be calculated directly. For crystalline, linear macromolecules, the derived thermodynamic functions, are reported previously [57, 69, 70]. For amorphous solids, the residual entropy at 0 K can be calculated by equating the entropy at the equilibrium melting temperature.

An analysis of many macromolecules indicates that the residual entropy of amorphous macromolecules at absolute zero (S_0^a) is 2.4 J K^{-1} mol^{-1} per mobile bead, as listed previously [57, 69, 70]. With the knowledge of S_0^a, the absolute

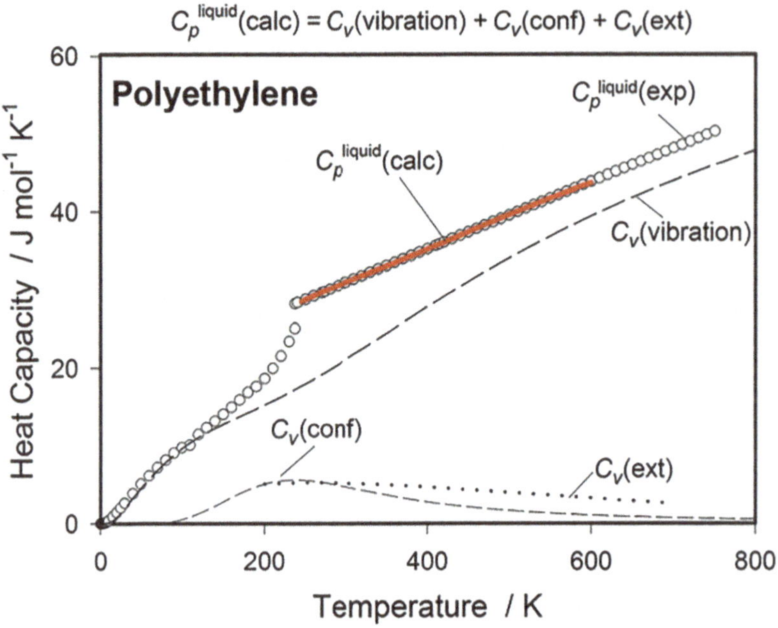

FIG. 2.35 Evaluation of liquid heat capacity of polyethylene based on vibration, C_v (vibration), conformation, C_v (conf), and anharmonic, C_v (ext) contributions [67, 68].

entropy of amorphous solids can be calculated. The crystalline state at 0 K is chosen as a standard state for all enthalpy and free enthalpy data [57]. The functions $(H_T^a \cdot H_0^c)$ and $(H_T^c \cdot H_0^c)$ are directly related to the heats of fusion at temperature T. From the absolute entropy and $H_0^a \cdot H_0^c$, the free enthalpy (Gibbs function) $G_0^a \cdot H_0^c$ can be calculated for amorphous solids. Fig. 2.36A–D show the changes in the thermodynamic functions: enthalpy, entropy, and Gibbs energy of PE, PNIPA, poly(3-hydroxybutyrate) (P3HB), and PLA with temperature [57].

7.10 Thermal characterization of stabilized amorphous selenium

Stabilized amorphous selenium is a prototype inorganic polymer-type glass that shows glass aging occurring over the time scale of hundreds of hours at room temperature. It thus allows the aging effects on the T_g measurements to be studied over a reasonable time scale. Fig. 2.37 shows the dependence of T_g as measured by TMDSC and DSC on the aging period. T_g measured using TMDSC was from the inflection point, and on the other hand, T_g measured by DSC was from the usual endothermic T_g peak onset definition that is common in the literature. The sample was first heated at 20°C min^{-1} to a temperature above T_g but below the crystallization temperature, e.g.,70–80°C, equilibrated, and then cooled down to room temperature at 5°C min^{-1}. It was then aged at room temperature

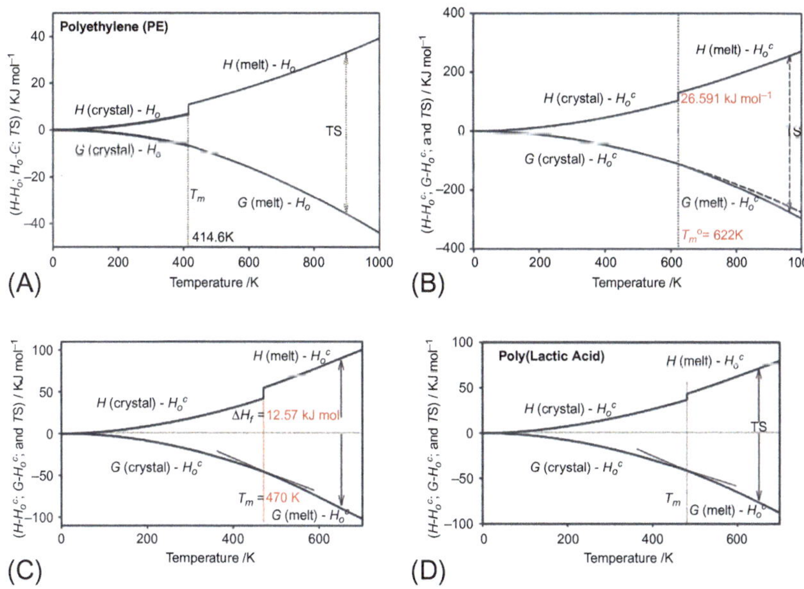

FIG. 2.36 (A) Enthalpy, entropy, and free enthalpy of polyethylene as derived from heat capacities. The origin of H and G has been set arbitrarily to zero at absolute zero in temperature, and the energy equivalent TS is plotted instead of S. (B) Enthalpy $(H\cdot H_{\mathrm{o}}^{\mathrm{c}})$, Gibbs energy $(G\cdot H_{\mathrm{o}}^{\mathrm{c}})$, and entropy contribution (TS) of PNIPA using the vibrational, solid heat capacity, C_{p} (vibration), and the liquid heat capacity $C_{\mathrm{p}}^{\mathrm{liquid}}$. $H_{\mathrm{o}}^{\mathrm{c}}$ is the constant enthalpy. (c) Enthalpy $(H\cdot H_{\mathrm{o}}^{\mathrm{c}})$, Gibbs free energy $(G\cdot H_{\mathrm{o}}^{\mathrm{c}})$, and entropy contribution (TS) for P3HB using the vibrational, solid heat capacity, C_{p} (vibration), and the liquid heat capacity $C_{\mathrm{p}}^{\mathrm{liquid}}$. (d) Enthalpy $(H\cdot H_{\mathrm{o}}^{\mathrm{c}})$, Gibbs free energy $(G\cdot H_{\mathrm{o}}^{\mathrm{c}})$, and entropy contribution (TS) for PLA using the vibrational, solid heat capacity, C_{p} (vibration), and the liquid heat capacity $C_{\mathrm{p}}^{\mathrm{liquid}}$ [57].

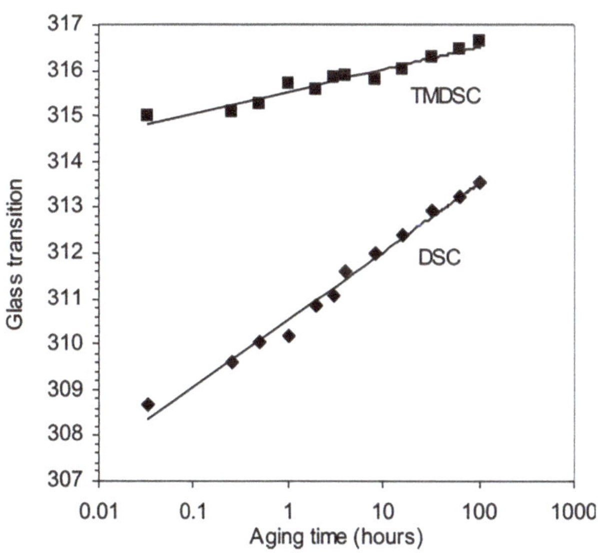

FIG. 2.37 T_{g} dependence on aging for DSC and TMDSC methods.

for the aging times shown in Fig. 2.37. The underlying heating scan was 1°C min^{-1} in both techniques. Over an aging time of 100h, TMDSC T_g changes by about 1°C whereas DSC T_g changes by about 5°C. The endothermic relaxation that usually occurs during heating through the glass transition normally becomes larger as the glass is aged longer. This endothermic relaxation peak in the conventional DSC makes the determination of T_g based on the onset definition highly sensitive to aging. Most significantly, there is a genuine relaxation effect on T_g itself in DSC heating scans, because the T_g definition in the DSC case depends on the starting structure and the initial enthalpy, $H(T_{o,t})$, which changes with aging time [71, 72].

7.11 Thermal characterization of arsenic triselenide (As_2Se_3)

Glass transition temperature of As_2Se_3 glass as observed in DSC scans is well known to depend on thermal history. The classical well-known example is the fact that the T_g value determined by DSC under a heating scan at a given rate r is different than that determined under a cooling scan at a rate $q = r$. Furthermore, T_g determined in a DSC heating scan also depends on aging. Fig. 2.38 shows the clear difference between T_g values observed in heating and cooling scans in DSC experiments; T_g values from $T_{g\text{-onset}}$ and $T_{g\text{-inflection}}$ definitions are different in heating and cooling scans. In contrast, T_g values determined in TMDSC scans, whether by a heating or cooling scan, as shown in Fig. 2.39, are the same for a given modulation period P, within experimental errors. T_g depends only on the modulation frequency ω ($= 2\pi/P$) provided that the underlying heating or cooling scan is "slow," usually less than 1°C min^{-1}.

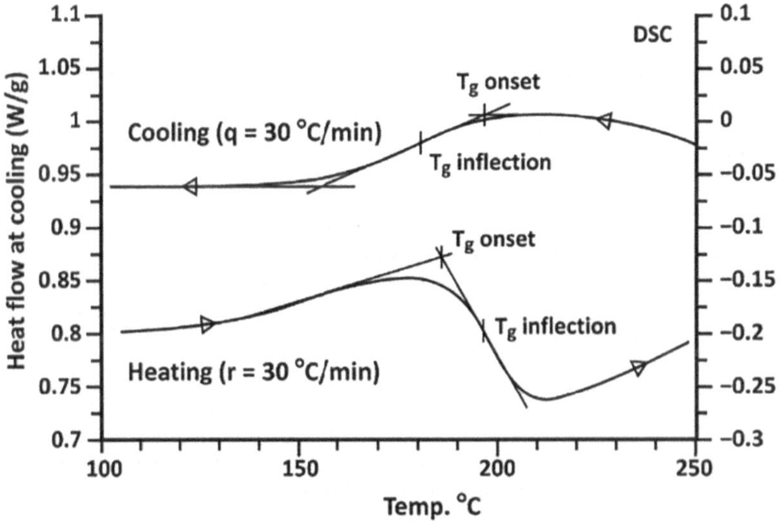

FIG. 2.38 Typical DSC heating and cooling scans on As_2Se_3 glass.

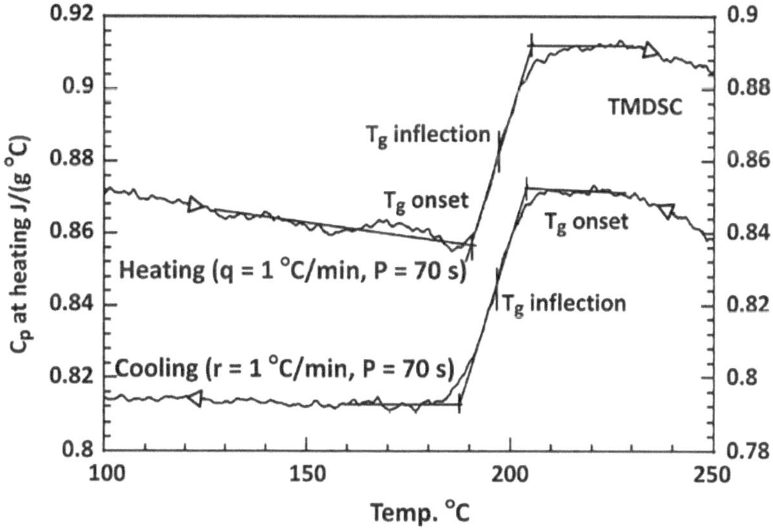

FIG. 2.39 Typical TMDSC heating and cooling scans on As_2Se_3 glass [71].

7.12 Thermal characterization of polyethylene terephthalate

The conventional DSC heating and cooling scans for PET samples are shown in Fig. 2.40 in which two heating schedules, labeled HA and HB, with different thermal histories are shown to highlight the dependence of T_g on the thermal history. In HA, the sample is equilibrated at 0°C and then heated at 10°C min^{-1} and in HB it is equilibrated at 50°C, and then heated at 10°C min^{-1}. The two T_{gs}

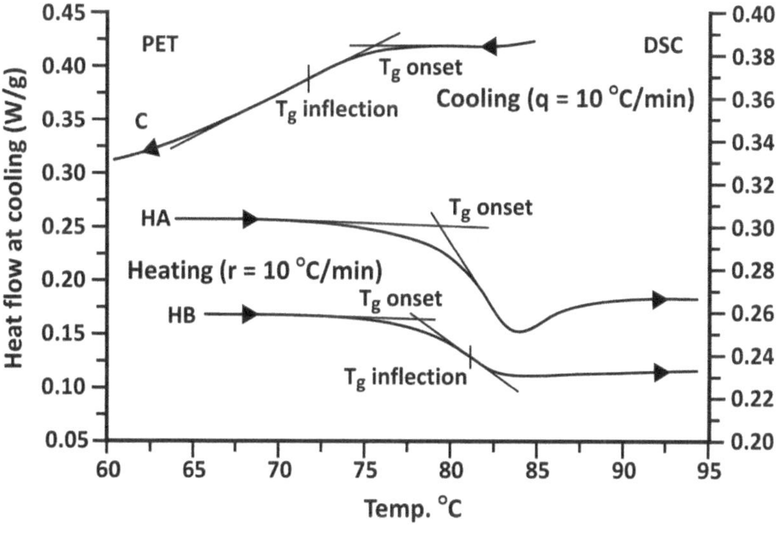

FIG. 2.40 DSC T_g measurements on PET (C: cooling; HA and HB: heating) [71].

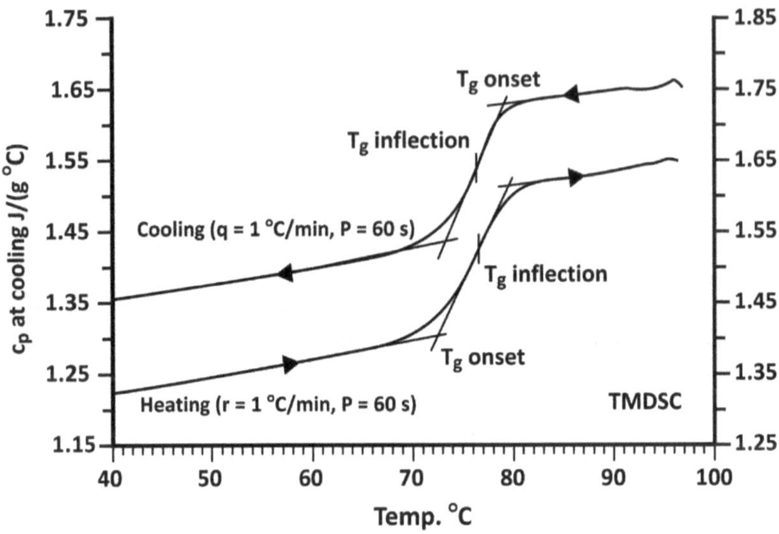

FIG. 2.41 TMDSC T_g measurements on PET [71].

from onset are different. HA exhibits an enthalpic relaxation. The TMDSC results are shown in Fig. 2.41 where c_p versus temperature scans in TMDSC are the same irrespective of whether the experiment is a heating or a cooling scan, and hence independent of the initial state [71].

Table 2.2 provides a useful summary of the T_g measurements from thermal analysis (TA) using DSC and TMDSC experiments. $T_g(I)$, corresponding to the inflection point has 10°C difference between heating and cooling DSC scans with $r = q$, but the difference is less than 0.5°C in the case of TMDSC heating and cooling scans. Clearly, the TMDSC provides a distinct advantage in providing T_g values that are independent of the underlying heating or cooling rate, and independent of the starting point of the scan, that is, thermal history [71].

TABLE 2.2 PET comparison of T_g measurements. $T_g(I)$ = T_g at inflection point; $T_g(O) = T_g$ at onset; r = heating and q = cooling, both in °C min^{-1} [71].

T_g	$T_g(I)$	$T_g(I)$	$T_g(I)$	$T_g(I)$	$T_g(O)$ HB	$T_g(O)$ HA	$T_g(O)$
TA	TMDSC	TMDSC	DSC	DSC	DSC	DSC	DSC
	$r = 1$	$q = 1$	$r = 10$	$q = 10$	$r = 10$	$r = 10$	$q = 10$
°C	76.16	76.05	81.41	71.51	77.80	80.73	75.29

7.13 Thermal characterization of polycarbonate (PC)

Both the DSC and TMDSC experiments for polycarbonate samples were like those on PET in terms of experimental conditions (r, q, P, and A). Fig. 2.42 shows the TMDSC results obtained with heating and cooling scans where it is apparent that the two C_P vs. T characteristics through glass transition are almost identical; the same T_g values as obtained from the inflection point, and independent of the starting state [71].

Table 2.3 provides a useful summary of the T_g measurements from DSC and TMDSC experiments. $T_g(I)$, corresponding to the inflection point, has 12°C difference between heating and cooling DSC scans with $r = q$, but the difference is less than 0.25°C in the case of TMDSC heating and cooling scans [71].

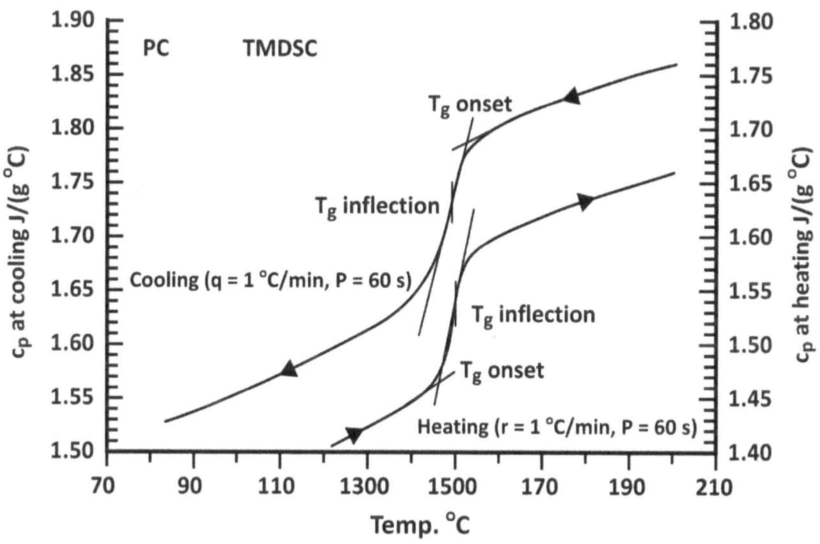

FIG. 2.42 TMDSC T_g measurements on polycarbonate [71].

TABLE 2.3 PC comparison of T_g measurements. $T_g(I) = T_g$ at inflection point; $T_g(O) = T_g$ at onset; r = heating and q = cooling rates, both in °C min^{-1} [71].

T_g	$T_g(I)$	$T_g(I)$	$T_g(I)$	$T_g(I)$	$T_g(O)$	$T_g(O)$
TA	TMDSC	TMDSC	DSC	DSC	DSC	DSC
	$r = 1$	$q = 10$	$r = 10$	$q = 10$	$r = 10$	$q = 10$
°C	149.47	149.23	154.00	142.12	149.23	150.11

7.14 Nonisothermal crystallization and stepwise quasiisothermal crystallization of polymers upon heating

Polymer crystallization that occurs above the glass transition temperature (T_g) upon heating is called "cold-crystallization" process. This type of crystallization occurs at temperatures well below the melting region, far from equilibrium, and due to high nucleation density, it is generally fast. For these reasons, this process may not be modulated, and commonly it is observed as a fully irreversible phenomenon in TMDSC curves. An example is reported in Fig. 2.43, which shows the comparison between the $c_{p,app}$ curve from standard DSC and the $c_{p,rev}$ curves at two different frequencies of temperature modulation for an initially amorphous poly(3-hydroxybutyrate) (P3HB) sample [73]. Above the glass transition temperature, which is located around 0°C, in the temperature range 25–45°C, the $c_{p,rev}$ curves overlap the apparent specific heat capacity ($c_{p,app}$) data, as well as the thermodynamic specific heat capacity of liquid P3HB, attesting that in this interval no processes involving latent heat take place. Around 30–45°C, simultaneously with the appearance of the cold crystallization peak in the $c_{p,app}$ curve, the $c_{p,rev}$ value suddenly decreases.

The $c_{p,rev}$ curves overlap in the interval 45–75°C, which proves that the reversing specific heat capacity is independent of the modulation frequency in this temperature range, and that no reversing latent heat is exchanged. This means that, in this temperature range, crystallization and melting do not take place consecutively in the two semiperiods, respectively, and that the $c_{p,rev}$

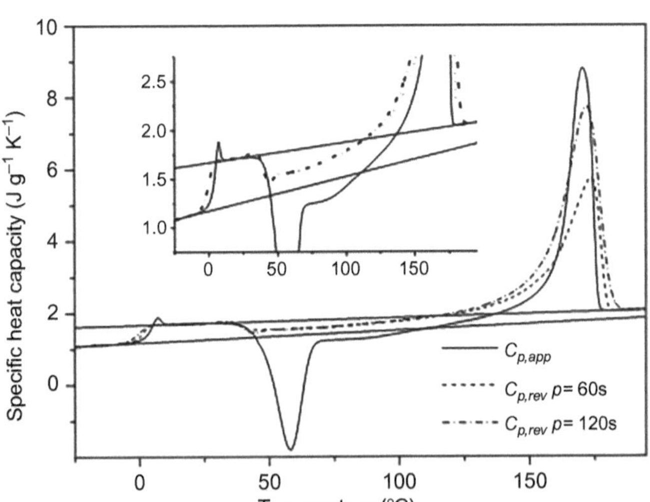

FIG. 2.43 Apparent specific heat capacity ($c_{p,app}$) by DSC and reversing specific heat capacity ($c_{p,rev}$) of poly(3-hydroxybutyrate) (PHB) measured by TMDSC after cooling from the melt at 200°C min^{-1} at modulation periods $p = 60$ and 120 s, temperature amplitude (AT) = 1.0°C, and cooling rate $q = 2$°C min^{-1}. The thin solid lines are the solid and liquid specific heat capacities, as taken from Ref. [73, 74]. The inset is an enlargement of the figure in the cold crystallization region.

curve corresponds to the baseline specific heat capacity. In fact, the general rule for the interpretation of the TMDSC curves is that when crystallization and melting latent heats do not contribute to the reversing specific heat capacity, i.e., when the excess specific heat capacity is zero, $c_{p,rev}$ is independent of the modulation frequency and corresponds to the baseline specific heat capacity of the crystallization process. The evolution of crystallinity can be thus followed by changes in $c_{p,rev}$, and the rapid decrease in $c_{p,rev}$ around 30–45°C has to be ascribed to an increased solid fraction of P3HB, which develops during an irreversible process that occurs continuously both in the heating and cooling semiperiods with the same rate. In contrast, above 75°C, the frequency dependence of the $c_{p,rev}$ curves attests that thermal processes involving reversing exchanges of latent heat, typical of polymer melting, take place in this temperature range.

A second example of cold crystallization studied by TMDSC is shown in Fig. 2.44. The cold crystallization was monitored during heating through a stepwise "quasiisothermal" method. Every 2°C, a quasiisothermal experiment of 20 min was performed, by oscillating the temperature around a fixed value. Also in Fig. 2.44, as in Fig. 2.43, the $c_{p,rev}$ curve of a melt-quenched, amorphous sample of PET exhibits a sharp decrease around 100°C, simultaneously with the cold-crystallization process. After cold crystallization, the crystallinity amount, determined by standard DSC, was 40%. For a comparison, Fig. 2.44 also illustrates the $c_{p,rev}$ curve of a semicrystalline PET sample with similar crystallinity degree, obtained with the same TMDSC experimental procedure.

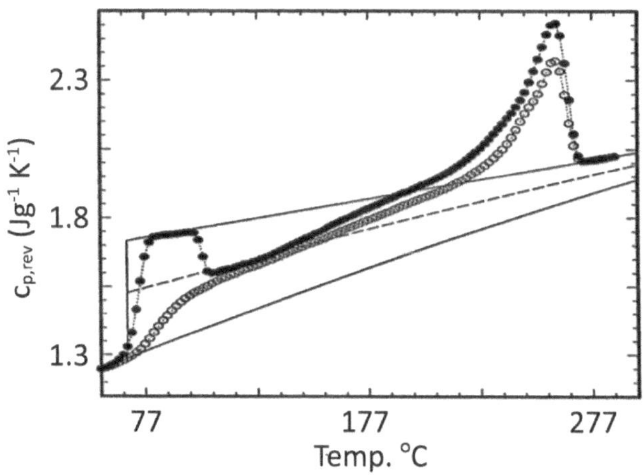

FIG. 2.44 Reversing specific heat capacity ($c_{p,rev}$) of a melt-quenched amorphous PET sample (filled circles) and a melt-crystallized PET sample (open circles) with crystallinity degree of 44%, measured by quasiisothermal TMDSC upon stepwise heating ($p = 60$ s, AT = 1.0°C, and oscillation time around each temperature: 20 min). The thin solid lines are the solid and liquid specific heat capacities, as taken from Refs. [3, 75]. The broken line is the calculated specific heat capacity for a 40% crystalline PET [76].

7.15 Thermal conductivity of polymers

Thermal conductivity is an intrinsic material property. Solid conductors (such as metals) typically have thermal conductivities in the range 10–400 W °C^{-1}, while insulators (such as polymers, glasses, and ceramics) have values in the range 0.1–2 W °C^{-1}. Furthermore, thermal conductivity changes as a weak function of temperature and rarely changes by a factor of 10 within a general class of materials. Determination of a material's thermal conductivity is important in evaluating its utility for specific applications so far. Thermal conductivity can be measured using several different instrumental techniques, one of these is based on DSC, which measures heat flow into or out of a material as a function of temperature or time. DSC is primarily used to measure transition temperatures and associated heats of reaction in materials, particularly polymers. Measurement of glass transition temperature, melting point, %crystallinity, degree of cure, decomposition temperature, and oxidative stability are specific examples of some of the more common DSC measurements for polymers as well. The most widely used approach for making DSC measurements is the heat flux DSC, in which the sample and reference materials (usually contained in metal pans) are placed on a thermoelectric disk inside a temperature programmed environment (Chapter 1). Heat flow in this approach is measured using the thermal equivalent of Ohm's Law where $dQ/dt = dT/R$ (Q is the heat, t is the time, T is the temperature, and R is the thermal resistance of thermoelectric disk).

Several researchers [77-80] have modified heat flux DSCs to measure the thermal conductivity of insulating materials such as thermoplastic solids, elastomers, thermoplastic melts, and pyrotechnics. Recently, a test specimen is placed in the DSC cell in contact with the sample platform. The DSC sensor measures both the temperature of one side of the specimen and the heat flow into it. A heat sink of known temperature is constructed to contact the opposite side of the test specimen. From the recorded heat flow and the temperature difference between the DSC cell and the heat sink (along with the test specimen dimensions), thermal conductivity can be calculated using the equation ($dQ/dt = -kA\ dT/dx$) where Q is the heat (J), t is the time (s), k is the thermal conductivity (W °C^{-1} m^{-1}), T is the temperature (°C), x is the height of test specimen (m), and A is the cross-sectional area of test specimen (m^2). This DSC measurement of thermal conductivity works well but requires modification of the commercially available DSC cell, as well as very careful attention to experimental details. TMDSC as a recent extension of traditional DSC minimizes those limitations.

Direct measurement of heat capacity by TMDSC is of primary interest in thermal conductivity measurement because heat capacity and thermal conductivity are related properties. TMDSC users have observed that the best heat capacity results are obtained when experimental conditions are selected to obtain maximum temperature uniformity across the test specimen. Small thin specimens, long oscillation periods, and complete encapsulation of the test specimen in sample pans of high conductivity (aluminum has a conductivity of about 235 W °C^{-1}) produce the best results. When test conditions lie outside of these guidelines, the value of the measured heat capacity declines. This is thought to be due to the thermal conductivity of the material preventing uniform temperature conditions across the test specimen. Alternatively, the effect of the specimen's thermal conductivity may be maximized through the use of thick test specimens and the use of open sample pans, which results in the application of the temperature oscillation to only one side of the test specimen [81].

For the determination of accurate heat capacity values from TMDSC, a precise calibration, which takes the falsification due to heat transfer problems into account, is needed. In contrast, knowing the influence of the heat transfer on the heat capacity data allows obtaining information about thermal conductivity of the sample and thermal contacts between sample and the measuring system. Therefore, it seems to be an attractive task to use TMDSC as a tool for thermal conductivity measurements of low thermal conductivity materials such as polymers. As reported first by Marcus et al. [16, 81], thermal conductivity can be measured without modification of the commercially available DSC cell. They calculated the thermal conductivity from the ratio of apparent and true heat capacities measured for a thick (about 3 mm) and a thin (about 0.5 mm) sample in a temperature modulated mode.

They used one additional calibration step to consider the heat losses through a surrounding purge gas. The method assumes that "the face of the specimen at the heat source follows the applied temperature modulation," which means no thermal resistance between the sample and the furnace. Marcus and Reading [16, 81] also mentioned that thermal conductivity could be easily obtained from a single sample run at several frequencies, i.e., it is not necessary to measure two samples. But what is necessary in any case is to know the heat transfer coefficient. If the heat transfer coefficient is not known, an accurate value of thermal conductivity cannot be obtained.

A new method reported recently describes the dynamic behavior of DSC and TMDSC systems under conditions of an appreciable temperature gradient inside the sample. A question about how to get all the parameters of the model arises. An algorithm was presented to determine the most important parameters: sample heat capacity, effective thermal contact between the sample and the furnace, and sample thermal conductivity for the case of real valued heat capacity and thermal conductivity. Furthermore, experimental results are presented with different solid samples and various experimental conditions. Measured thermal conductivity data are compared with expected values. Finally, possible reasons for some systematic bias between measured and expected thermal conductivity values were discussed [82]. The method for measuring thermal conductivity was applied to composites of low-density polyethylene containing 15 wt% of particulate aluminum (LDPE/Al), resulting from the recycling of Tetra Pak aseptic packaging, and postconsumer PET [82].

The heat capacity (c_p) and thermal conductivity (k) of the composites were measured in a modulated DSC (MDSC 2910, TA Instruments) at five

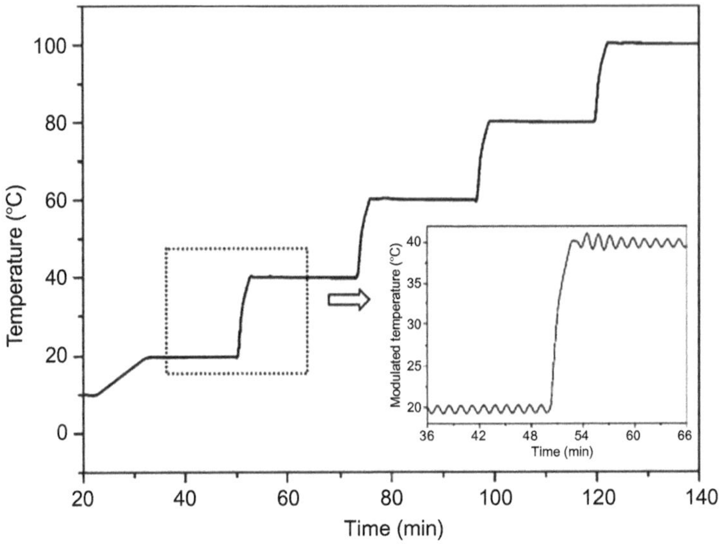

FIG. 2.45 Temperature program used for c_p and k measurements [82].

temperatures: 20°C, 40°C, 60°C, 80°C, and 100°C. Temperature, heat capacity, and thermal conductivity are first calibrated using indium, sapphire, and polystyrene. Cylindrical specimens of polystyrene were also obtained from injection molded samples with the same dimensions as the composites. The experiment consisted in keeping the specimen for 20 min at each temperature of interest, with oscillation amplitude of ±0.5°C within a period of 80 s. The measurements at all the temperatures were attempted in a single DSC run using 5 °C min^{-1} as the heating rate as shown in Fig. 2.45. The experiments and calculations were carried out according to Lopes and Felisberti calculations [82].

The MHF is one of the responses obtained in a TMDSC experiment. The heat flow depends on mass and, therefore, on the dimensions of the specimen. Fig. 2.46 shows how the variation in the specimen thickness affects the heat flow. The heat capacity is obtained from the relationship between the resultant heat flow amplitude and the amplitude of the oscillatory temperature that produces it. The heat capacity is determined using sapphire as standard. The ratio of experimental value and the c_p value of the literature for sapphire is the calibration constant (Kc_p) that is determined for a range of temperatures of interest, assuming a constant value. However, because c_p, and consequently Kc_p, changes with temperature, it is proposed to determine Kc_p for each temperature. Table 2.4 presents Kc_p for the temperatures for which heat capacity and thermal conductivity were determined. The heat capacity of the sample is obtained using thin specimens. These results, as a function of composition, are shown in Fig. 2.47 and Table 2.4. The c_p values correspond to the average of the measurements in the last 10 min of isotherm for each temperature.

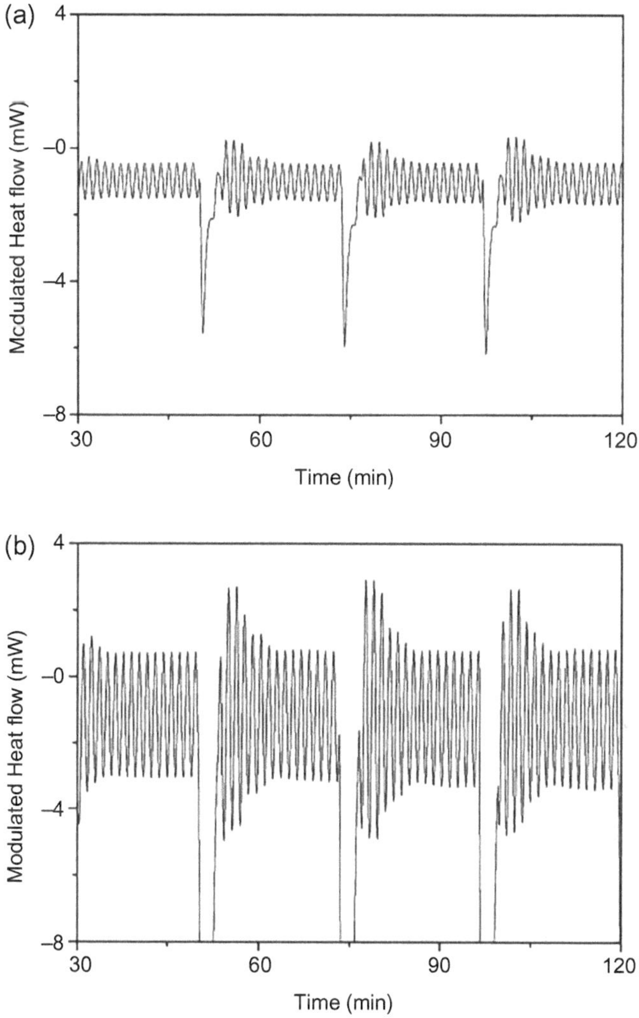

FIG. 2.46 Effect of specimen thickness on the modulated heat flow amplitude: (A) 0.4 mm; (B) 3 mm. Each block corresponds to modulation in a different temperature [82].

The dependence of thermal conductivity on temperature is shown in Fig. 2.48. The values are also listed in Table 2.4. For all the samples, the thermal conductivity decreases with the increase in temperature. This behavior is expected for semicrystalline polymers in the temperature range between glass transition temperature and melting.

The influence of composition on the thermal conductivity can be better visualized as shown in Fig. 2.49. The composites present a complex dependence of thermal conductivity on composition. The thermal conductivity is higher for

TABLE 2.4 Calibration constant (Kc_p), heat capacity (c_p), and thermal conductivity (k) of the composites as a function of temperature and composition.

		c_p ($J g^{-1} °C^{-1}$)					λ ($W m^{-1} °C^{-1}$)				
		% PET					*% PET*				
Temperature (°C)	Kc_p	*0*	*30*	*50*	*70*	*100*	*0*	*30*	*50*	*70*	*100*
20	1.26	2.6567	1.8660	1.7587	1.5326	1.1438	0.24 ± 0.02	0.23 ± 0.02	0.27 ± 0.02	0.15 ± 0.02	0.19 ± 0.01
40	1.35	2.9687	2.0636	1.9035	1.6680	1.2194	0.22 ± 0.02	0.22 ± 0.02	0.26 ± 0.02	0.14 ± 0.02	0.18 ± 0.01
60	1.44	3.3349	2.3010	2.0802	1.8179	1.3038	0.20 ± 0.02	0.20 ± 0.01	0.25 ± 0.02	0.133 ± 0.009	0.18 ± 0.01
80	1.51	3.7716	2.6188	2.3178	2.0969	1.6014	0.19 ± 0.01	0.19 ± 0.01	0.24 ± 0.01	0.106 ± 0.007	0.16 ± 0.01
100	1.63	4.4140	3.0074	2.6097	2.2849	1.6449	0.17 ± 0.01	0.17 ± 0.01	0.21 ± 0.01	0.088 ± 0.006	0.15 ± 0.01

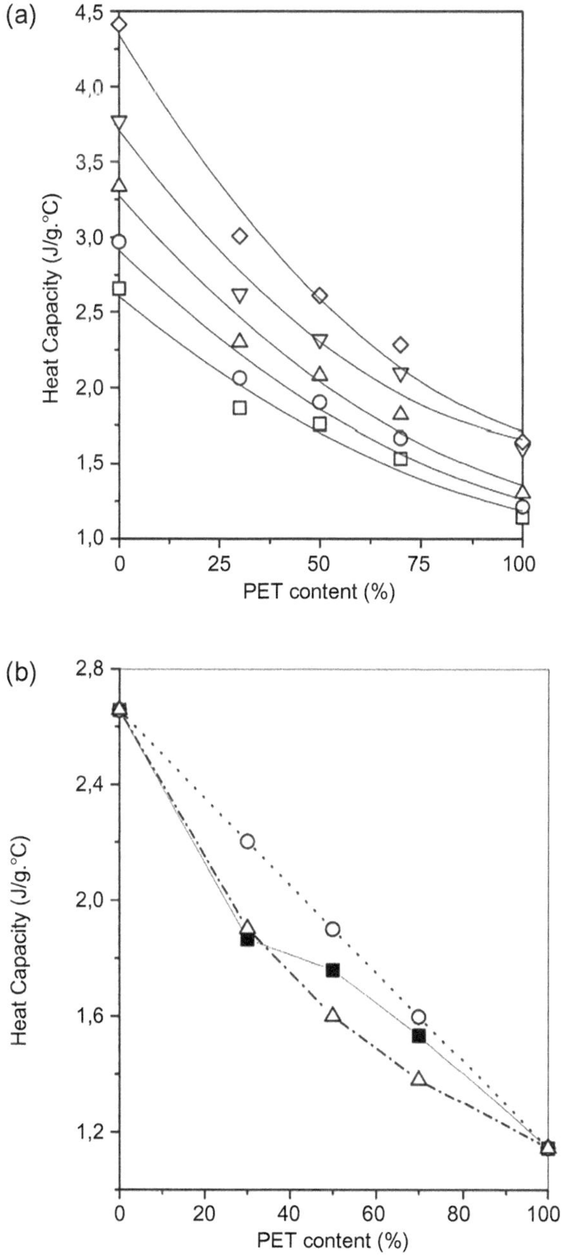

FIG. 2.47 Influence of composition on the heat capacity of the PET/(LDPE/Al) composites [82].

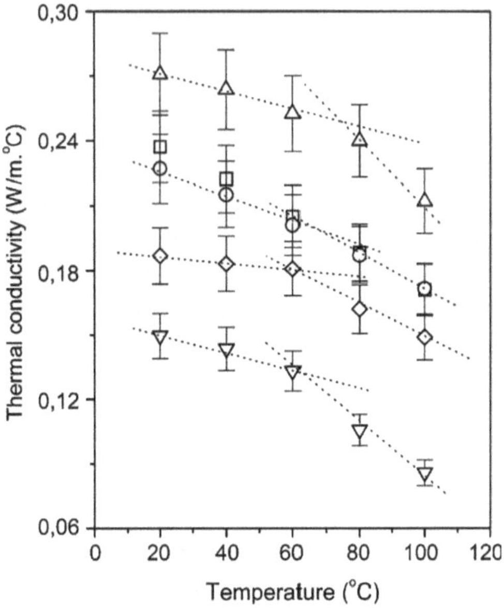

FIG. 2.48 Influence of temperature on thermal conductivity of the PET/(LDPE/Al) composites [82].

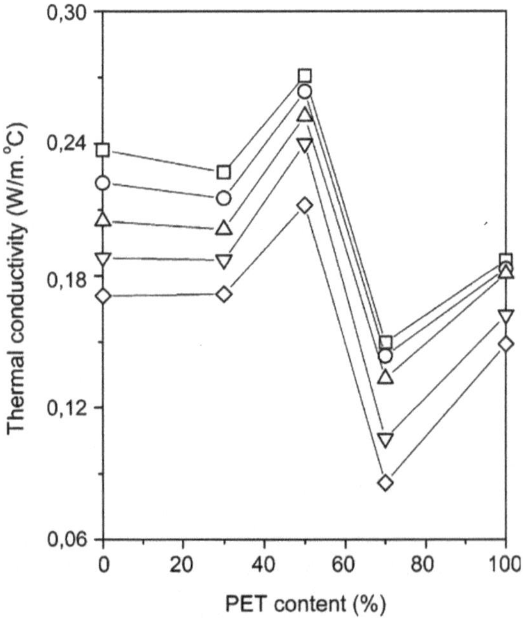

FIG. 2.49 Influence of composition on the thermal conductivity of the PET/(LDPE/Al) composites [82].

LDPE/Al than for PET. In comparison with the constituent components, composites containing 50 and 70 wt% of PET have thermal conductivity higher and lower than both, respectively, while the one with 30 wt% of PET presents an intermediate value. This complex behavior, as well as the profile of the curves of thermal conductivity versus temperature, reflects the morphological differences and the presence of different phases in the composite. The composites PET/(LDPE/Al) present five phases: amorphous and crystalline phase of PET, amorphous and crystalline phase of polyethylene, and aluminum dispersed in polyethylene [82].

8 More applications of the TOPEM technique

8.1 Solid-solid transition in sodium nitrate

Phase transitions are classified into first-order and second-order ones. The difference between first- and second-order phase transitions is that first-order phase transition involves a latent heat with the entropy, volume, and energy of the thermodynamic system change abruptly. Familiar examples are the melting of ice and the boiling of water. In contrast, second-order phase transition is a phenomenon in which effects due to latent heat do not occur and changes occur continuously. Schawe et al. [5] applied TOPEM technique to study the solid-solid transition in sodium nitrate at 275°C.

Sodium nitrate shows a second-order phase transition around 275°C [83] (Fig. 2.50). The transition was investigated by Schawe et al. in their classical

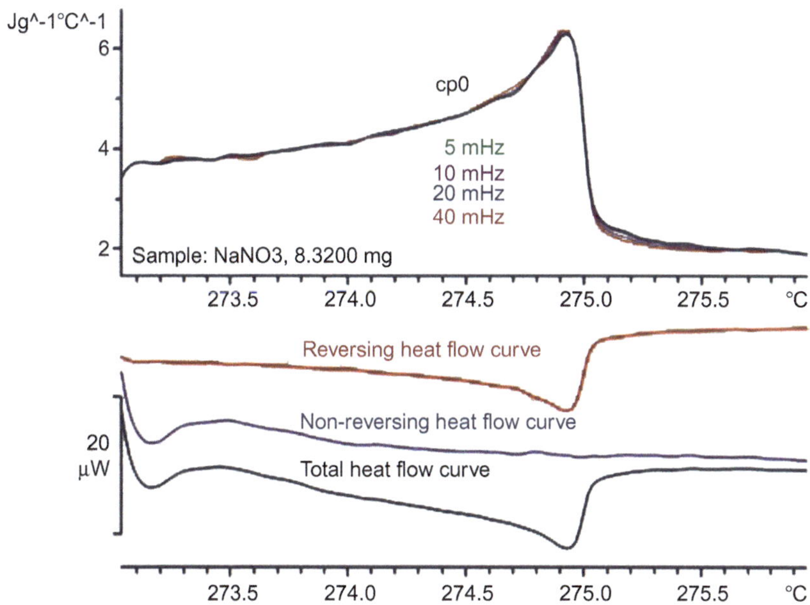

FIG. 2.50 Quasistatic heat capacity $c_{p,0}$ at different frequencies (top) and the total, reversing, and nonreversing (bottom) heat flow curves of the solid-solid transition of sodium nitrate.

paper using TOPEM technique and the frequency dependence of quasistatic heat capacity $c_{p,0}$ was later reported by Mettler-Toledo GmbH [44].

During the phase transition of sodium nitrate, the heat capacity first increases with increasing temperature and then suddenly decreases within 100m $°C^{-1}$ at the critical temperature of about 275°C. To gain information about the physical nature of such transitions, measurements were performed at very low heating rates (20m $°C^{-1}$ min^{-1}) and very small pulse heights (5m $°C^{-1}$). In the nonreversing heat flow curve, the transition occurs without the release of latent heat. This behavior is expected for a pure second-order phase transition.

This proves that TOPEM is clearly able to differentiate between effects that exhibit sensible and latent heat flows. The frequency independence of the transition temperature indicates close molecular dynamics of both sodium nitrate phases.

8.2 Glass transition (T_g) and its frequency dependence

At glass transition temperature, a step-like increase in heat capacity of a polymer is observed since the heat capacity values are affected by disorder in a sample. At T_g, the content of mobile amorphous material increases. Alternatively, the heat capacity decreases on crystallization because the content of mobile amorphous material decreases.

The temperatures of a sample undergoing step-like c_p change at T_g are sensible in nature and show reversible fluctuations depending on the modulation program. Moreover, T_g is a frequency-dependent relaxation effect because of the different molecular dynamics associated with these two c_p steps [84].

The different molecular dynamics associated with these two c_p steps is apparent in their frequency dependence. The glass transition step is shifted to higher temperatures as the modulation frequency increases [84].

The glass transition is thus a relaxation process. Knowledge of its frequency dependence allows information to be obtained about molecular dynamics and permits predictions to be made regarding the stability of the material's structure.

Information about the frequency dependence facilitates the interpretation of unknown events. For example, a glass transition shifts to higher temperatures at higher frequencies.

A sample of technical polystyrene (PS) was measured at a cooling rate of 0.2°C min^{-1}. The frequency-dependent heat capacity at six frequencies between 200 and 5 mHz was then calculated from this single measurement (Fig. 2.51) [44].

One important result is the dependence of the relaxation frequency on temperature as shown in the activation diagram ($\log f$ vs. $1/T$). The frequency range can be extended to low frequencies by increasing the maximum switching time in the TOPEM temperature program. In PET, the glass transition shifts to higher temperatures at higher frequencies. This can be seen in the measurement of PET

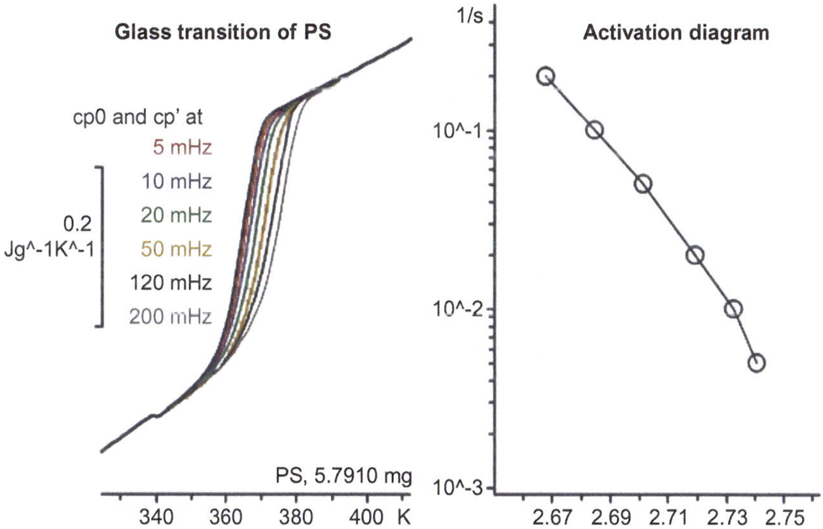

FIG. 2.51 Frequency dependence of heat capacity in PS.

shown in Figs. 2.52 and 2.53. In contrast, with cold crystallization, the jump in the heat capacity is clearly independent of frequency.

Fraga et al. applied TOPEM to study the glass transition of polycarbonate [33]. Typical experimental parameters applied in their TOPEM data collection were as follows:

- The temperature range for the scan was 180–80°C to cover the glass transition interval ($T_g \approx 145$°C).
- The underlying rate, was almost invariably cooling to avoid problems associated with structural relaxation, ranged from −2°C to −0.02°C min^{-1}.
- The amplitude of the temperature pulse took values of −0.5°C, −0.25°C, and −0.1°C.
- The switching time range, which limits the duration of the pulses, had a minimum of 15 s and a maximum of 30 s or any longer time as considered appropriate.

To calculate the response function for the system (other than the sample), there must be a region of the data in which there is no transition; it is for this reason the authors applied cooling curves starting at 180°C, well above the nominal glass transition which is centered at ≈145°C. The TOPEM evaluation is made within a calculation window, the width of which is set by the user. It is recommended that it be less than or equal to one third of the width of the transition interval, with larger windows giving better signals, and has a default value of 120 s. Once the calculation window has been selected, the TOPEM evaluation yields the curve of $c_{p,0}$, the "quasistatic" specific heat capacity. Fig. 2.54

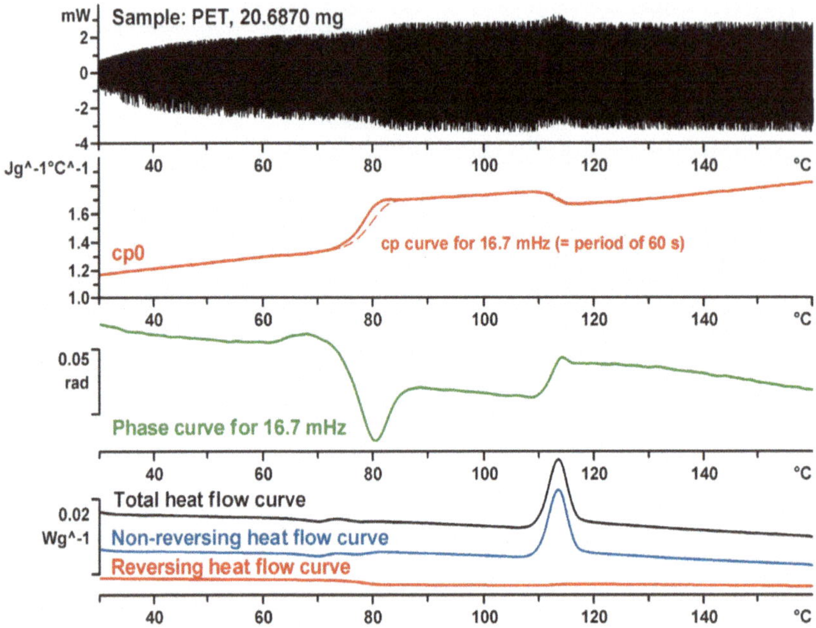

FIG. 2.52 TOPEM curves for PET thermal behavior.

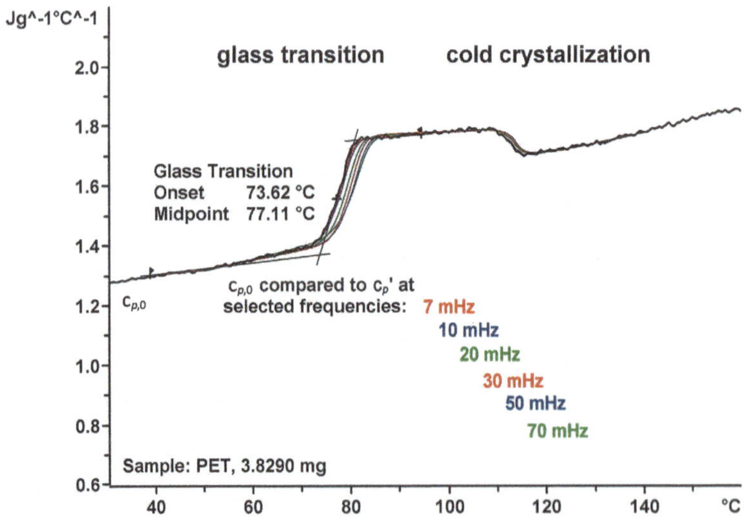

FIG. 2.53 Glass transition and frequency dependence of heat capacity in PET.

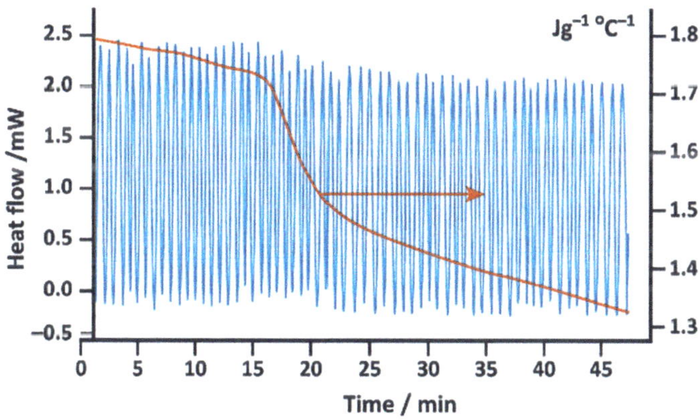

FIG. 2.54 Heat flow response to stochastic temperature pulses of magnitude ±0.25°C during cooling of polycarbonate at −2°C min^{-1} from 180°C to 80°C with switching time range of 15–30 s, from which the "quasistatic" specific heat capacity curve shown is derived. Left-hand axis refers to the randomly alternating heat flow signal; right-hand axis refers to the sigmoidal-shaped specific heat capacity curve [33].

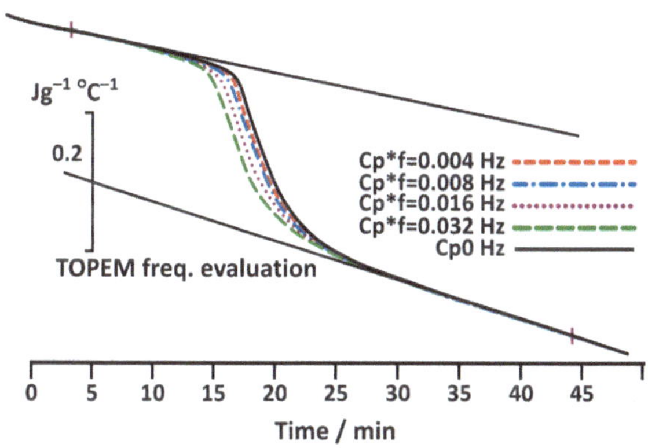

FIG. 2.55 Quasistatic specific heat capacity from Fig. 2.54 (full line) with asymptotes defined by transition interval selected. Lines show the frequency evaluation at selected frequencies of 32, 16, 8, and 4 mHz [33].

illustrates this with the original heat flow modulations from which the $c_{p,0}$ curve has been obtained.

The TOPEM calculated complex specific heat capacity curves at different frequencies of 32, 16, 8, and 4 mHz are shown in Fig. 2.55. The effect of increasing frequency is to reduce the time (increase the temperature) at which the complex specific heat capacity displays its transition. The curve for the low frequency value of 4 mHz is barely visible as it is superposed almost exactly onto the quasistatic specific heat capacity curve. It should be taken into account the quasistatic specific heat capacity is the complex heat as the frequency tends to zero, $c_{p,0} = c_p^*(\omega \to 0)$. From the set of complex specific heat capacity curves at different frequencies, it would be possible to determine, for example, an activation energy from the temperature shift of the curves as a function of the frequency. The curves for all frequencies less than or equal to 4 mHz were reported to be superposed onto the curve for $c_{p,0}$, the quasistatic specific heat capacity. The authors concluded that 4 mHz appears as a kind of limiting frequency for TOPEM [33].

From a practical point of view, this ability of TOPEM to extract multifrequency information from a single experiment is of great importance. It means that thermal events that depend on the frequency, such as the glass transition, can be separated by this criterion from those that are not frequency dependent, such as crystallization and melting. This can be a very useful attribute in situations in which it is difficult to identify unequivocally a certain transition.

Fraga et al. [33] suggested that the so-called "quasistatic" specific heat capacity, $c_{p,0}$, is not equivalent to the dynamic specific heat capacity $c_p(\omega)$ as ω tends to 0 but corresponds to a limiting frequency of about 4 mHz. They suggested that this limit is related to the heating and cooling rates of the stochastic temperature pulses. Comparison of the frequency dependence of the dynamic glass transition as determined by TOPEM with that found by ADSC indicates differences that are attributed to the same effect.

Vitrification of isothermally cross-linked epoxy resins

Vitrification is the transformation of a substance into a glass, that is, a noncrystalline amorphous solid. It is usually achieved by heating materials until they liquidize, then cooling the liquid, often rapidly, so that it passes through the glass transition to form a glassy solid. Certain chemical reactions also result in glasses [44, 85].

In polymerization reactions, the glass transition temperature of the reaction products increases with increasing degree of polymerization. If the glass transition temperature of the reaction product is greater than the reaction temperature, vitrification can occur during the reaction. The viscosity of the reaction mixture then increases markedly, and the reaction practically stops. Vitrification is observed as a step in the heat capacity (sensible heat flow) curve. In conventional DSC, this step is overlapped by the contribution from the reaction enthalpy (latent heat flow).

TOPEM was applied to study the isothermal curing (at 80°C) of diglycidyl ether of bisphenol A (DGEBA) epoxy resin using 4,4′-diaminodiphenyl methane (DDM) as a cross-linking agent (Fig. 2.56). The total heat flow curve shows an exothermic reaction peak. The quasistatic heat capacity ($c_{p,0}$) curve can be simultaneously determined. As the reaction proceeds, $c_{p,0}$ first increases and then decreases in a distinct step due to vitrification. In vitrified material,

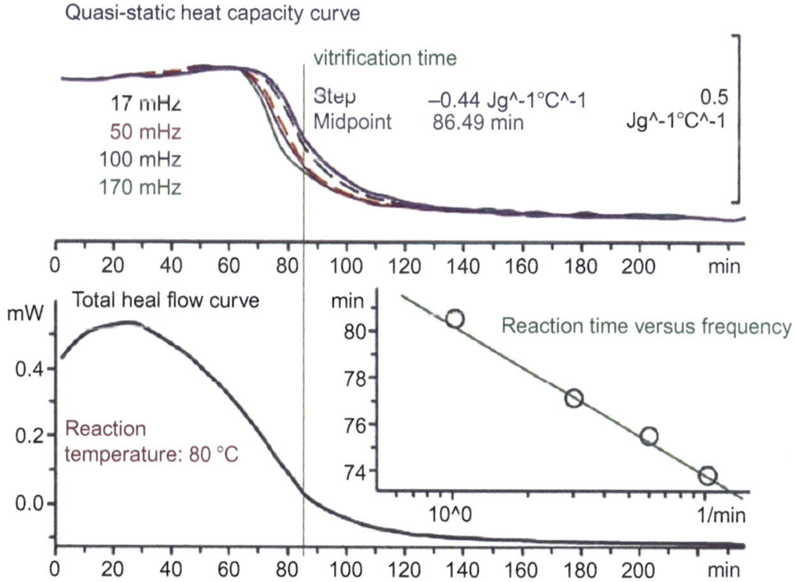

FIG. 2.56 Isothermal curing (at 80°C) of diglycidyl ether of bisphenol A (DGEBA) epoxy resin using 4,4′-diaminodiphenyl methane (DDM) as a cross-linking agent.

diffusion is greatly hindered, and the reaction rate slows down until the reaction almost stops. The vitrification time (86.5 min) characterizes the curing reaction.

Since vitrification is in fact a chemically induced glass transition, the step is also frequency dependent. The multifrequency evaluation shows that the step shifts to shorter times at higher frequencies. An important advantage of this type of evaluation is that all the curves originate from a single measurement of the same sample.

When powder coatings are applied to sheet metal and other surfaces, the particles first soften, then coalesce, and finally cure. The glass transition at 62°C is accompanied by a large relaxation peak (Fig. 2.57). In a DSC measurement, the effect could be interpreted as "melting" and is difficult to evaluate as a glass transition. The reversing and nonreversing heat flow curves in the TOPEM analysis, however, very clearly show that two effects overlap. The phase curve at 110°C indicates the coalescence of the powder particles. This can be simultaneously observed using DSC microscopy. This effect causes a change in the heat transfer in the sample. The total heat flow curve is practically the same as the DSC curve. The enthalpy of curing obtained from the conventional DSC curve is also the same as that from the TOPEM measurement [44].

Recently, Leszczak et al. analyzed the curing reaction of liquid-crystalline epoxy compositions by using TOPEM [86]. Liquid-crystalline epoxy resins belong to a relatively new group of thermosetting materials, and their curing

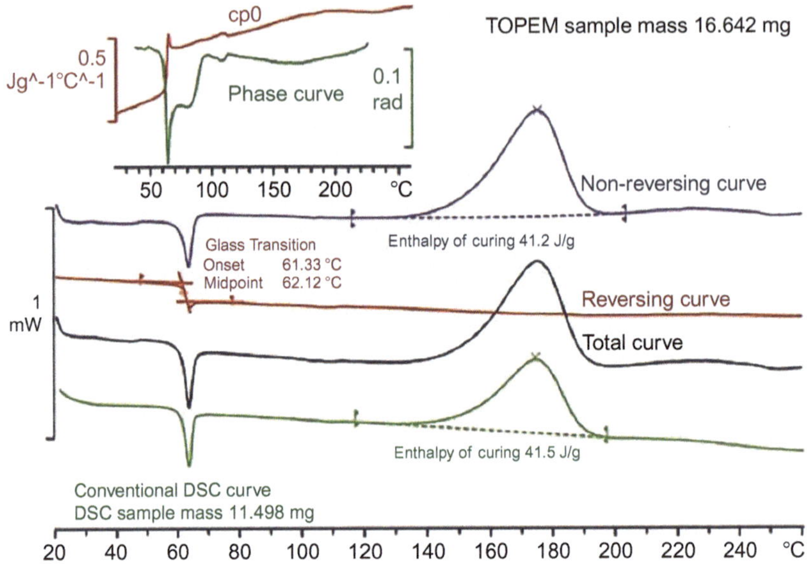

FIG. 2.57 The reversing and nonreversing heat flow curves in the TOPEM analysis of powder coating curing.

process is not yet completely understood. The curing behavior of the composition of a liquid-crystalline diepoxy monomer (LCEM) with the central triaromatic mesogenic group was studied using conventional DSC and TOPEM together with ^{1}H NMR spectroscopy. The stoichiometric amount of primary aromatic diamine, 4,4′-diaminodiphenylmethane (DDM), was used as a curing agent. TOPEM allowed the separation of the thermal effects related to reversible processes (e.g., phase transitions) from irreversible processes (cross-linking reaction). Irreversible curing reaction was shown to occur at the same time as reversible transition of the LCEM from crystal to the liquid-crystalline nematic phase. TOPEM data was substantiated by ^{1}H-NMR analysis of the LCEM/DDM mixture preheated to a temperature from the temperature range of complex thermal changes. The obtained spectrum clearly showed the presence of partially cured products in the investigated samples and additionally proved that during the endothermic phase transition of the LCEM monomer, an exothermic cross-linking reaction occurs. TOPEM method proved very useful for the optimization of the cross-linking conditions of liquid crystalline polymer thermosets and the analysis of the polymer compositions that, for example, require fusion before the proper curing process. Complex heat capacity measured by ADSC provides proof of the different reactivity of primary and secondary amines in amine-epoxy systems [87].

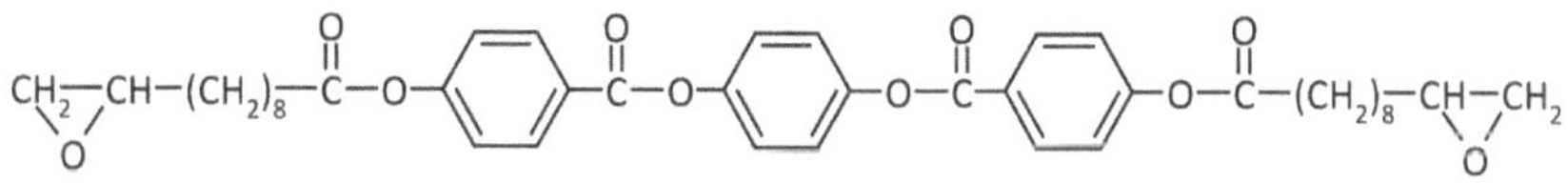

LCEM

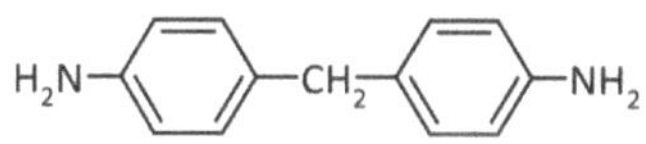

DDM

The conventional DSC analysis (Fig. 2.58) showed that there is a deformation in the liquid-crystalline transition of LC monomer (i.e., from crystal to liquid-crystalline phase transition) during heating of the analyzed composition, which suggested that the partial exothermic cross-linking reaction with epoxy groups involvement occurs simultaneously with the endothermic phase transition of liquid-crystalline epoxy monomer (LCEM).

The endothermic changes registered on the curve of the LCEM/DDM composition in the temperature range 70–92°C (Fig. 2.58) result from the polymorphic transition of the LCEM monomer and the melting of the amine DDM. The difference between areas of DDM melting peaks for pure hardener and mixed composition curves is caused by small mass fraction of amine in the composition (about 12%). Enthalpies are calculated in $J\,g^{-1}$, so it is expected that DDM melting enthalpy in the mixture with epoxy is about 10 times lower than that in hardener alone. The distinct broad exothermic effect accompanying the cross-linking reaction appears above 135°C, just above the endothermic peak assigned to the melting of the LCEM to the liquid crystalline nematic phase. Temperature peaks from mixed composition are slightly shifted compared to pure LCEM and hardener, because of mutual interactions. The shape, enthalpy and temperature of the endothermic peak assigned to melting of LCEM in the LCEM/DDM composition indicate that in this temperature range, simultaneous partial exothermic cross-linking takes place.

TMDSC has been used before to study the kinetics of the curing process. It has been shown by Hutchinson group that TOPEM provides a great advantage over traditional DSC when it comes to determining the glass transition temperature for a fully cured system of reactive trifunctional epoxy-triglycidyl-paminophenol (TGAP) [88].

Results of the analyses of LCEM/DDM compositions for different heating rates of 2°C, 5°C, 10°C, and 20°C$\,min^{-1}$ (Fig. 2.59) show that the enthalpy of both transitions, endothermic at about 130°C and exothermic related to cross-linking above 135°C, increases with increasing heating rate during measurements. The total (summed) thermal effect of both changes also increases, as shown in Fig. 2.59.

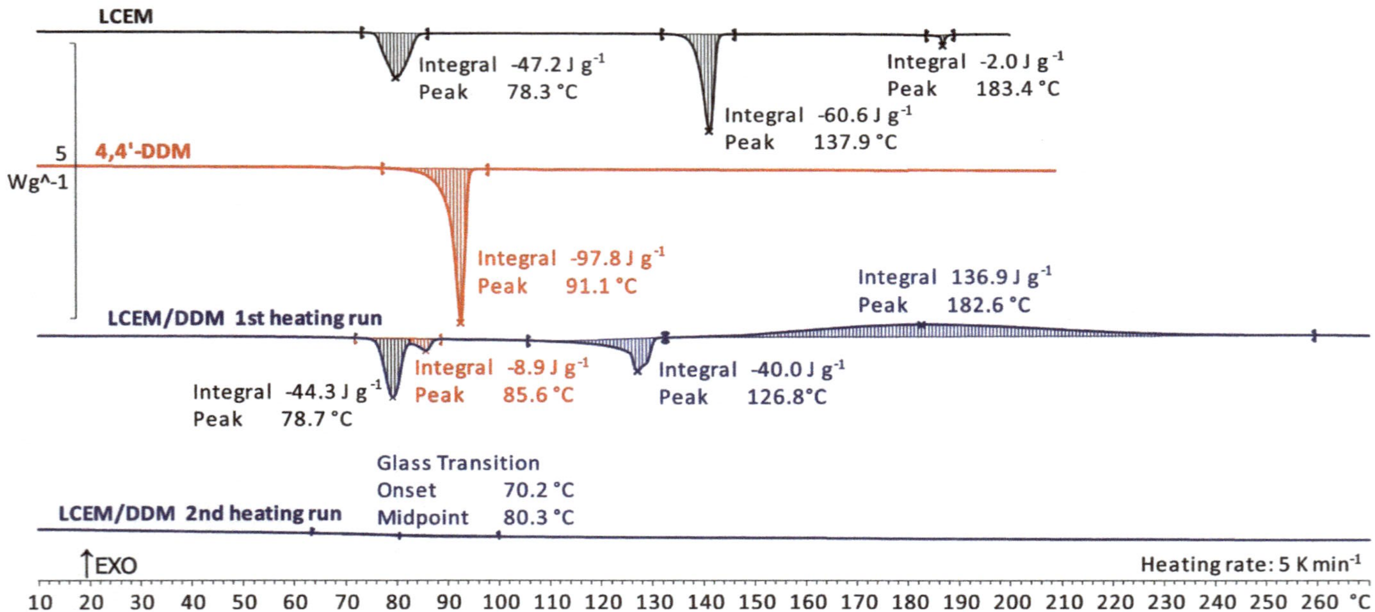

FIG. 2.58 Conventional DSC curves of the monomer LCEM, amine DDM, and LCEM/DDM stoichiometric mixture; heating rate of 5°C min^{-1}.

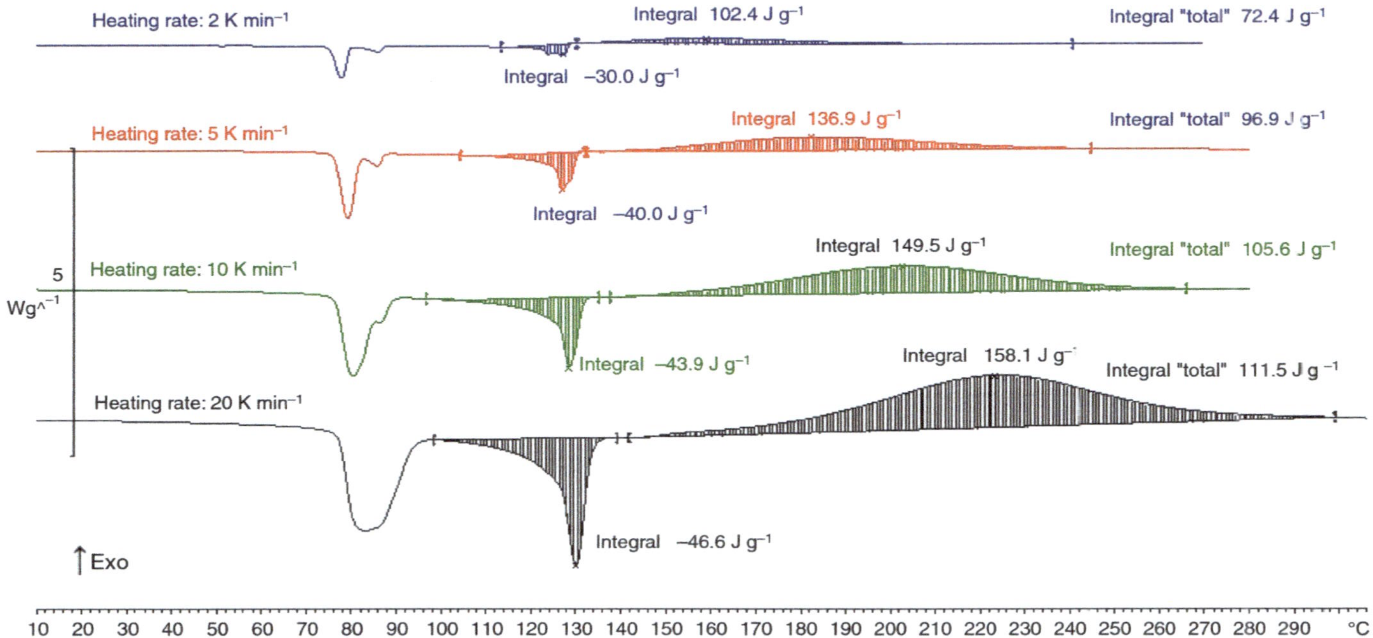

FIG. 2.59 Dynamic DSC curing curves of the LCEM/DDM composition at different heating rates of 2, 5, 10, and 20°C min^{-1}.

Discrepancies in the enthalpy values between samples heated at different rates may be caused by the appearance or nonappearance of the liquid-crystalline phase during the cross-linking reaction. At work [89], it was found that the formation of the LC phase reduces the viscosity of the reaction mixture, which results in an increase in enthalpy of reaction and degree of curing. However, it is somewhat unexpected that viscosity can have influence not only on the rate of curing, but also on the cross-linking density, and in case of liquid-crystalline epoxies, the presence of mesophase during curing affects the activation energy dependent on the degree of cure, what can be the main cause of this phenomenon [89]. Anisotropic mesophase also affects spatial possibilities of contacting reactive functional groups, thereby changing the network density.

This complicated system of simultaneously occurring factors is the reason why it is almost impossible to accurately estimate the degree of cure. The authors think, however, that reaction enthalpy 158.1 J g^{-1} achieved with 20°C min^{-1} heating rate is close to maximum possible, because of inability to precure in the range of endothermic transition (110–140°C) caused both by short period of time in case of 20°C min^{-1} heating (about 1.5 min) and difficulties with LC formation, as presented in Ref. [89].

Next, to separate the possible reversible effects (e.g., phase transitions of the mixture components) from irreversible (curing the resin), TMDSC measurements were performed. The TMDSC curves for the LCEM/DDM composition are shown in Fig. 2.60. These are heat flow curves marked as nonreversing (red line), reversing (green line), and total heat flow (black line). In Fig. 2.60, the curve of the composition recorded with the conventional DSC method at a heating rate of 2°C min^{-1} (blue line) is also presented.

The courses of the total heat flow curves obtained from both DSC analyses, TOPEM and conventional, are very similar, with similar enthalpy values of endothermic and exothermic phase transitions. On the reversing heat flow curve, the endothermic peak was only recorded, with enthalpy of 19.5 J g^{-1} higher than the enthalpy of transition in the same temperature range determined from the total heat flow curve. Of note, on the nonreversing heat flow curve, two exothermic effects can be clearly observed. The enthalpy value of the second peak recorded on that curve above 100 J g^{-1} coincides with the results from the conventional DSC and total heat flow curve of TMDSC. This clearly confirms the irreversibility of the cross-linking process. However, the presence of the first exothermic effect with maximum at about 125°C and enthalpy of 19 J g^{-1} suggests a cross-linking reaction at 110–130°C and reduction in the enthalpy of the endothermic peak (49.1 J g^{-1}) on the reversing heat flow curve. The obtained dependences indicate that the total effect of endothermic transformation from a temperature range of 110–130°C is the sum of the endothermic phase transformation of the monomer and the partial exothermic cross-linking of the LCEM/DDM composition, which was distinctly confirmed by the results of the experiments described below. The difference in shapes of first DSC peak and first total heat flow TOPEM peak may be due to temperature fluctuations present in temperature-modulated technique. Moreover, LC transition during

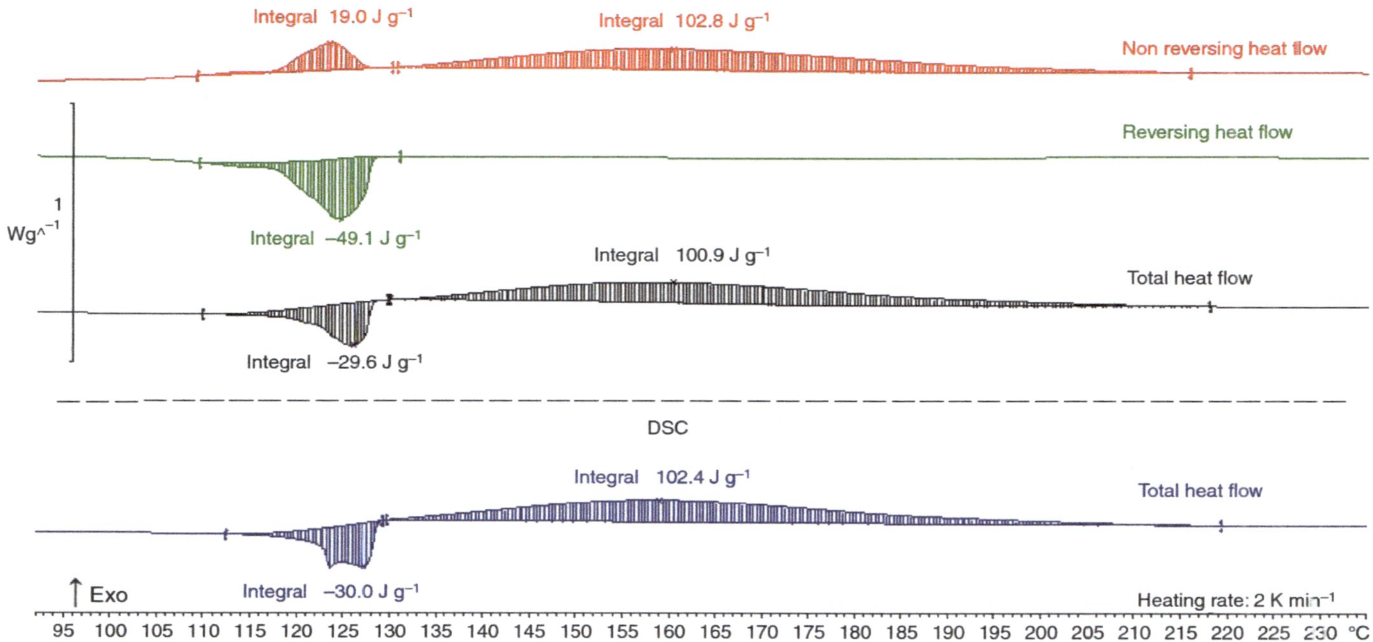

FIG. 2.60 DSC curves of the LCEM/DDM composition; TOPEM and DSC analysis recorded at the heating rate of 2°C min^{-1}, pulse height 1°C (±0.500°C), pulse width 15–30 s. Heat flow curves marked as nonreversing (red line), reversing (green line), and total (black line) heat flow. Conventional DSC thermogram at a heating rate of 2°C min^{-1} (blue line) is also presented.

hardening should be theoretically visible on reversing heat flow curve affecting the total enthalpy of curing, but the authors have observed the flat baseline in the curing temperature range. The reason for the lack of this reversible transition lies probably in constantly changing and growing of macromolecule during hardening. It is impossible to have reversible effect, when the polymer or pre-polymer chain being the source of LC is no longer present in the composition, because it has transformed into polymer molecule; therefore, the mesophase creation can still be predicted.

Nonreversing results from 70°C to 90°C temperature range did not provide proof of any irreversible process, so this fragment of curve is not presented on the picture for more clarity of crucial part of experiment. Two samples were taken from the temperature range of complex thermal changes—meaning that samples of the fresh compositions were heated to 127°C and 130°C in the DSC apparatus with a heating rate of $2°C\,min^{-1}$, and then after rapid cooling, the samples were subjected to further DSC analysis with a heating rate of $10°C\,min^{-1}$ and ^{1}H NMR analysis. The results of DSC analyses of LCEM/DDM samples, which in the first cycle were heated to 127°C and 130°C and in the second to 270°C, are shown in Fig. 2.61 together with the result of the DSC analysis of the fresh mixture.

As was expected, the compositions were heated first to 127°C and 130°C in the next, and second heating cycle was further cross-linking with exothermic effect. Additionally, on the curves from the second run, there are no phase transitions of components. The thermal effect of the cross-linking process is lower than the enthalpy of curing the fresh mixture, which means that during heating to 127°C and 130°C, a partial cross-linking of the composition occurs.

Styrene butadiene rubber (SBR) elastomer containing a low-molecular-weight oil

The conventional DSC curve of SBR containing a low-molecular-weight oil shows overlapping thermal effects that are difficult to identify and properly assign (Fig. 2.62). In contrast, a TOPEM measurement performed at $2°C\,min^{-1}$ allowed the transitions to be quickly and unambiguously identified [44].

In the reversing heat flow curve, the glass transition of the elastomer is observed as a characteristic step at −34°C. In addition, the reversing heat flow curve exhibits an endothermic effect at 4°C. This can be assigned to the melting of the oil. The nonreversing heat flow shows an exothermic effect at about −16°C. This is due to the cold crystallization of the low-molecular-weight constituents.

8.3 Composite of carbon fibers and epoxy resin

The glass transition and degree of cure of the polymer matrix are important quality criteria for composite materials. In case of high fiber content, the glass

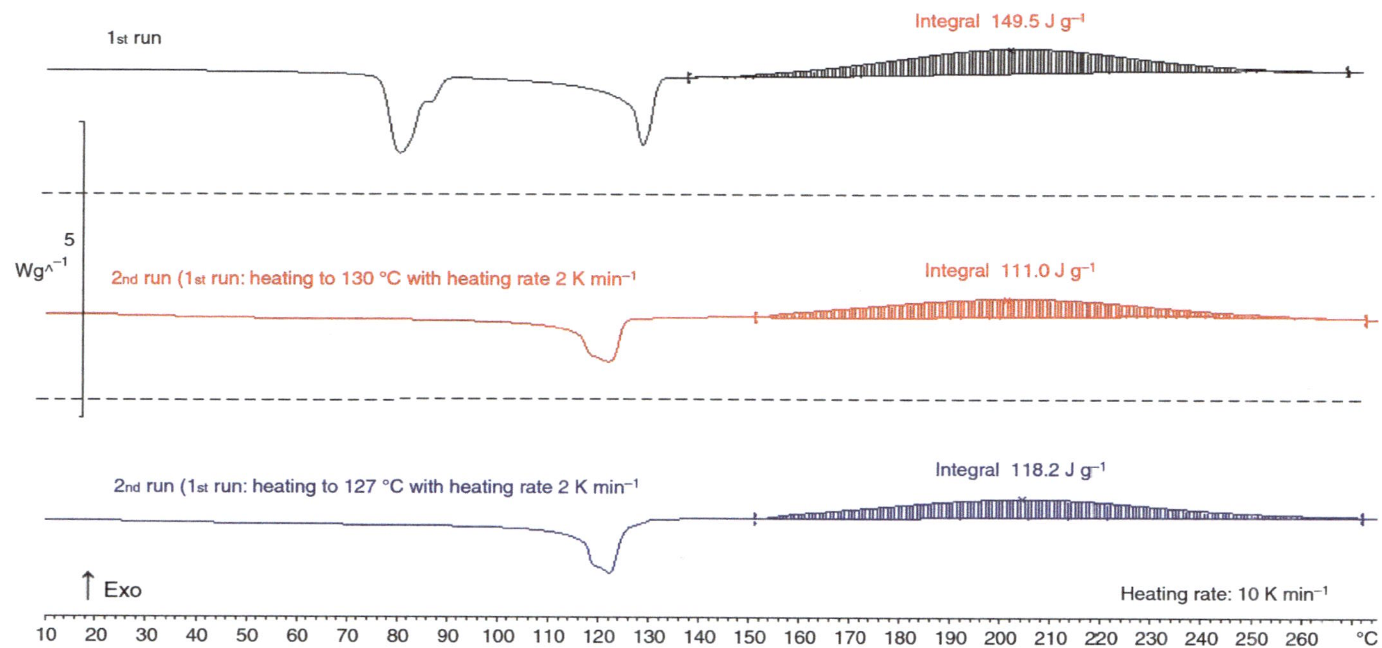

FIG. 2.61 Conventional DSC curves of the LCEM/DDM composition: analysis of a fresh sample (top) and preheated samples to 130°C and 127°C (second run); heating rate $10°C\,min^{-1}$.

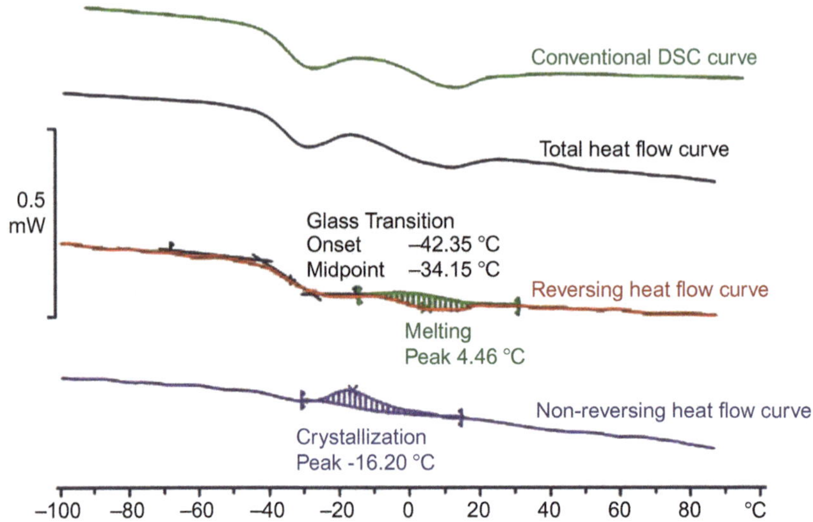

FIG. 2.62 TOPEM measurement performed at 2°C min^{-1} showing overlapping thermal effects on SBR elastomer containing a low-molecular-weight oil [44].

transition is usually broad and is accompanied by only a small change in c_p. In addition, it may be overlapped by a postcuring reaction [44, 90].

In the example, the postcuring reaction and onset of decomposition mask the expected glass transition. Conventional DSC therefore shows only an exothermic postcuring reaction followed by decomposition (Fig. 2.63). Using TOPEM, the glass transition of the material can be clearly seen in the reversing heat flow curve at 216°C. The nonreversing heat flow shows the postcuring reaction and the onset of decomposition. TOPEM thus enables the glass transition to be clearly separated from the two other processes. The total heat flow corresponds to the result of a conventional DSC experiment.

Separation of overlapping events in pharmaceutical formulations

The DSC curves of pharmaceutical formulations often exhibit several overlapping thermal effects. In this example, the total heat flow curve (analogous to the conventional DSC curve) shows a broad vaporization peak that overlaps two sharp peaks at 60°C and 125°C (Fig. 2.64). The quasistatic heat capacity curve shows that the first peak at 61°C is due to a glass transition. The peak is therefore the result of enthalpy relaxation [44].

The second event is a phase transition that also appears as a peak in the $c_{p,0}$ curve. The frequency evaluation shows that the peak temperature is independent of frequency. Evaluation of the nonreversing heat flow curve yields a moisture loss of about 6.5%, which is the reason for a slight reduction of 0.28 J g^{-1} °C^{-1} in the heat capacity.

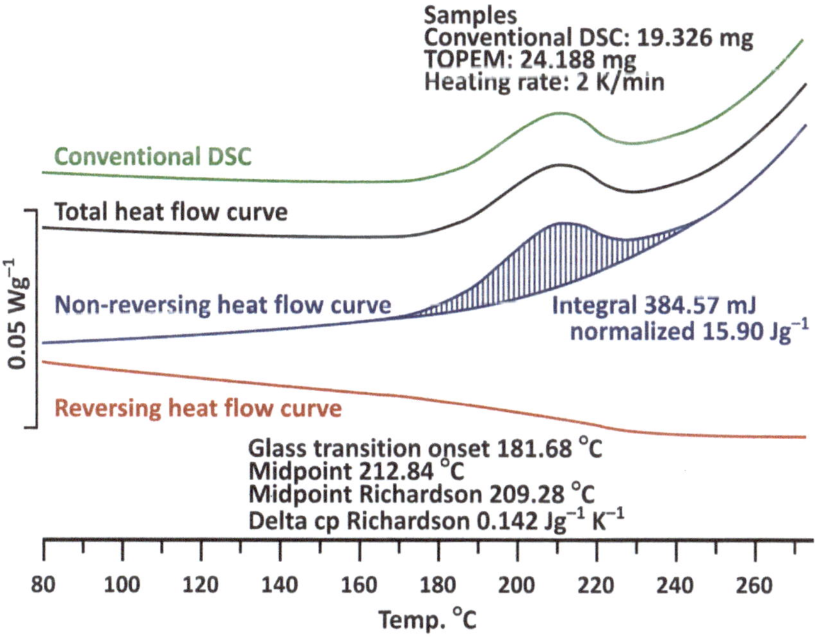

FIG. 2.63 Conventional DSC and TOPEM curves of a composite of carbon fibers and epoxy resin.

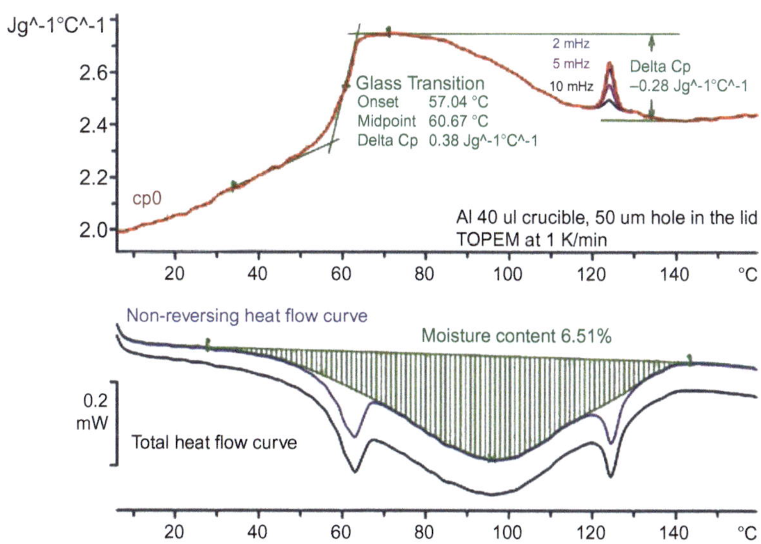

FIG. 2.64 The DSC curves of pharmaceutical formulations showing a broad vaporization peak that overlaps two sharp peaks at 60°C and 125°C.

8.4 Investigation of the phase behavior of a lactose solution

The phase behavior of aqueous lactose solution has gained importance because lactose is a main constituent acid whey, a by-product in dairy industry that cannot be simply disposed of due to its toxicity during decomposition, robbing oxygen from rivers and streams creating health and environmental concerns. The failure of lactose to crystallize in the presence of lactic acid and calcium ions and remain in its amorphous form adds to the problem.

The concentration and crystallization behavior of lactose and its interactions with other constituents of acid whey especially lactic acid and calcium ions have been reported using DSC measurements. Knowledge of the thermal behavior of lactose-water solutions is also important in the development freeze-drying processes [91]. In an earlier study, Mettler-Toledo [44] reported the thermal behavior of lactose-water solutions using TOPEM measurements. There is a very large melting peak of water in a 4.5% lactose solution together with a thermal event at −30°C, which could be interpreted as a melting process or as a glass transition.

The TOPEM measurement under quasiequilibrium conditions (heating rate of 0.1°C$\,$min^{-1} with temperature pulse heights of 0.005°C) yields a nonreversing heat flow curve with an endothermic peak (Fig. 2.65). This information indicates that the effect at −30°C is a melting process that is accompanied by a small c_p change, as seen in the reversing heat flow curve.

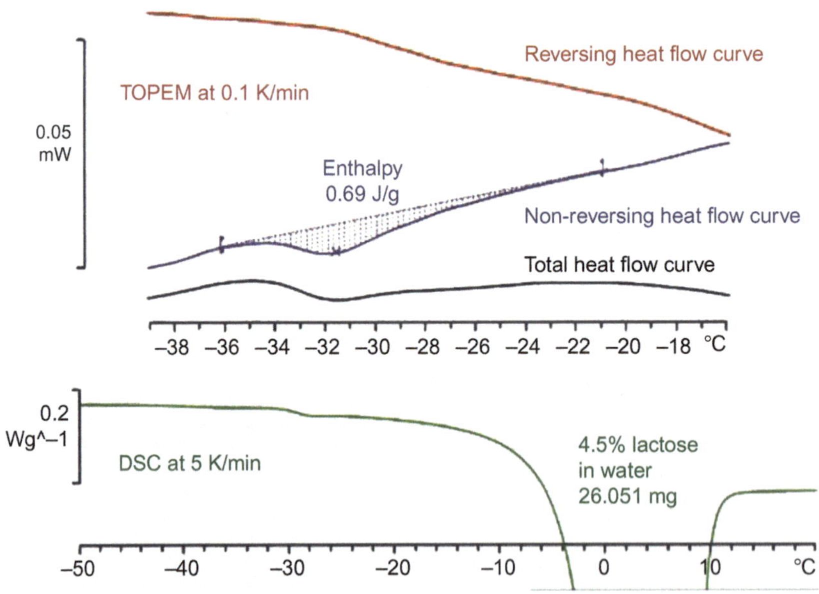

FIG. 2.65 TOPEM measurements under quasiequilibrium conditions (heating rate of 0.1°C$\,$min^{-1} with temperature pulse heights of 0.005°C) of 4.5% lactose aqueous solution yielding a nonreversing heat flow curve with an endothermic peak.

The use of TOPEM in studying the first heating run in disposable cups

Mettler-Toledo Team draw conclusions about the material obtained from the first heating run in disposable cups using TOPEM. They showed how overlapping effects in the first heating run can be separated and interpreted. This allowed the correct enthalpy of melting of the original sample to be determined with conclusions about the morphology of the sample [92].

The cups were manufactured using a deep drawing process at temperatures around the glass transition. In the production process, the poly-lactic acid (PLA) material experienced high and much localized cooling rates. Conventional DSC measurements were performed in the temperature range 25–180°C at a heating rate of 10°C min^{-1} [93]. At the end of the first heating run, the sample was held at 180°C for 5 min to ensure melting was complete and then cooled at 10°C min^{-1}. A thin layer of about 3 mg fine aluminum oxide powder (Al_2O_3) was spread over the bottom of each crucible. The purpose of this was to reduce disturbances due to mechanical stresses in the sample during the first heating run.

The TOPEM measurement was performed in the temperature range 25–180°C using a pulse height of ±0.5°C and a pulse width between 15 and 30 s. The underlying heating rate was 2°C min^{-1}. The evaluation was performed using the default parameters.

Inspection of the first heating curve shown in black in Fig. 2.66 using conventional DSC measurements shows a high sample crystallinity that might be expected because the enthalpy of melting is larger than that in the second measurement. Moreover, the glass transition appears at a higher temperature and exhibits a large endothermic relaxation peak. The height of the Δc_p step of 0.08 J g^{-1} $°C^{-1}$ at the glass transition indicates that the mobile amorphous phase content is low, which is consistent with a high degree of crystallinity. Cold crystallization is not detected.

In the second heating run (red curve in Fig. 2.66), the sample exhibits a glass transition at about 59°C with a Δc_p step height of 0.51 J g^{-1} $°C^{-1}$ together with a cold crystallization peak at about 125°C and a peak due to melting at about 153°C. The enthalpies of crystallization and melting of about 22 J g^{-1} are the same. The sample was therefore completely amorphous. The thermal effects in the second heating run are easy to interpret. The curve shows a glass transition with a Δc_p typical for amorphous polymers and the mass fraction of the melting crystallites corresponds to that of the crystallites formed during cold crystallization.

A thorough investigation carried out by Mettler-Toledo Team using TOPEM technique shows that the above interpretation is rather superficial. Fig. 2.67 shows the quasistatic heat capacity $c_{p,0}$ curves obtained from TOPEM measurements recorded as a function of temperature. The c_p values of the three curves agree well above 170°C in the melt. In the first heating run below 50°C, the values are slightly lower than those of the following two measurement curves because of the initial poorer thermal contact of the sample.

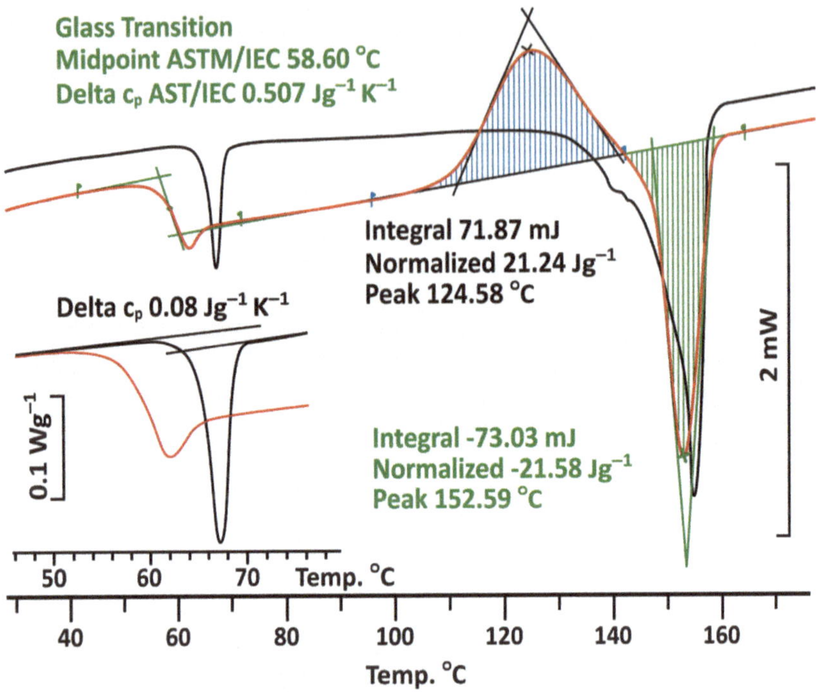

FIG. 2.66 Evaluation of the first (black) and second (red) conventional DSC heating runs measured at 10°C min^{-1} for disposable cups sample. The glass transition region is shown on an expanded scale.

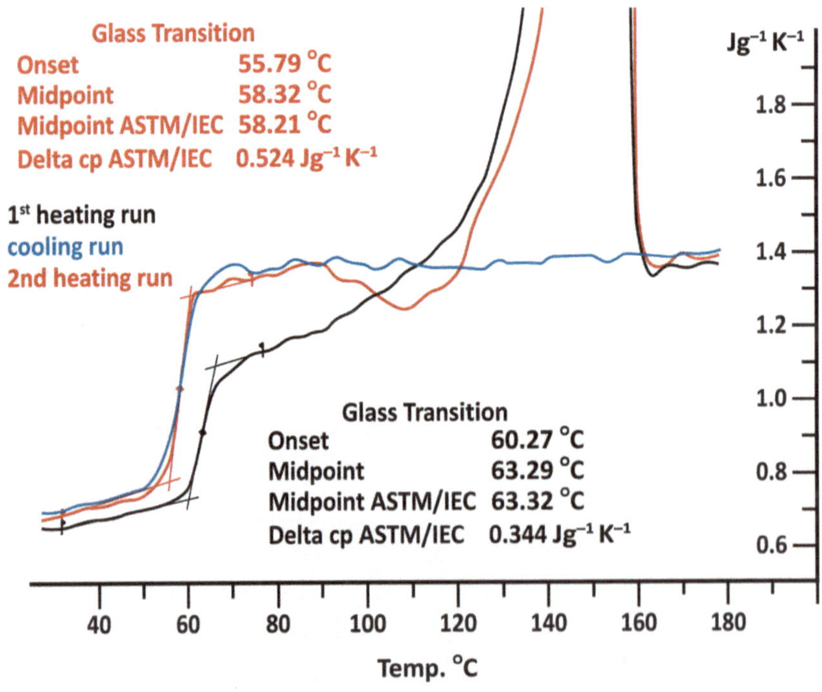

FIG. 2.67 TOPEM measurements showing the evaluation of the specific heat capacity of the first (black) and second (red) heating runs and the cooling curve (blue) for disposable cups sample.

The height of the Δc_p step in the first heating measurement is appreciably larger than would be expected from the evaluation of the conventional DSC measurement: $T_g = 63°C$, $\Delta c_p = 0.34\,J\,g^{-1}\,°C^{-1}$. This means that the content of mobile amorphous phase in the original sample was larger than that evaluated in Fig. 2.66. Crystallization does not occur in the cooling run despite the relatively low cooling rate of $2°C\,min^{-1}$. The Δc_p step of the glass transition in the cooling run and in the second heating measurement are practically the same and agree very well with the result from the conventional DSC measurement: $T_g = 58°C$, $\Delta c_p = 0.52\,J\,g^{-1}\,°C^{-1}$. The decrease in the specific heat capacity between 90°C and 110°C in the second heating run is due to cold crystallization.

Fig. 2.68 shows the total heat flow curves. These curves correspond to conventional DSC measurements performed at a heating rate of $2°C\,min^{-1}$. They allow the enthalpy of cold crystallization to be determined, which begins at the glass transition. Extrapolation from the melt provides a good baseline for this determination and yields a value of about $30\,J\,g^{-1}$ for the second heating run (red curve). The evaluation of the first heating run in the same way yields a value of $15\,J\,g^{-1}$, which is just half of that obtained in the second heating run. This shows that cold crystallization also occurs in the first heating run.

Reevaluation of the first and second conventional DSC heating runs at $10°C\,min^{-1}$ is illustrated in Fig. 2.69. The baseline construction used in

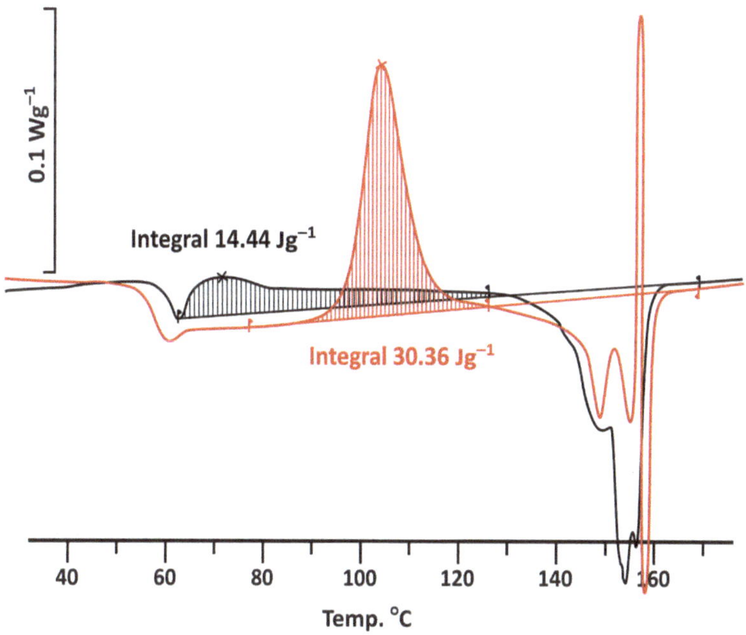

FIG. 2.68 Total heat flow curves for the first (black) and second (red) heating runs of the TOPEM measurement for disposable cups sample.

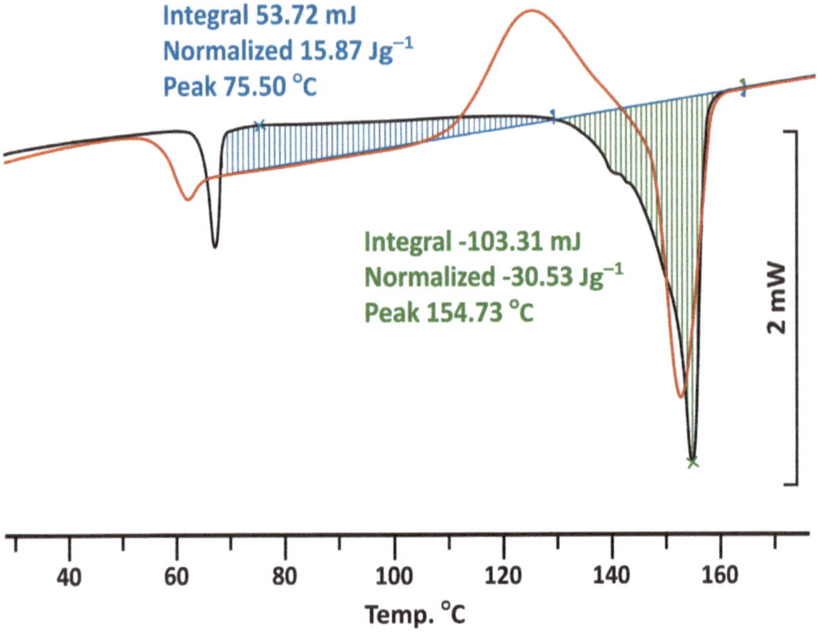

FIG. 2.69 Reevaluation of the first (black) and second (red) conventional DSC heating runs at 10°C min^{-1}.

Fig. 2.69 is applied to the first heating run of the DSC measurement in Fig. 2.66. Reevaluation yields a value of 16 J g^{-1} for the enthalpy of cold crystallization, which is in good agreement with the result from the TOPEM measurement. The enthalpy of melting for the first heating run is almost 31 J g^{-1} and is thus about 50% larger than that in the second heating run. Furthermore, it now becomes clear that 50% of the crystallites that melt in the first heating run are formed through cold crystallization during the measurement. Contrary to the initial results obtained from the evaluation of the first heating run using conventional DSC, the degree of crystallinity turns out to be significantly smaller.

The correct value for the enthalpy of melting is about 15 J g^{-1} [92]. An evaluation that does not take cold crystallization into account therefore yields a value for the degree of crystallinity that is much too high for the original sample.

A thermal behavior of the biodegradable polylactide (PLA) disposable cups was concluded in which cold crystallization begins immediately at the glass transition. The mobility of the polymer chains increases, and mechanical stresses produced in the deep drawing production process of the cup decrease. The sample crystallizes during the heating run, reorganizes, and melts. The total heat flow curve, which shows the behavior of the sample at the underlying heating rate of 2°C min^{-1}, agrees very well with a conventional DSC measurement at 2°C min^{-1}.

Meanwhile due to the favorable mechanical properties of poly(L-lactic acid) (PLLA), it is one of the most frequently used bio-based polyesters. However, some properties such as slow crystallization rate and low crystallinity restrict its commercial uses. Pandey et al. were able to reduce its induction period, the acceleration of crystallization, the increase of the final value of the degree of crystallinity, and the acceleration of lamellar thickening through 10% doping of silk, which were studied using DSC [93].

9 Summary

This chapter reports the definition, theory, principles, and operating conditions (e.g., sample size, modulation period, measuring transitions, and modulation amplitude) of TMDSC method and its applications, especially its use for studying the thermodynamic properties of polymers and pharmaceuticals. We provided detailed comparison between DSC and TMDSC methods to show the advantages of TMDSC over traditional DSC in resolving interfering thermal events, providing the best sensitivity for otherwise weak transitions, and improving the precision of heat capacity measurements. Moreover, we talked in this chapter about all techniques of TMDSC, including the isothermal step (IsoStep) method, the ADSC method, and, the most advanced TOPEM method and their unique applications such as studying the thermal response of samples at different frequencies.

References

[1] L.I. Pooley, A.S. Abu-, M.J. Bakar, R.W. Cran, K.A.M. Moinuddin, Measurements of specific heat capacity of common building materials at elevated temperatures: a comparison of DSC and HAD, J. Therm. Anal. Calorim. 141 (2020) 1279–1289. C. Potter, Advanced Techniques and Strategies in Thermal Analysis and Calorimetry, TA Instruments-Waters LLC. V. Tkachenko, L. Josien, G. Schrodj, S. Hajjar-Garreau, S. Urbaniak, J. Poly, A. Chemtob, A DSC and XPS characterization of core-shell morphology of block copolymer nanoparticles, Colloid Polym. Sci. 298 (2020) 1095–1105.

[2] N.H. Concord, N.Y. Albany, C.T. Meriden, Thermal Analysis & Rheology, TA Instruments, Inc., 109 Lukens Drive, New Castle, DE 19720, 2018. 10–11 April.

[3] M.C. Righetti, Crystallization of polymers investigated by temperature-modulated DSC, Materials 10 (2017) 442.

[4] M. Wagner, Thermal Analysis in Practice: Fundamental Aspects, Carl Hanser Verlag, Munich, Cincinnati, 2018.

[5] J.E.K. Schawe, T.H. Hutter, C. Heitz, I. Alig, D. Lellinger, Stochastic temperature modulation: a new technique in temperature-modulated DSC, Thermochim. Acta 446 (2006) 147–155.

[6] H. Gobrecht, K. Hamann, G. Willers, Complex plane analysis of heat capacity of polymers in the glass transition region, J. Phys. E: Sci. Instrum. 4 (1971) 21–23.

[7] M. Reading, D. Elliott and V. Hill, Proceedings of the 21st North Am. Therm. Anal. Soc. Conf, 1992, 145-150.

[8] M. Reading, D. Elliott, V.J. Hill, A new approach to the calorimetric investigation of physical and chemical transitions, J. Therm. Anal. 40 (1993) 949–955.

[9] P.S. Gill, S.R. Sauerbrunn, M. Reading, Modulated differential scanning calorimetry, J. Therm. Anal. 40 (1993) 931–939.
[10] Introducing the Temperature Modulated DSC, Hitachi High-Tech, Tokyo, 1993.
[11] M. Reading, A. Luget, R. Wilson, Modulated differential scanning calorimetry, Thermochim. Acta 238 (1994) 295–307.
[12] M. Reading, Method and Apparatus for Gas Flow Modulated Differential Scanning Calorimetry, US Patent 005 624 187 A, 1997.
[13] A.A. Lacey, C. Nikolopoulos, M. Reading, A mathematical model for modulated differential scanning calorimetry, J. Therm. Anal. 50 (1997) 279–333.
[14] M. Reading, B.K. Hahn, Method and Apparatus for Modulated Differential Analysis, US Patent 522 477 5, 1993.
[15] M. Reading, Method and Apparatus for Spatially Resolved Modulated Differential Analysis, US Patent 5 248 199, 1993.
[16] S.M. Marcus, M. Reading, Method and Apparatus for Thermal Conductivity Measurements, US Patent 5 335 993, 1994.
[17] M. Reading, B.K. Hahn, Method and Apparatus for Modulated Differential Analysis, US Patent 5 346 306, 1994.
[18] M. Reading, Method and Apparatus for AC Differential Thermal analysis, US Patent 5 439 291, 1995.
[19] M. Pyda, Temperature-modulated differential scanning calorimetry, Encycl. Polym. Sci. Technol. (2014), https://doi.org/10.1002/0471440264.pst623.
[20] P. Claudy, J.C. Commercon, J.M. Letoffe, Quasi-static study of the glass transition of glycerol by DSC, Thermochim. Acta 128 (1988) 251–260.
[21] S.C. Mraw, D.F. Naas, the heat capacity of stoichiometric titanium disulfide from 100 to 700 k: absence of the previously reported anomaly at 420 K, J. Chem. Thermodyn. 11 (1979) 585–592.
[22] W.F. Hemminger, H.K. Cammenga, Methoden der Thermischen Analyse, Springer-Verlag/ Wiley-VCH Verlag GmbH&Co. KGaA, Berlin/Weinheim, 1990, p. 260.
[23] R. Riesen, G. Widmann, R. Truttmann, Alternating thermal analysis techniques, Themochim. Acta 272 (1996) 27–39.
[24] J. Cao, Y. Long, R.A. Shanks, Experimental investigation into the heat capacity measurement using an modulated DSC, J. Therm. Anal. 50 (1997) 365–373.
[25] M. Cassettari, G. Salvetti, E. Tombari, S. Veronesi, G.P. Johari, Thermodynamics of isothermal vitrification by chemical reactions in a liquid, Il Nuovo Cimento 14D (1992) 763–766.
[26] H. Gobrecht, K. Hamann, G. Willers, Complex plane analysis of heat capacity of polymers in the glass transition region, J. Phys. E: Sci. Instrum. 4 (1971) 21.
[27] U. Jorimann, G. Widmann, R. Riesen, Temperature modulated DSC (TMDSC) applications and limits of phase information, c_p determination and effect separation, J. Therm. Anal. Calorim. 56 (1999) 639–647.
[28] J.E.K. Schawe, Modulated temperature DSC measurements: the influence of the experimental conditions, Thermochim. Acta 271 (1996) 127–140.
[29] TA Instruments, Modulated DSC Workshop 10/94, Melbourne, Oct. 1994.
[30] A. Hensel, J. Dobbertin, J.E.K. Schawe, A. Boller, C. Schick, Temperature modulated calorimetry and dielectric spectroscopy in the glass transition region of polymers, J. Therm. Anal. 46 (1996) 935–954.
[31] S. Weyer, A. Hensel, C. Schick, Phase angle correction for TMDSC in the glass-transition region, Thermochim. Acta 304-305 (1997) 267–275.
[32] B. Wunderlich, R. Androsch, M. Pyda, Y.K. Kwon, Heat capacity by multi-frequencies Sawtooth modulation, Thermochim. Acta 348 (2000) 181–190.

[33] I. Fraga, S. Montserrat, J.M. Hutchinson, TOPEM, a new temperature modulated DSC technique, application to the glass transition of polymers, J. Therm. Anal. Calorim. 87 (2007) 119–124.

[34] J. Schawe, I. Alig, D. Lellinger, Method and Apparatus for Thermally Analyzing A Material, EP 1091208, 2001.

[35] J. Schawe, I. Alig, D. Lellinger, US Patent 6,551,835, 2001.

[36] Thermal Analysis Application No. UC 71 Application published in Mettler-Toledo Thermal Analysis UserCom 7. D.A. Ditmars, S. Ishihara, S.S. Chang, G. Bernstein, E.D. West, Enthalpy and heat-capacity standard reference material: synthetic sapphire (a-$A1_2O_3$) from 10 to 2250 K, J. Res. Nat. Bureau Stand. 87 (1982) 159–163. M. Yokota1, K. Sugane, I. Tsukushi, M. Shibata, Evaluation of the heat capacity of amorphous polymers composed of a carbon backbone below their glass transition temperature, Polym. J. (2020), https://doi.org/10.1038/s41428-020-0317-x.

[37] K. Kanaria, T. Ozaw, Simulation of temperature modulated DSC of temperature dependent heat capacity, Thermochim. Acta 304-305 (1997) 201–207.

[38] B. Cassel, W.J. Sichina, New Software for Sample Characterization: Stepscan DSC, Thermal Analysis Application Note, PerkinElmer Instruments, 2000.

[39] M.S. Lopes, T.A. Catelani, A.L.C.S. Nascimento, J.S. Garcia, M.G. Trevisan, Ketoconazole: compatibility with pharmaceutical excipients using DSC and TG techniques, J. Therm. Anal. Calorim. 141 (2020) 1371–1378. L. C. Thomas, Why Modulated DSC®?: An Overview and Summary of Advantages and Disadvantages Relative to Traditional DSC Modulated DSC®, Paper #1, TA Instruments, 109 Lukens Drive, New Castle, DE 19720, USA. L. C. Thomas, Modulated DSC® Paper #2, Modulated DSC® Basics; Calculation and Calibration of MDSC® Signals, TA Instruments Technical Paper TP 007, TA Instruments, 109 Lukens Drive, New Castle, DE 19720, USA.

[40] S.R. Rabel, J.A. Jona, M.B. Maurin, Applications of modulated differential scanning calorimetry in preformulation studies, J. Pharm. Biomed. Anal. 21 (1999) 339–345.

[41] M. Wagner, Thermal Analysis in Practice: Fundamental Aspects, Hanser Publishers, Munich, Cincinnati, 2018, p. 287.

[42] J. Leng, Analysis of Poly(L-lactide) Nanocomposites Based on Layered Double Hydroxides: Comparison of NiAl and MgAl LDH, MSc Thesis, Fakultät III—Prozesswissenschaften der Technischen Universität Berlin, 2016.

[43] L. Parmentier, TOPEM®, A New Temperature Modulated DSC Technique: A critical Review, First Master of Pharmacy, Pharmaceutical Care, Institute for Pharmaceutical Technology and Biopharmacy, University of Duesseldorf, 2008–2009.

[44] TOPEM®—The New Advanced Multi-Frequency TMDSC Technique, Mettler-Toledo GmbH, Analytical, Schwerzenbach, Internet: www.mt.com/ta, 08/2005.

[45] U.S. Syamala, Calculation of MTDSC signals, factors effecting the signals and applications in drug development, MOJ Bioequiv. Bioavail. 5 (2018) 141–149.

[46] Y. Kraftmakher, Modulation Calorimetry: Theory and Application, Springer-Verlag Berlin Heidelberg, New York, 2004.

[47] J.E.K. Schawe, T. Hütter, TOPEM®: The latest innovation in temperature modulated DSC, in: Proceedings of the NATAS 32nd Annual Conference, Williamburg, VA, 2005.

[48] J.E.K. Schawe, W. Winter, The influence of heat transfer on temperature-modulated DSC measurements, Thermochim. Acta 298 (1997) 9–16. I. Moon, R. Androsch, B. Wunderlich, A calibration of The various heat-conduction paths for a heat-flux-type temperature-modulated DSC, Thermochim. Acta 357–358 (2000) 285–291.

[49] N.O. Birge, P.K. Dixon, N. Menon, Specific heat spectroscopy: origins, status and applications of the 3ω method, Thermochim. Acta 304–305 (1997) 51–66.

[50] E. Donth, J. Korus, E. Hempel, M. Beiner, Comparison of DSC heating rate and HCS frequency at the glass transition, Thermochim. Acta 304–305 (1997) 239–249.
[51] J.E.K. Schawe, E. Bergmann, Investigation of polymer melting by temperature modulated differential scanning calorimetry and it's description using kinetic models, Thermochim. Acta 304-305 (1997) 179–186.
[52] J.E.K. Schawe, Description of thermal relaxation of polystyrene close to the thermal glass transition, J. Polym. Sci. B: Polym. Phys. 36 (1998) 2165–2175.
[53] H. Huth, M. Beiner, S. Weyer, M. Merzlyakov, C. Schick, E. Donth, Glass transition cooperativity from heat capacity spectroscopy-temperature dependence and experimental uncertainty, Thermochim. Acta 377 (2001) 113–124.
[54] J.E.K. Schawe, G.R. Strobl, Superheating effects during the melting of crystallites of syndiotactic polypropylene analysed by temperature-modulated differential scanning calorimetry, Polymer 39 (1998) 3745–3751.
[55] J.E.K. Schawe, A comparison of different evaluation methods in modulated temperature DSC, Thermochim. Acta 260 (1995) 1–16.
[56] M. Schubnell, C. Heitz, T. Hütter, S. Sauerbrunn, J.E.K. Schawe, A multifrequency temperature-modulated technique For DSC, Am. Lab. 38 (2005) 18–21.
[57] M. Pyda, A.C. Kubicka, Thermodynamic Properties of Polymers, Encyclopedia of Polymer Science and Technology, John Wiley & Sons, Inc, 2018, https://doi.org/10.1002/0471440264.pst369.pub2.
[58] B. Wunderlich, Reversible crystallization and the rigid-amorphous phase in semicrystalline macromolecules, Prog. Polym. Sci. 28 (2003) 283–450.
[59] M. Pyda, Temperature-Modulated Differential Scanning Calorimetry, Encyclopedia of Polymer Science and Technology, John Wiley & Sons, Inc, 2014, https://doi.org/10.1002/0471440264.pst623.
[60] B. Wunderlich, J. Grebowicz, Thermotropic mesophases and mesophase transitions of linear, flexible macromolecules, in: N.A. Platé (Ed.), Liquid Crystal Polymers II/III, Adv. Polym. Sci., vol. 60/61, 2014, pp. 1–59. B. Wunderlich, M. Moeller, J. Grebowicz, H. Baur, Conformational Motion and Disorder in Low and High Molecular Mass Crystals, Adv. Polym. Sci., vol. 87, Springer, Berlin, 1988.
[61] Hitachi High-Tech, Application Brief, Introducing the Temperature Modulated DSC, TA N0.64, MAR, 1993. I.D.S. Silva, H. Schäfer, N.G. Jaques, D.D. Siqueira, A. Ries, D.S. Morais, K. Haag, K. Koschek, L.H. Carvalho, R.M.R. Wellen, An investigation of PLA/Babassu cold crystallization kinetics, J. Therm. Anal. Calorim. 141 (2020) 1389–1397.
[62] C.A. Kennedy, M. Stancescu, R.A. Marriott, M.A. White, Recommendations for accurate heat capacity measurements using a Quantum Design physical property measurement system, Cryogenics 47 (2007) 107–112. A. Scheie, LongHCPulse: Long Pulse Heat Capacity on a Quantum Design PPMS, Institute for Quantum Matter and Department of Physics and Astronomy Johns Hopkins University, Baltimore, MD, 2017. K. Umeo, Improved AC-calorimeter for measuring the absolute value of specific heat up to 10 GPa, J. Phys.: Conf. Ser. 950 (2017), 042022. J. Pak, B. Wunderlich, Heat capacity by Sawtooth-modulated, standard heat-flux differential scanning calorimeter with close control of the heater temperature, Thermochim. Acta 367–368 (2001) 229–238.
[63] B. Wunderlich, H. Baur, Heat capacities of linear high polymers, Adv. Polym. Sci. 7 (1970) 151–368.
[64] M. Pyda, B. Wunderlich, Reversing and nonreversing heat capacity of poly(lactic acid) in the glass transition region by TMDSC, Macromolecules 38 (2005) 10472–10479.

[65] B. Wunderlich, Y. Jin, A. Boller, Mathematical description of differential scanning calorimetry based on periodic temperature modulation, Thermochim. Acta 238 (1994) 277–293.
[66] A. Czerniecka-Kubicka, J. Schliesser, B.F. Woodfield, I. Zarzyka, M. Pyda, Vibrational heat capacity of poly(N-isopropylacrylamide), Polymer 63 (2015) 108–115.
[67] M. Pyda, in: E. Piorkowska, G. Rutledge (Eds.), Handbook of Polymer Crystallization, John Wiley & Sons, Inc, 2013, pp. 265–286.
[68] G. Strobl, The Physics of Polymers, Concepts for Understanding Their Structures and Behavior, Springer-Verlag, Berlin, 2007.
[69] M. Varma-Nair, B. Wunderlich, Heat capacity and other thermodynamic properties of linear macromolecules X. Update of the ATHAS 1980 Data Bank, J. Phys. Chem. Ref. Data 20 (1991) 349.
[70] D.E. Kirkpatrick, B. Wunderlich, Thermal analysis of the phase transitions of poly(p-xylylene), Die Makromol. Chem. 186 (1985) 2595–2607.
[71] D. Tonchev, S.O. Kasap, Thermal characterization of glasses and polymers by temperature modulated differential scanning calorimetry: glass transition temperature, in: C.A. Brebbia, W.P. De Wilde (Eds.), High Performance Structures and Materials II, WIT Press, 2004, pp. 223–232.
[72] A. Peled, E. Hadziioannou, Thermal characterization of amorphous selenium films obtained by low-temperature photodeposition, J. Mater. Sci. 26 (1991) 1769–1774.
[73] M.L. Di Lorenzo, M. Gazzano, M.C. Righetti, The role of the rigid amorphous fraction on cold crystallization of poly(3-hydroxybutyrate), Macromolecules 45 (2012) 5684–5691.
[74] C. Schick, A. Wurm, A. Mohamed, Vitrification and devitrification of the rigid amorphous fraction of semicrystalline polymers revealed from frequency-dependent heat capacity, Colloid Polym. Sci. 279 (2001) 800–806.
[75] Z. Qiu, T. Ikehara, T. Nishi, Crystallization behaviour of biodegradable poly(ethylene succinate) from the amorphous state, Polymer 44 (2003) 5429–5437.
[76] I. Okazaki, B. Wunderlich, Reversible melting in polymer crystals detected by temperature-modulated differential scanning calorimetry, Macromolecules 30 (1997) 1758–1764.
[77] J. Chiu, P.G. Fair, Determination of thermal conductivity by differential scanning calorimetry, Thermochim. Acta 34 (1979) 267–273.
[78] A.K. Sircar, J.L. Wells, Thermal conductivity of elastomer vulcanizates by differential scanning calorimetry, Rubber Chem. Technol. 54 (1982) 191–207.
[79] K.Y. Keating, C.S. McLaren, Thermal conductivity of polymer melts, Thermochim. Acta 166 (1990) 69–76.
[80] A. A. Duswalt, 48th Calorimeter Conference, Durham NC, (1993) 10.
[81] S.M. Marcus, R.L. Blaine, Thermal conductivity of polymers, glasses and ceramics by modulated DSC, Thermochim. Acta 243 (1994) 231–239. S.Z.D. Cheng, Handbook of Thermal Analysis and Calorimetry. Vol. 3: Applications to Polymers and Plastics, Elsevier Science B. V., 2002.
[82] C.M.A. Lopes, M.I. Felisberti, Thermal conductivity of PET/(LDPE/AI) composites determined by MDSC, Polym. Test. 23 (2004) 637–643.
[83] G.L. Janz, F.J. Kelly, J.L. Perano, Melting and pre-melting phenomena in alkali metal nitrates, J. Chem. Eng. Data 9 (1964) 133.
[84] M. Wagner, Thermal Analysis in Practice: Fundamental Aspects, Hanser Publications, Munich, 2018, p. 287.
[85] C.B. Thompson, Analysis of the Mechanical Properties of DGEBF/4,4′-DDS, Thermoset Systems as a Function of Cure Rate, Thesis, The University of Southern Mississippi, 2015.

[86] B. M-Leszczak, M. Kisiel, J.B. Lechowicz, N. Buszta, R. Ostatek, M. Włodarska, Analysis of curing reaction of liquid-crystalline epoxy compositions by using temperature-modulated DSC TOPEM®, J. Therm. Anal. Calorim. 138 (2019) 2435–2444.

[87] S. Montserrat, I. Cima, isothermal curing of an epoxy resin by alternating differential scanning calorimetry, Thermochim. Acta 330 (1999) 189–200.

[88] J.M. Hutchinson, F. Shiravand, Y. Calventus, I. Fraga, Isothermal and non-isothermal cure of a tri-functional epoxy resin (TGAP): a stochastic TMDSC study, Thermochim. Acta 529 (2012) 14–21.

[89] Y. Li, M.R. Kessler, Cure kinetics of liquid crystalline epoxy resins based on biphenyl mesogens, J. Therm. Anal. Calorim. 117 (2014) 481–488.

[90] S. Sprenger, M.H. Kothmann, V. Altstaedt, Carbon fiber-reinforced composites using an epoxy resin matrix modified with reactive liquid rubber and silica nanoparticles, Compos. Sci. Technol. 105 (2014) 86–95.

[91] R.S.N. Wijayasingh, Phase Behavior of Lactose as Affected by Presence of Lactic Acid and Minerals, M.Sc. Thesis, College of Health and Biomedicine, Faculty of Health, Engineering and Science, Victoria University, 2015.

[92] The Use of TOPEM® to Analyze the First Heating Run of Biodegradable Disposable Cups, Thermal Analysis Application No. UC384, METTLER TOLEDO Thermal Analysis UserCom 38.

[93] A.K. Pandey, P.T.N. Diep, R. Patwa, V. Katiyar, S. Sasaki, S. Sakurai, in: V. Katiyar, A. Kumar, N. Mulchandani (Eds.), DSC and SWAXS Studies on the Effects of Silk Nanocrystals on Crystallization of Poly(L-Lactic Acid), Springer Nature Singapore Pte Ltd, 2020, https://doi.org/10.1007/978-981-15-1251-3 (Chapter 14).

Chapter 3

Thermal analysis in materials science

Various ways have been proposed to construct nanostructured inorganic materials from metal-organic frameworks (MOFs) and porous coordination polymers (PCPs) [1,2]. These include porous carbons [3], metal oxides [4,5], metal sulfides [6,7], metal phosphides [8], metal carbides [9], and metallic nanoparticles [10]. They also include new multifunctional hybrid materials (composites) due to their highly controllable composition, structures, and porosity [11], with potential in catalytic, biological, and energy-related applications.

For example, the thermal decomposition of these materials in different environments yields functional nanomaterials. The regulated thermal treatment of their nanostructures as precursors, in air or inert atmosphere, yields nanoporous metal oxides, sulfides, phosphides, carbides, and metal or alloy nanoparticles [4,8]. The resulting metal frameworks participate as metal sources, and the organic components, which can be removed by a simple process of calcination, produce pores. The original morphologies are almost retained, even after the thermal treatments. This strategy has proven to be a promising solid-state chemistry route for the fabrication of nanoporous metal oxides, carbides, and metal and alloy nanoparticles with similar morphologies. Thermal analysis methods are applied for studying, tracking, and understanding the thermal decomposition of MOFs and PCPs in different environments, as well as for calculating the weight loss or gain by heating [12,13]. In addition, TGA is applied to estimate the interlatticed water molecules in the crystal structure.

1 Production of nanostructured metal oxides

The urgent need for nanoporous metal oxides with well-defined compositions, fine morphologies, and highly crystallized frameworks is motivating scientists to discover novel and facile synthetic methods for their potential in catalytic, biological, and energy-related applications. A solid-state strategy for preparing nanoporous inorganic materials with a predictable nanostructure is highly promising and necessary for large-scale production. By using, for example,

Thermal Analysis. https://doi.org/10.1016/B978-0-323-90191-8.00003-8

cyano-bridged coordination polymers (CPs), many nanoporous metal oxides can be prepared by annealing in ambient conditions. MOFs and PCPs have a large fraction of metal ions among the organic ligands that can be used as metal sources, while the organic components can be removed by simple calcination and produce abundant pores. Thus, during thermal treatment, the metal content is retained and it takes part in the formation of the framework.

The thermal degradation of MOF and PCP nanostructures is mainly studied using TGA method nowadays. It is well known that a multi-step transformation often happens during heating of MOF and PCP nanostructures (Fig. 3.1). In the first step, water molecules are removed by heating from room temperature up to ~150°C. The second weight loss can be assigned to the organic groups or ligands bridges which start to release at ~300°C. During this stage, the metal framework is oxidized concurrently. Finally, a recrystallization process occurs without any weight loss producing nanostructured metal oxides with the same morphology as that of the starting MOFs and PCPs.

Pang et al. [14] have produced $Mn_3[Co(CN)_6]_2{\cdot}nH_2O$ with a uniform corner truncated nanocubic precursor and, by controlling the calcination conditions, have obtained porous corner-truncated nanocubic Mn_3O_4-Co_3O_4 composite nanomaterials. As shown in the TGA thermogram (Fig. 3.2), the decomposition process passes through two steps. The first step represents the loss of water molecules as the temperature is increased from ambient to 160°C. The second step represents the removal of the cyanide groups after heating up to 370°C. This is the usual decomposition behavior of compounds that contain CN groups. The composite reflects different degrees of porosity and surface roughness due to heating at different intervals. The surface fuses after heating for 3 h, thus heating for a short time is preferred.

Moreover, the regular thermal decomposition of the as-prepared Mn-Ru CP nanocubes by using TGA are shown in Fig. 3.3 [15]. In the first stage of heating from room temperature to ~300°C, there is a small weight loss (~5%). This

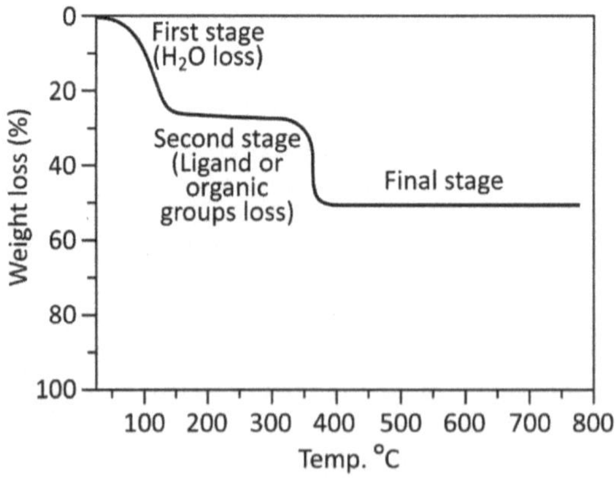

FIG. 3.1 TGA thermogram for MOFS and PCPs nanostructures in air.

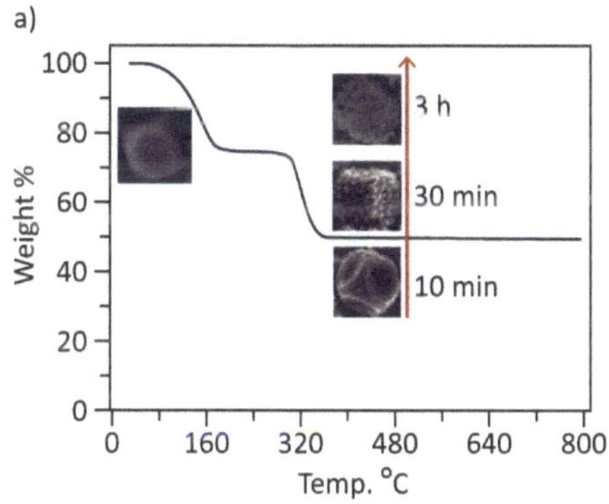

b)

$$3Mn_2[Co(CN)_6]_2 \cdot nH_2O + 46O_2 \longrightarrow 3Mn_3O_4 + Co_3O_4 + 36CO_2 + nH_2O$$

FIG. 3.2 (A) TGA curve and (B) the decomposition equation of $Mn_3[Co(CN)_6]_2 \cdot nH_2O$ [14].

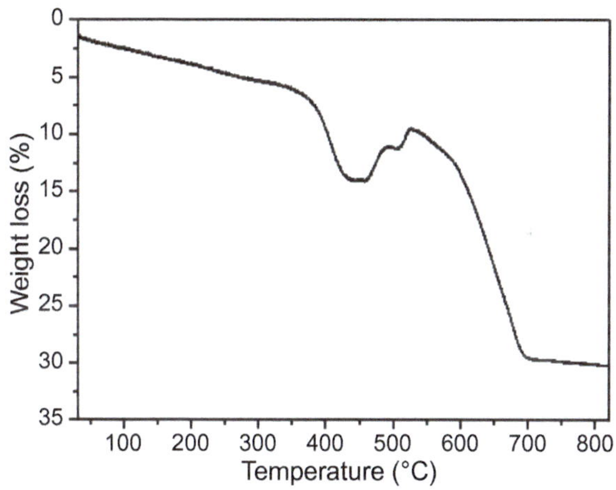

FIG. 3.3 TGA curve of the as-prepared Mn-Ru CP nanocubes at a heating rate of $5°C\,min^{-1}$ from room temperature to around 800°C [15].

weight loss is expected to correspond to the removal of interlatticed water molecules, which reveals a finite contribution of water molecules in the as-prepared Mn-Ru CP. On continuous heating, a sharp weight loss occurs from ~300°C to ~400°C (~11%). This weight loss corresponds to the removal of CN groups and at the same time, recrystallization of metal oxide starts. Then, with further increase in the temperature up to ~500°C, a weight gain is observed (~4%). This reveals the presence of metallic phase (expected to be Ru metal) at the second stage, which converts into oxides by further heating. During this stage, the weight gain corresponds to the addition of oxygen. Since there are two transition metals at this stage (Mn and Ru), the crystallization temperature of oxides are different. In the last stage, all Mn-Ru oxides are completely degraded.

The thermal degradation of $Fe_4[Fe(CN)_6]_3 \cdot xH_2O$ at different temperatures was also reported in an effort for the preparation of various nanoporous iron oxides [16]. The morphology of the yields after calcination at 250°C and 400°C retained the cubic shape, but their sizes were slightly reduced and roughness of their surface increased. The main cause of the reduced particle size is the removal of the cyanide groups and the interstitial water molecules during heating, which leads to the formation of magnetic γ-Fe_2O_3. Other spinel metal oxides were obtained (e.g., Co_2O_3 and $FeCo_2O_4$) by annealing $Co_3[Co(CN)_6]_2 \cdot nH_2O$ and $Fe_3[Co(CN)_6]_2 \cdot nH_2O$ in air under controlled conditions [17]. Hu et al. have demonstrated a strategy for the fabrication of binary $Fe_xCo_{3-x}O_4$ porous nanostructures, involving the retained morphology and the thermal transformation of $Fe_3[Co(CN)_6]_2 \cdot nH_2O$ nanoparticles (Fig. 3.4) [18]. After pyrolysis of sphere-like $Fe_3[Co(CN)_6]_2 \cdot nH_2O$ nanoparticles, binary nanoporous $Fe_xCo_{3-x}O_4$ is formed. The $Fe_xCo_{3-x}O_4$ product is composed of many nanoparticles, without significant aggregation, and its morphology has undergone negligible change as compared to the precursor.

Yan et al. have prepared Co_3O_4 nanocages, nanoparticles, and a solid dendritic structure by using a simple strategy that involves the thermal degradation of nanoparticles of anhydrous $Co_3[Co(CN)_6]_2$ at various temperatures [19]. TGA thermograms of the as-prepared $Co_3[Co(CN)_6]_2$ and $Co_3[Co(CN)_6]_2 \cdot nH_2O$ are shown in Figs. 3.5 and 3.6. $Co_3[Co(CN)_6]_2$ nanocubes calcined at 450°C and 650°C yielded nanocages of porous Co_3O_4. This porous structure formation phenomenon can be explained in the terms of "Kirkendall effect" phenomenon. This phenomenon is based on a nonequilibrium interdiffusion process, which leads to

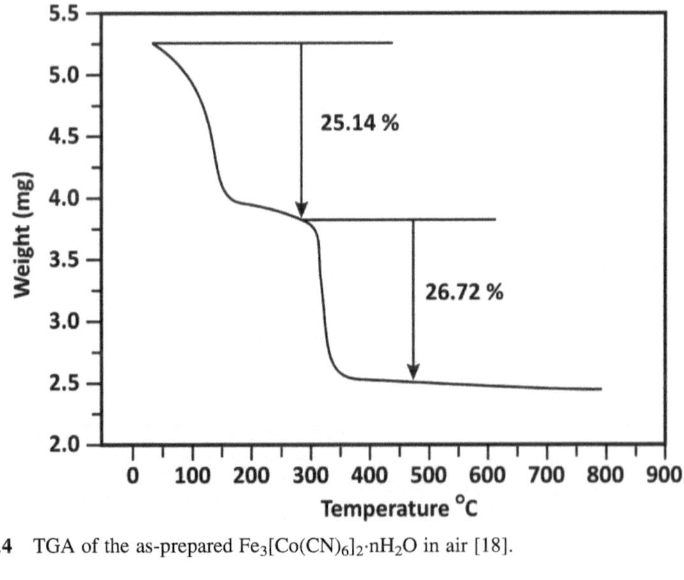

FIG. 3.4 TGA of the as-prepared $Fe_3[Co(CN)_6]_2 \cdot nH_2O$ in air [18].

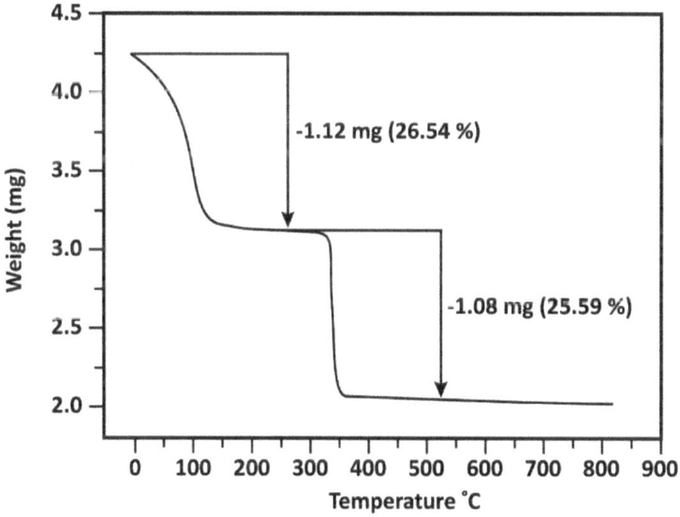

FIG. 3.5 TGA of the as-prepared $Co_3[Co(CN)_6]_2$ [19].

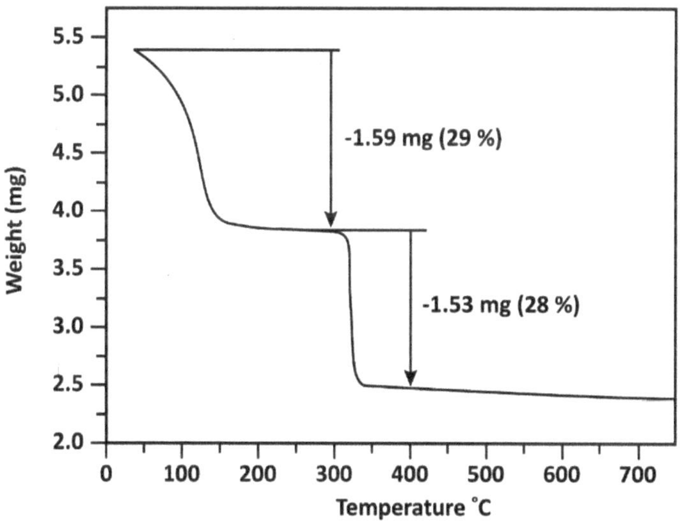

FIG. 3.6 TGA of the as-prepared $Co_3[Co(CN)_6]_2 \cdot nH_2O$ [20].

the formation of nanocages. The morphology starts to change upon heating at 650°C because the nanocubes consist of primary particles that fall into smaller grains and the initial structures have collapsed. This change becomes apparent after heating at 750°C, at which time new dendritic structures replace the nanocubes. As the calcination temperature increased up to 850°C, the grains gather, recrystallize, and rearrange into large irregular nanoparticles. This kind of mechanism is explained by "Ostwald ripening" (i.e., a template-free strategy), which can be described as the growth of large crystals from small ones. The overall thermal reaction is shown in Fig. 3.7A.

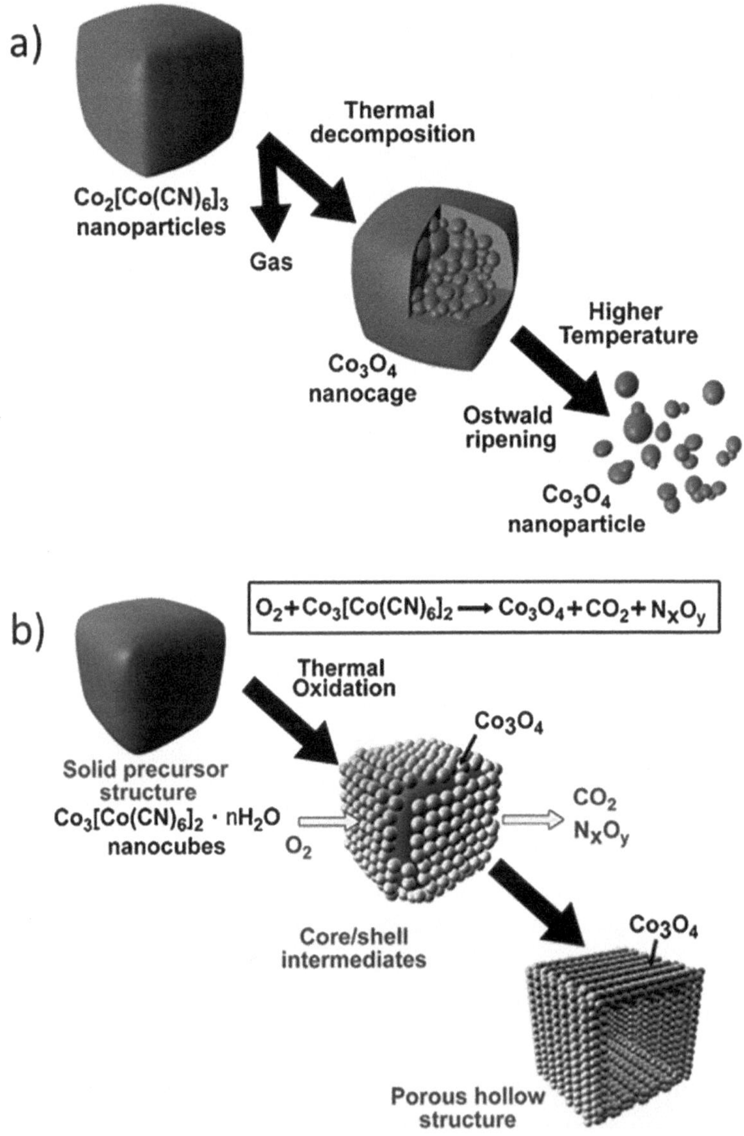

FIG. 3.7 (A) Synthetic protocol of Co_3O_4 nanocages and nanoparticles via thermal decomposition of $Co_3[Co(CN)_6]_2$. (B) Illustration of the synthetic procedures used to fabricate Co_3O_4 nanocages by a thermal treatment process of $Co_3[Co(CN)_6]_2{\cdot}nH_2O$.

The term "Kirkendall Effect" is explained by different atomic diffusive rates of elements under heating. Due to the difference in their diffusion rates, the generation of pores in the low-melting component side of the diffusion couple near the interface can create hollow nanostructures. For example, solid $Co_3[Co(CN)_6]_2{\cdot}H_2O$ is converted into Co_3O_4 nanocages by heating in air [20]. Their thermal decomposition in air is studied by TGA (Fig. 3.6). During the thermal oxidation stage, a thin intermediate shell is formed on the surface of $Co_3[Co(CN)_6]_2{\cdot}nH_2O$ nanocubes. Then, it acts as a barrier that separates the inner part of $Co_3[Co(CN)_6]_2{\cdot}nH_2O$ from aerobic oxygen. There are many voids/holes that allow oxygen to penetrate, and a subsequent outward diffusion of Co cations occurs. During oxidation processes, the diffusion rate of Co is greater than that of the atmospheric oxygen, which eventually generates hollow cavities in the Co_3O_4 nanocages (Fig. 3.7B).

Starting with $Mn_3[Co(CN)_6]_2{\cdot}nH_2O$ nanocubes, Zhang et al. have reported controlled thermally derived $Mn_xCo_{3-x}O_4$ nanocages with thin porous shells [21]. The thermal decomposition of $Mn_3[Co(CN)_6]_2{\cdot}nH_2O$ was tracked by TGA (Fig. 3.8), which shows a behavior similar to that of cyano-bridged CPs. The fabrication process for the $Mn_xCo_{3-x}O_4$ nanocages involves two steps (Fig. 3.9A). Shape-controlled $Mn_3[Co(CN)_6]_2{\cdot}nH_2O$ precursor nanocubes, with well-defined crystal structures, were synthesized. Polyvinylpyrrolidone (PVP) was used as the capping agent during this step to direct the anisotropic growth of $Mn_3[Co(CN)_6]_2{\cdot}nH_2O$ through a preferential adsorption on specific crystal facets. Thus, the PVP drives Mn^{2+} and $[Co(CN)_6]^{3-}$ to self-assemble into cube-shaped nanocrystals. Then, the nanocrystals increase in size, rearrange, and homogenize, and the crystalline $Mn_3[Co(CN)_6]_2{\cdot}nH_2O$ nanocubes are further enhanced with less defects. Repeated washing and recrystallization

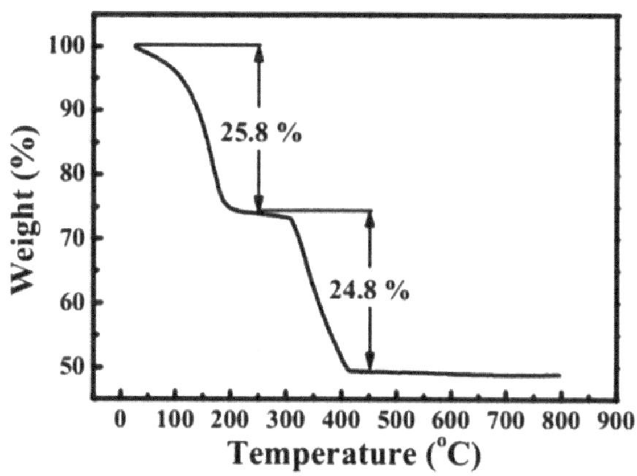

FIG. 3.8 TGA curves of the thermal decomposition of $Mn_3[Co(CN)_6]_2{\cdot}nH_2O$ precursor nanocubes in air flow with a heating rate of $10°C\,min^{-1}$.

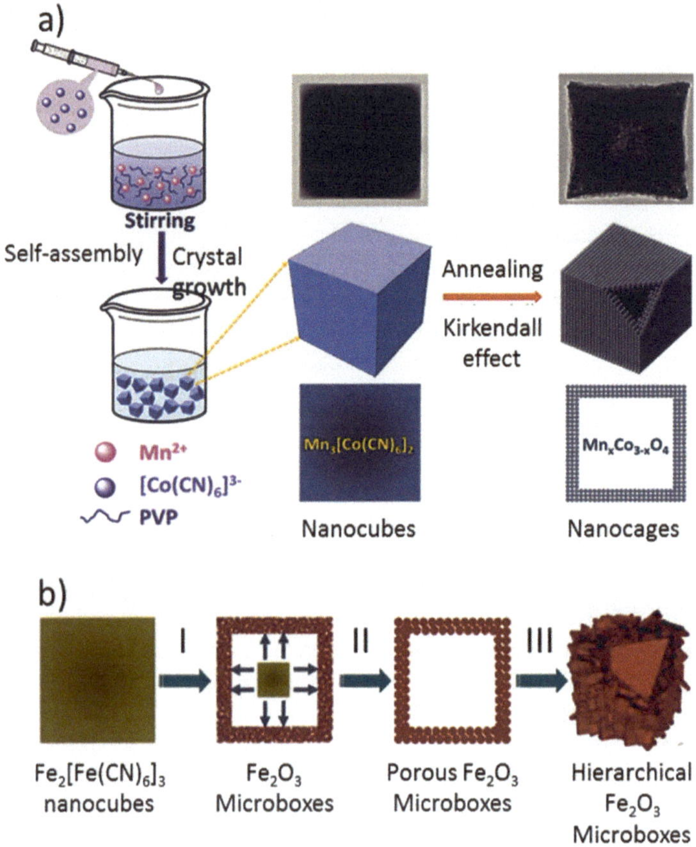

FIG. 3.9 (A) Schematic illustration of the formation of $Mn_xCo_{3-x}O_4$ nanocages from $Mn_3[Co(CN)_6]_2{\cdot}nH_2O$ nanocubes. (B) Illustration of the fabrication process of hollow Fe_2O_3 microboxes, and their shell evolution at elevated temperatures. *((A: Reproduced from L. Zhang, L. Shi, L. Huang, J. Zhang, R. Gao, D. Zhang, Rational design of high-performance DeNOx catalysts based on $Mn_xCo_{3-x}O_4$ nanocages derived from metal-organic frameworks, ACS Catal. 4 (2014) 1753–1763. Copyright© 2014, American Chemical Society. B: Reproduced from L. Zhang, H.B. Wu, S. Madhavi, H.H. Hng, X.W. Lou, Formation of Fe_2O_3 microboxes with hierarchical Shell structures from metal-organic frameworks and their lithium storage properties, J. Am. Chem. Soc., 134 (2012) 17388–17391. Copyright© 2012, American Chemical Society.))*

decrease the defects on the surfaces of the nanocubes. Then, $Mn_xCo_{3-x}O_4$ nanocages are formed via annealing $Mn_3[Co(CN)_6]_2{\cdot}nH_2O$ nanocubes. A variety of organic and inorganic species in the as-prepared $Mn_3[Co(CN)_6]_2{\cdot}nH_2O$ nanocubes, including PVP, crystal water, and cyanide ligands, can be decomposed and volatilized by thermal treatment in air to generate hollow and porous structures.

Additionally, Fe_2O_3 microboxes with hierarchically nanoporous shells have been synthesized by thermal degradation of $Fe_4[Fe(CN)_6]_3$ microcubes in

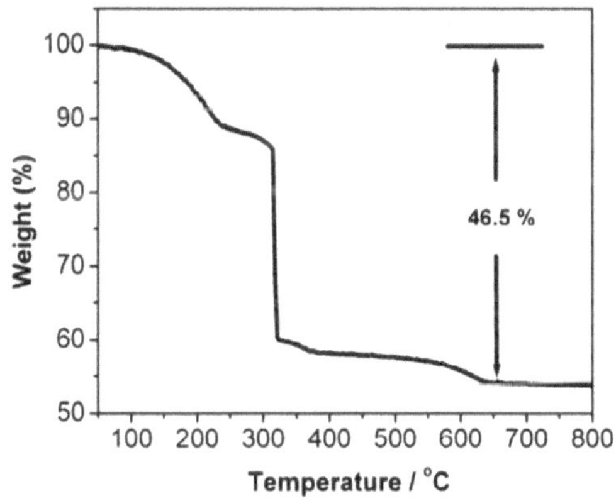

FIG. 3.10 TGA curve of $Fe_4[Fe(CN)_6]_3$ microcubes thermal degradation in an air flow.

aerobic conditions at various temperatures [22]. Through a simultaneous decomposition and oxidation of $Fe_4[Fe(CN)_6]_3$ microcubes accompanied by a crystallization of iron oxide shells, a scalable synthesis of anisotropic hollow structures could be realized. The various shell structures were monitored by tuning the annealing temperature (Fig. 3.9B). Uniform $Fe_4[Fe(CN)_6]_3$ microcubes were synthesized, and Fe_2O_3 microboxes were obtained after subsequent thermal treatment in air.

The reaction takes place in various stages according to TGA curve (Fig. 3.10). A first stage occurs below 350°C leading to the transformation of $Fe_4[Fe(CN)_6]_3$ into iron oxide in the near-surface region. By continuous heating of $Fe_4[Fe(CN)_6]_3$ microcubes, a large cavity is formed and continuous gas evolution with weight loss occurs. When the annealing temperature is increased up to 550°C (stage II), Fe_2O_3 microboxes, with relatively smooth and dense shells, are transformed into highly porous microboxes derived from enlarged Fe_2O_3 nanoparticles.

Hu et al. have announced the use of Prussian blue (PB) $Fe_4[Fe(CN)_6]_3{\cdot}xH_2O$ nanocubes for the synthesis of multiscale, in a controllable way, hollow superparamagnetic iron oxide particles with nanopores (Fig. 3.11) [23]. Calcination of small solid PB nanoparticles (SPB, around 110nm) at 250°C and 400°C generates amorphous iron oxide. Meanwhile, starting with correspondingly small hollow PB nanoparticles (SHPB, around 110nm) results in the formation of a mixed phase of crystalline α- and γ-Fe_2O_3 at higher temperature and amorphous γ-Fe_2O_3 is formed at lower temperature that shows superparamagnetism. A pure phase of crystalline γ-Fe_2O_3 is obtained by annealing large hollow PB (LHPB, around 190nm) particles at low temperature and α-Fe_2O_3 is obtained at higher temperature. This reflects the critical effect of size and hollow structure on the final thermal-derived iron products. In the same context, Ariga et al. [24] and Azhar et al. [25] have synthesized various PB analogues and used them for the synthesis of various hollow inorganic materials by annealing technique.

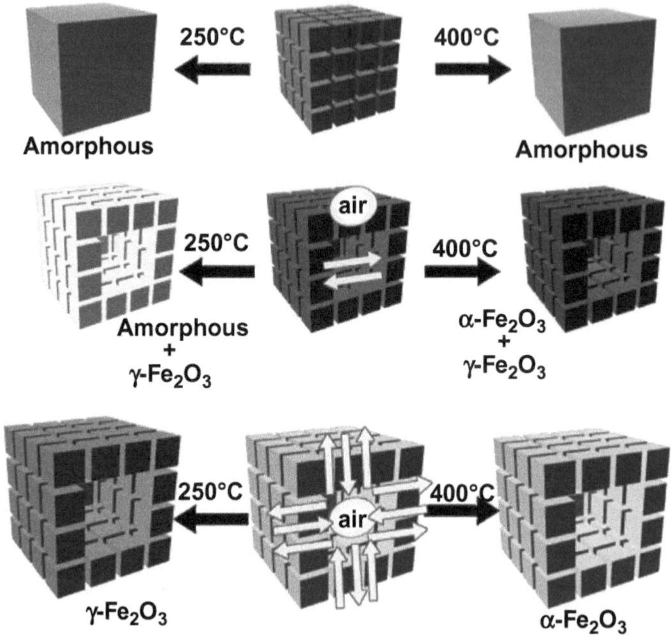

FIG. 3.11 Schematic illustrations of the decomposition of various PB nanocubes [SPB (110nm), SHPB (110nm), and LHPB (190nm)] into iron oxides through annealing in air at 250°C and 400°C.

The thermal decomposition process for original (PB) precursors was examined by TGA. TGA curves (Fig. 3.12) indicate that all three samples encountered a drastic weight loss around 250°C. The rates of weight loss of PB precursors were different from each other. LHPB showed the highest rate of weight loss, while SPB showed the lowest one. As the weight loss in TGA is related to thermal decomposition during the solid-state reaction, it was concluded that the thermal decomposition rates of the PB precursors varied by hollow cavity sizes of the nanocubes.

2 Estimation of carbon content

Kaneti et al. have reported the synthesis of highly microporous Ni-doped Co/CoO/NC hybrid particles with a rhombic dodecahedral morphology by calcination of bimetallic Ni-Co-zeolitic imidazolate framework (Ni-Co-ZIF) MOFs particles at 500°C under an argon atmosphere [26]. They used TGA to estimate carbon content. To determine the carbon content in the Ni-doped Co/CoO/NC hybrid, TGA measurement was carried out under an air atmosphere with a heating rate of $10°C\,min^{-1}$ (Fig. 3.13). The weight loss between 250°C and 320°C is mainly attributed to the combustion of N-doped carbon (NC). The amount of carbon in the hybrid material can be approximated to be 51 wt%.

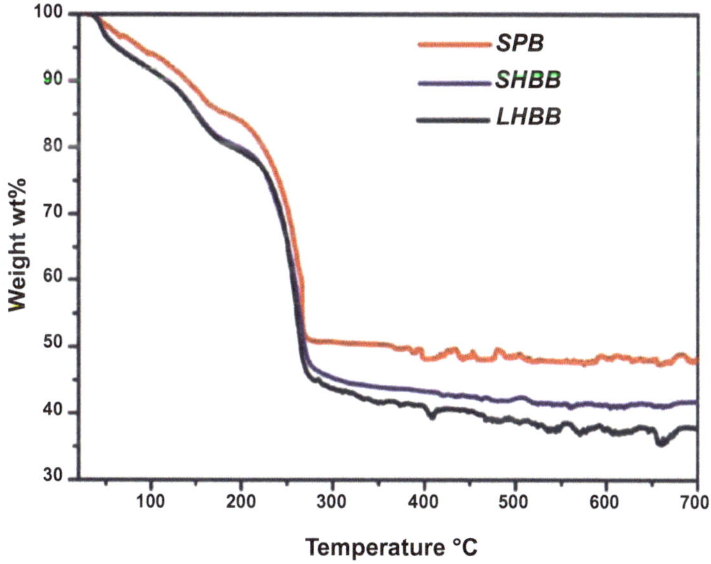

FIG. 3.12 TGA curves of thermal decomposition of SPB, SHPB, and LHPB of various PB nanocubes.

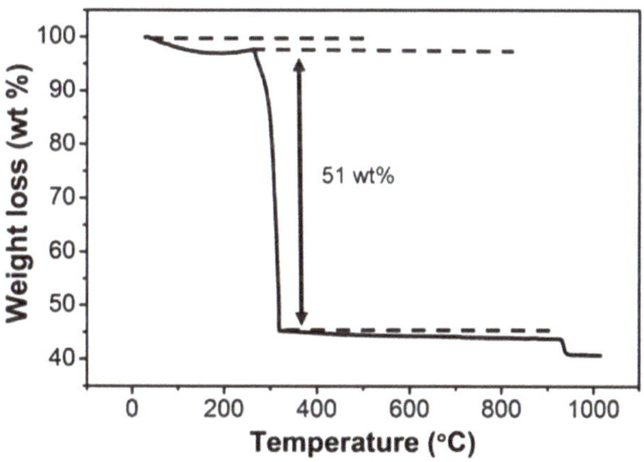

FIG. 3.13 TGA curve of the Ni-doped Co/CoO/N-doped carbon (NC); TGA was carried out under air atmosphere at a heating rate of $10^\circ C\,min^{-1}$.

Co_3O_4@NC was synthesized by Wang et al. [27] using ZIF-67 as the precursor and a dynamic template to obtain the desired morphology. During a controlled two-step calcination process, the central metal ions were transformed into Co_3O_4 NPs (around 5 nm). The organic linkers were carbonized to form interconnected nitrogen-doped carbon layers on these Co_3O_4 NPs [27].

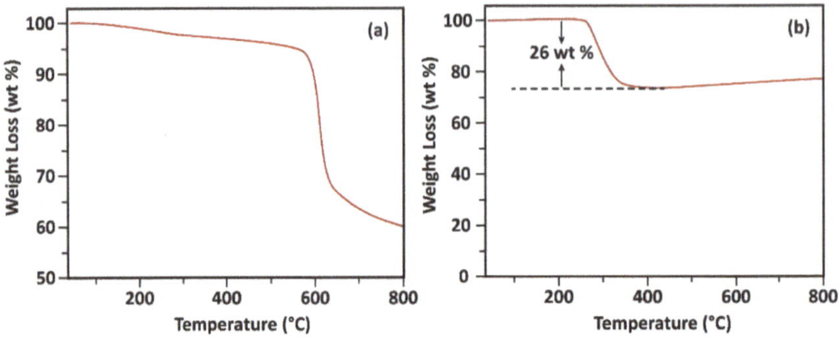

FIG. 3.14 TGA plot of (A) the ZIF-67 precursor; TGA was carried out in Ar with a heating rate of 10°C min^{-1} and (B) the as-obtained Co_3O_4@NC; TGA was carried out in air with a heating rate of 10°C min^{-1}.

By annealing ZIF-67 in Ar at 550°C, a dramatic weight loss was demonstrated between 500°C and 650°C according to TGA curves (Fig. 3.14A). This step of annealing led to the formation of metallic Co, which was subsequently converted to Co_3O_4 by heating at 150°C in air for 12h.

The carbon content of Co_3O_4@NC was determined by using TGA performed in air (Fig. 3.14B). The weight loss between 250°C and 370°C is associated with the combustion of NC. Therefore, the carbon content in Co_3O_4@NC was evaluated to be around 26wt%.

3 Production of metal carbides and metal and alloy nanoparticles

Hitherto, the thermal decomposition of MOF and PCP nanostructures has been reported to produce nanostructured metal carbides, metals nanoparticles, and alloys. For example, the thermal decomposition of hexacyanoferrate compounds in vacuum has been reported. The thermal decomposition reactions of europium hexacyanoferrate(III) and ammonium europium hexacyanoferrate(II) compounds occur through three steps [28]. The first step, heating at a low temperature (~370°C), which results in the formation of Eu hydroxide and $Fe(CN)_2$, as well as the evolution of cyanogen $(CN)_2$, hydrogen cyanide (HCN), and ammonia (NH_3) as normal products of the decomposition. In the second step, with continuous heating between 600°C and 700°C, $Fe(CN)_2$ further decomposes to Fe_3C. By heating at elevated temperatures, the iron carbide is transformed finally into metallic iron and carbon. Then, europium hydroxide loses water to recrystallize and forms EuO.

Aparicio et al. reported the thermal decomposition of PB in an inert atmosphere [29]. The decomposition reaction is based on the successive loss of water molecules at low temperature, followed by the release of cyano groups as normal decomposition components. Three principal stages were observed, including the loss of water, change of crystal structure, and the decomposition of PB. A monoclinic PB was obtained by heating at 400°C, while the formation of various polymorphs of iron carbides (e.g., Fe_2C) was observed by continuous heating at elevated temperatures. Increasing the temperature to above 700°C induces degradation of the primarily formed iron carbides into metallic iron, cementite, and graphite.

The thermal degradation of triple-shell PB hollow spheres was studied at 450°C, 550°C, and 700°C under nitrogen by Zakaria [30]. Well-retained nanoporous spheres (~270 nm) of iron carbides (Fe_2C) were observed after heating at 450°C. Graphite, free carbon, and metallic iron contaminants were present in the Fe_2C phase. Surprisingly, graphite and metallic iron contamination decreased after heating at high temperatures (i.e., 550°C and 700°C) (Fig. 3.15).

The XRD patterns (obtained at 450°C) are assignable to an orthorhombic Fe_2C as a main product (PDF card no. 03-065-1456). Some peaks corresponding to graphite, free carbon, and metallic iron are also present. Surprisingly, after heating at elevated temperatures (i.e., 700°C), the peaks corresponding to metallic iron and graphite improved and became sharper. This suggests that heating at elevated temperature in an inert atmosphere leads to the transformation of iron carbide particles into metallic iron (Fig. 3.15).

The in situ crystallization of $Ni(H_2O)_2[Ni(CN)_4]{\cdot}4H_2O$ nanoflakes on the surface of graphene oxide (GO) sheets was recently demonstrated [31,32]. A nucleation process of $Ni(H_2O)_2[Ni(CN)_4]{\cdot}4H_2O$ started from the surface of graphene oxide sheets, and then $Ni(H_2O)_2[Ni(CN)_4]{\cdot}4H_2O$-coated graphene sheets self-assembled to form ordered lamellar nanomaterials (layer-by-layer structures). The regulated thermal treatment process in nitrogen produces an

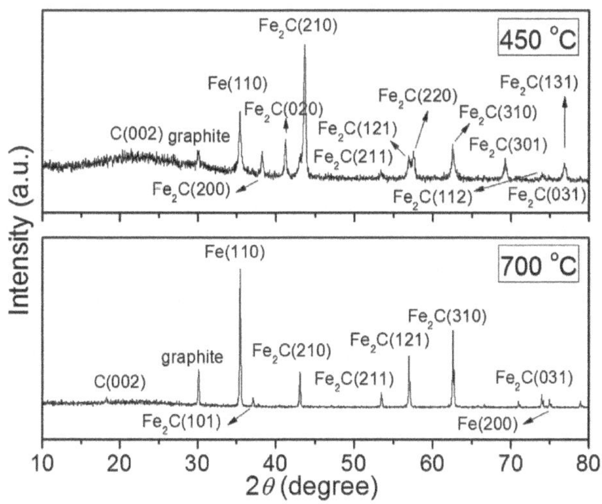

FIG. 3.15 Wide-angle XRD patterns of nanoporous iron oxide spheres obtained after calcination of triple-shelled PB spheres under nitrogen at 450°C and 700°C for 1 h.

Ni_3C-graphene oxide composite with similar layer-by-layer structure (Fig. 3.16). At the same time, in a similar synthetic approach, $Cu(H_2O)_2[Ni(CN)_4].4H_2O$ nanoflakes were successfully prepared on the surface of GO sheets, which are converted into CuNi-oxide/rGO composite by annealing in air [33].

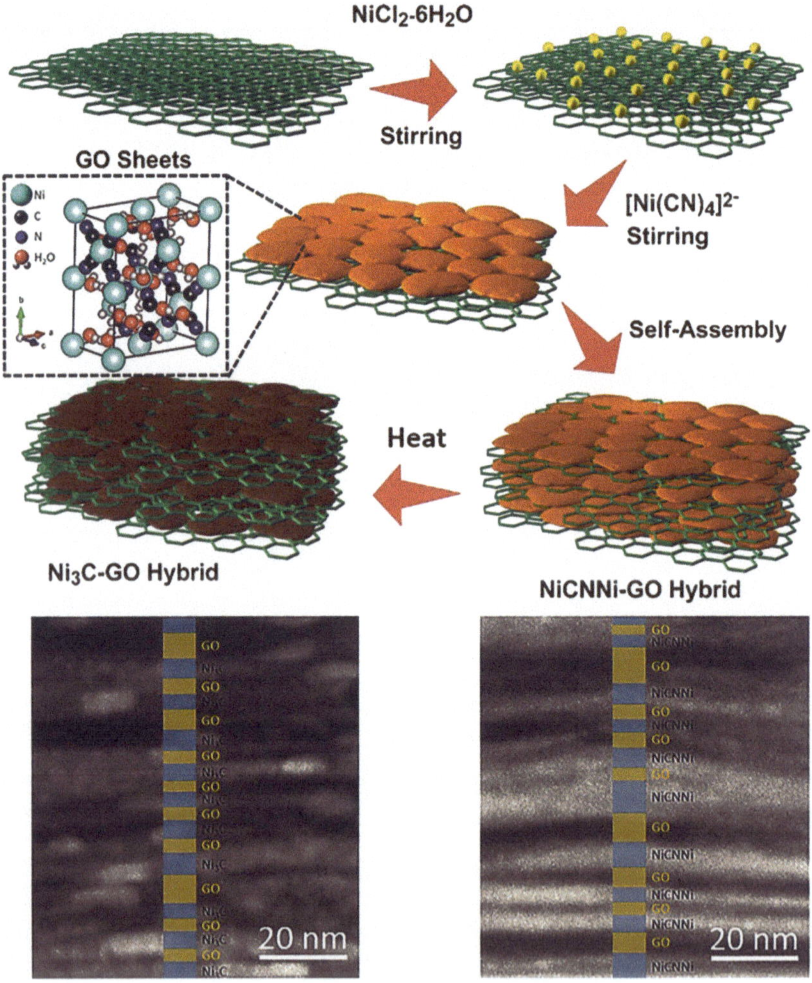

FIG. 3.16 Assembly of NiCNNi-GO hybrid layer-by-layer structure, and its thermal conversion to Ni_3C-GO hybrid. The cross-sectional TEM images of NiCNNi-GO and Ni_3C-GO hybrids and the crystal structure $Ni(H_2O)_2[Ni(CN)_4]{\cdot}4H_2O$ between the GO sheets are displayed. *((Reproduced from M.B. Zakaria, C. Li, Q. Ji, B. Jiang, S. Tominaka, Y. Ide, J.P. Hill, K. Ariga, Y. Yamauchi, Self-construction from 2D to 3D: one-pot layer-by-layer assembly of graphene oxide sheets held together by coordination polymers, Angew. Chem. Int. Ed. 55 (2016) 8426–8430. Copyrights© 2016 WILEY-VCH Verlag GmbH & Co. KGaA, Weinheim.))*

The thermal decomposition of $Ni(H_2O)_2[Ni(CN)_4].4H_2O$ flakes in an inert atmosphere (e.g., nitrogen) is expected to produce Ni_3C with a two-dimensional morphology. Thus, $Ni(H_2O)_2[Ni(CN)_4].4H_2O$ nanoflakes, without graphene oxide sheets, were heated in nitrogen at 350°C, 450°C, 550°C, and 650°C. The sample calcined at 350°C underwent incomplete removal of the organic components. The sample calcined at 450°C generated diffraction patterns that are assignable to pure crystalline Ni_3C (a trigonal *R-3c* structure). On further heating at 500°C and 650°C, Ni_3C is completely transformed into the cubic nickel metal (fcc-type). Although pure crystalline Ni_3C is obtained at 450°C, the original 2D morphology completely collapsed and destroyed because of the crystallization and fusion of the metal framework. On the other hand, the NiCNNi-graphene oxide layer-by-layer structure is relatively stable, even with thermal treatment at elevated temperatures, leading to the formation of Ni_3C-graphene oxide with morphology similar to NiCNNi-graphene oxide layer-by-layer structure of the starting precursor.

The thermal decomposition of $Ni(H_2O)_2[Ni(CN)_4]\cdot 4H_2O$ CP nanoflakes under nitrogen was further studied by TGA as presented in Fig. 3.17 [34]. It shows the same decomposition behavior as the CN-containing compounds. It is well known that a multi-step transformation often occurs during heating. In the first step, water molecules are removed (weight loss of ~32.75%) by heating from room temperature up to ~160°C. There are $4H_2O$ molecules calculated from TG curve, thus the final structure would be $Ni(H_2O)_2[Ni(CN)_4]$, which matches Hoffman type. Then, the obtained product was thermally stable until ~300°C and the second weight loss started from 300°C until ~520°C (~16.46%). This weight loss is anticipated to correspond to the loss of CN groups, which are released as CN_2 and N_2 gases. The third weight loss, which started from ~520°C, corresponds to the organic residue (e.g., graphitic carbon); in the final stage, fcc-Ni metal is obtained.

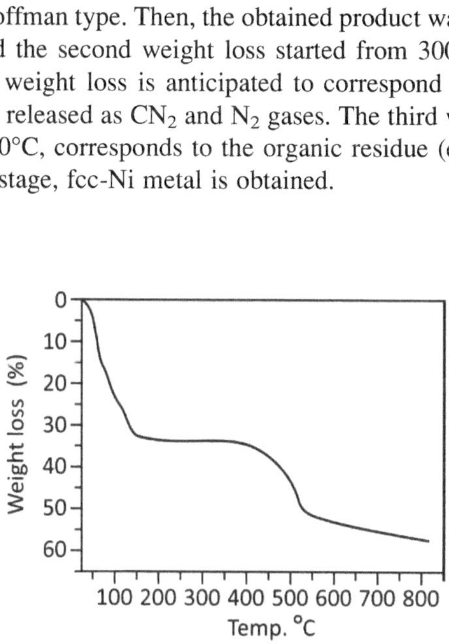

FIG. 3.17 TGA curve of the as-prepared $Ni(H_2O)_2[Ni(CN)_4]\cdot 4H_2O$ CP nanoflakes at a heating rate of $5°C\,min^{-1}$ under nitrogen gas flow from room temperature to 800°C.

The theoretical calculation of nickel contribution in $Ni(H_2O)_2[Ni(CN)_4]\cdot 4H_2O$ CP is around 31.44%, but the weight residue after the thermal decomposition process is around 43.1%. Hence, it was expected that Ni is contaminated by carbon, which is in accordance with the reports indicating that nickel metal together with solid amorphous carbon are obtained by thermal decomposition of Ni_3C in inert atmosphere [34].

4 Production of metal sulfides, selenides, and phosphides

The nanostructured metal sulfide fabrications have attracted extensive interests because of their potential in various applications, including solar cells, fuel cells, optoelectronics, thermoelectrics, memories, supercapacitors, and laser-induced breakdown spectroscopy (LIBS) devices. Among the investigated metal sulfides, zinc sulfide (ZnS), copper sulfide (Cu_2S), nickel sulfides (Ni_xS_y), and cobalt sulfides (Co_xS_y) are the most studied transition metal sulfides recently [35].

Sennu et al. reported the synthesis of two-dimensional (2D) Co_3S_4 in a nanosheet-like morphology via simple hydrothermal process [36]. TGA and DTA for the as-synthesized Co_3S_4 NS are performed (Fig. 3.18). The TGA curve has five distinct stages of weight loss. The first weight loss stage ended at 240°C, stemming from evaporation of trace water with organic content (i.e., thioacetamide, corresponding to the formation of CoS_2 phase), it is an endothermic reaction. The second weight loss began at 240°C until 440°C and is attributed to the desulfurization of CoS_2 phase in the exothermic backbone with subsequent formation of Co_3S_4 phase. Following this, desulfurization was

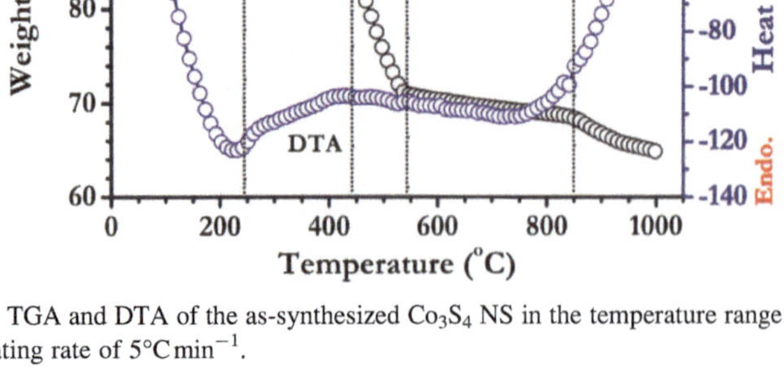

FIG. 3.18 TGA and DTA of the as-synthesized Co_3S_4 NS in the temperature range: 20–1000°C and at a heating rate of $5°C\,min^{-1}$.

caused by the formation of different crystalline CoS (440–550°C) and Co_9S_8 (550–840°C) phases occurring at different temperatures from 450°C to 1000°C.

Apart from metal sulfides, scarce reports demonstrate that MOFs have also been successfully used to derive metal selenide nanostructures [35]. For example, Hu et al. demonstrated a facile formation of hybrid nanoboxes with complex shell architecture where a CoSe-enriched inner shell is intimately confined within a carbon-enriched outer shell (abbreviated as CoSe@carbon nanoboxes). The synthesis is realized through the manipulation of the template-engaged reaction between Co-based zeolitic imidazolate framework (ZIF-67) nanocubes and Se powder at elevated temperatures. The content of CoSe in the CoSe@carbon nanoboxes was determined by TGA (Fig. 3.19) to be about 69 wt% [37]. During the TGA measurement in air flow, carbon would be burned out, while CoSe is converted into Co_3O_4 with 39 wt% of original mass retained. Based on the following chemical reaction: $3CoSe_{(s)} + 2O_{2(g)} = Co_3O_{4(s)} + 3Se_{(g)}$, the contents of CoSe and carbon are calculated to be 69 wt% and 31 wt%, respectively.

Sun et al. developed a facile method for synthesizing $CoSe_2$ nanoparticles uniformly anchored on carbon fiber paper ($CoSe_2$/CF) via pyrolysis and selenization of in situ grown zeolitic imidazolate framework-67 (ZIF-67), as shown in Fig. 3.20 [38]. First, the CF paper was treated with nitric acid to introduce

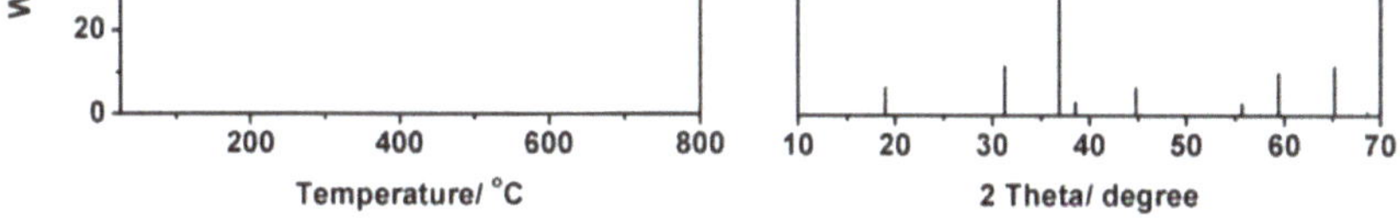

FIG. 3.19 (A) TGA curve of the CoSe@carbon nanoboxes at a ramping rate of $10°C\,min^{-1}$ from room temperature to 800°C, and (B) XRD pattern of the product after TGA measurement.

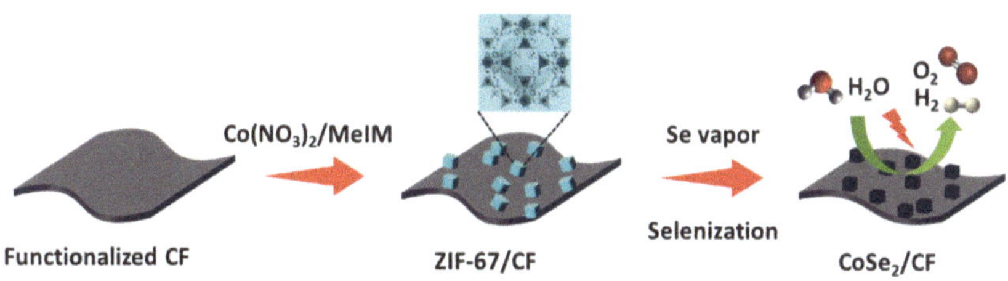

FIG. 3.20 Schematic representation of the synthesis of $CoSe_2$ nanoparticles on CFs.

abundant carboxylic (—COOH) groups, which can serve as nucleation centers for the growth of self-assembled MOFs. Several cycles of alternate introduction of Co^{2+} ions and an organic linker (2-methylimidazole) onto the CF paper led to the formation of ZIF-67 nanocrystals. Co^{2+} ions were readily adsorbed via electrostatic interactions with the negatively charged —COOH groups on the CFs. Pyrolysis and selenization in the tube furnace in an Ar atmosphere containing Se vapor converted the MOF nanocrystals to $CoSe_2$ nanoparticles.

The nanosized transition metal phosphides have undergone extensive studies recently because of their potential applications as electrocatalysts for hydrogen and oxygen evolution reactions, anode materials for lithium and sodium ion batteries, catalysts for hydrodenitrogenation and hydrodesulfidation [35]. Very recently, MOFs have been demonstrated as excellent precursors for the fabrication of porous or hollow transition metal phosphides nanomaterials.

For example, Liu and Li have reported the synthesis of hollow CoP polyhedrons using ZIF-67 polyhedrons as precursors. In their proposed strategy, ZIF-67 polyhedrons with sizes of around 700nm were initially calcined in air at 350°C to yield Co_3O_4 polyhedrons, followed by subsequent phosphidation with hydrated sodium hypophosphite ($NaH_2PO_2{\cdot}H_2O$) (ratio = 1:20) at 250°C in a N_2 atmosphere to obtain hollow CoP polyhedrons with uniform distribution of P element. When heated at temperatures ≥250°C, hypophosphites decomposed to release PH_3, which could directly react with metal oxides or MOFs to form TMPs ($2NaH_2PO_2 = PH_3 + Na_2HPO_4$) [39,40].

5 TGA in sol-gel science

Soft templating method is a promising way to fabricate mesoporous inorganic materials using a large structure directing agents as templates which are then removed by thermal treatments (Fig. 3.21). Concerning the soft-templating method, it is very common to use various structure directing agents (e.g., CTAB, P123, and F127) to form mesoporous metal oxides. Soft template-based methods gained more attention because the templates could be easily removed by calcination and present opportunities for achieving various porous morphologies [41].

In the solution synthesis route, the key issue is a controllable hydrolysis-condensation behavior of the metal species. A strong interaction between templates and precursors is necessary to form ordered mesostructure. The coassembly process is driven by weak noncovalent bonds such as hydrogen bonds and electrostatic bonds between surfactants and inorganic species. Cooperative assembly between inorganic precursors and surfactants is generally involved simultaneously with the hydrolysis and condensation process, leading to mesostructured composites. After the removal of templates (e.g., surfactants or block copolymers) by thermal treatments (Table 3.1), mesoporous metal oxides frameworks are obtained (Fig. 3.21). The removal of templates and degradation of polymers can be studied by TGA [61].

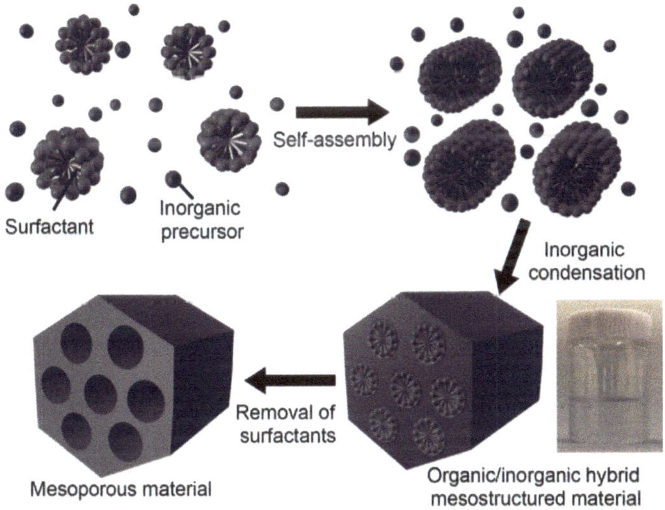

FIG. 3.21 Schematic representation of the soft-templating pathway leading to mesoporous non-siliceous materials.

TABLE 3.1 Degradation temperatures of some surfactants and block copolymers as determined by TGA.

Template	Degradation temp (°C)	Reference
P123	~250	[42]
HmSEBmS	~490	[42]
PMMAN-*b*-PAN	365–490	[43]
SDS	160–380	[44]
CTAB	180–340	[44]
F127	305–481	[45,46]
PS-*b*-PMMA	~450	[47]
PEO-*b*-PS	430–800	[48]
KLE	~415	[49]
Polyacrylate latex	343.22	[50]
Fluorinated polyacrylate latex	363.21	[50]
PEO-PPO	203–210	[51]
Brij 56	350–400	[52,53]
Brij 76	320–370	[52]

Continued

TABLE 3.1 Degradation temperatures of some surfactants and block copolymers as determined by TGA.—cont'd

Template	Degradation temp (°C)	Reference
CTAC	590–620	[52]
PEO_{20}-*b*-PPO_{70}-*b*-PEO_{20}	~350	[54]
Brij	400–460	[55]
PS_{230}-*b*-PAA_{156}	430–460	[56]
PAN_{450}-*b*-$PMMA_{89}$	400–450	[57]
PEO-*b*-PCL	350–430	[58]
PS-*b*-P2VP	430–450	[59]
PEO-*b*-PMMA	~400	[60]

TGA was used to determine the degradation temperature of Pluronic P123 and sulfonated polystyrene-*block*-polybutadiene-*block*-polystyrene (HmSEBmS) triblock copolymers and its effect on TiO_2 crystallization (Fig. 3.22) [42]. In the TGA curve for P123, weight loss started sharply around 150°C and was complete at 250°C, a much lower temperature than the crystallization temperature of titania (~350°C). Thus, at temperatures higher than 250°C, there were no templates for the films. Therefore, the crystal growth was not suppressed. In fact, the crystallites grew, and the original mesoporous structure easily collapsed with increase in the calcination temperature. However, in the sulfonated HmSEBmS system, only 35 wt% was lost at 400°C, and the template was totally removed around 500°C. Therefore, the remaining polymer template can act as a support to prevent rapid crystal growth of TiO_2 at higher temperatures resulting in the retention of the original mesoporous structure in this system.

Palanisamy et al. studied the thermal stability of the poly(methyl methacrylate-b-polyacrylonitrile) (PMMAN) block copolymer and their blends with homopolymer polyacrylonitrile (PAN) at various blend ratios using TGA (Fig. 3.23) [43]. The weight loss can be expressed in four distinct steps: weight loss due to moisture removal, weight loss attributed to nitrile oligomerization and loss of other volatile gaseous products, weight loss between 365°C and 490°C assigned to thermal degradation of polymers, and finally weight loss due to pyrolytic degradation of PAN. From the thermograms, it is evident that the samples with higher PAN content have higher thermal stability compared to pure PMMAN.

He et al. [62] have reported a sol-gel method for the derivation of mesoporous $GaAlPO_4$ glass. They studied the thermal degradation behavior of

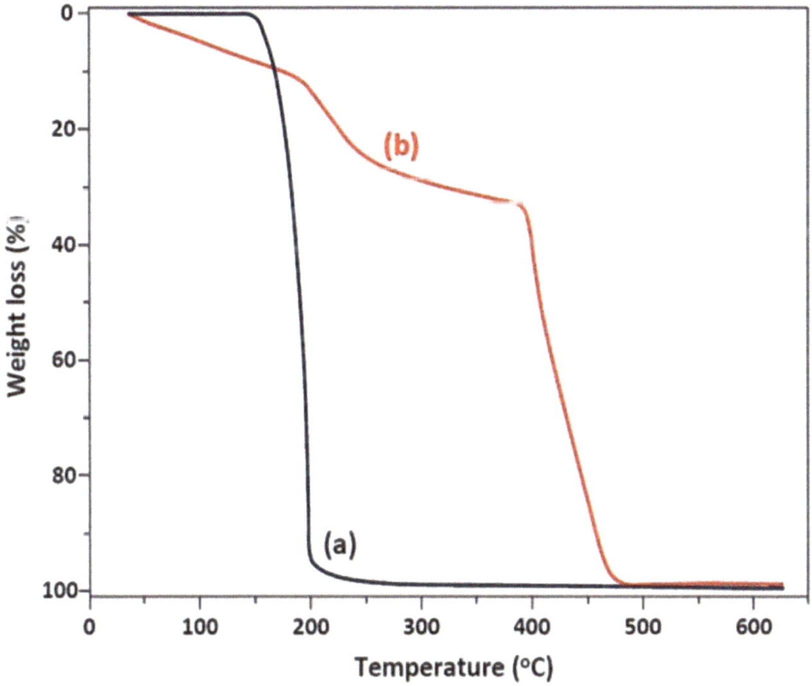

FIG. 3.22 TGA curves of (a) Pluronic P123 and (b) sulfonated HmSEBmS triblock copolymers at a heating rate of 1°C min^{-1}.

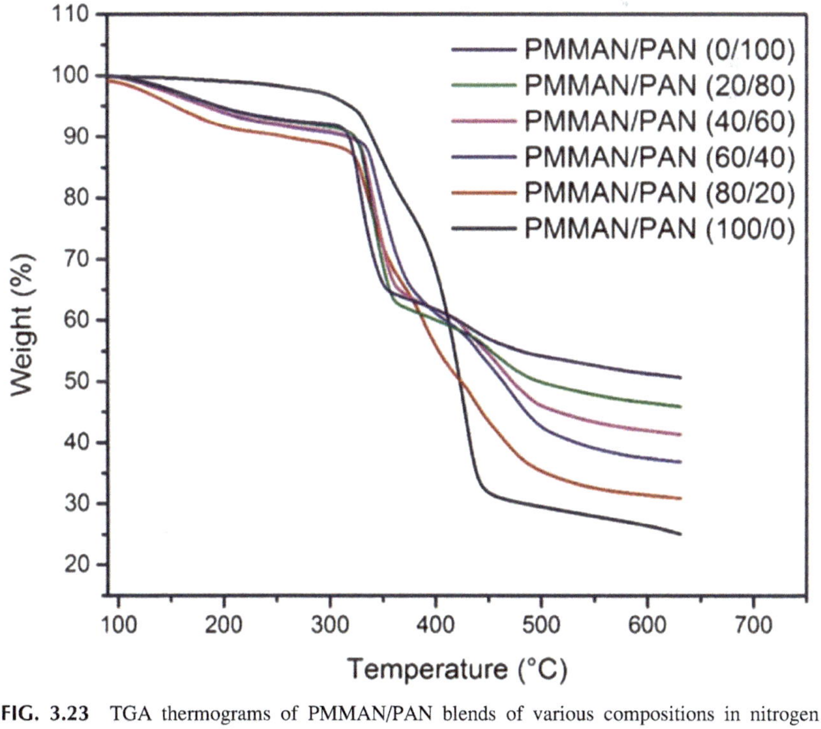

FIG. 3.23 TGA thermograms of PMMAN/PAN blends of various compositions in nitrogen atmosphere.

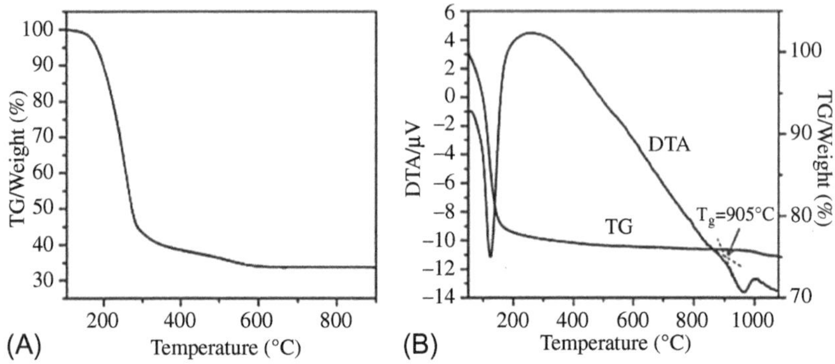

FIG. 3.24 (A) TGA trace of $Ga_{0.1}Al_{0.9}PO_4$ xerogel formed at room temperature. (B) Combined TGA/DTA of $Ga_{0.1}Al_{0.9}PO_4$ glass after annealing at 600°C for 6h.

$Ga_{0.1}Al_{0.9}PO_4$ xerogel as a precursor of mesoporous $GaAlPO_4$ glass by TGA/DTA (Fig. 3.24). The TGA trace of $Ga_{0.1}Al_{0.9}PO_4$ xerogel indicates that the volatile components (e.g., organics and water) are not completely removed below 400°C, and a final weight loss was found around 700°C (Fig. 3.24A). Fig. 3.24B illustrates the TGA and DTA curves for $Ga_{0.1}Al_{0.9}PO_4$ glass, in which the minor weight loss observed below 200°C is most likely due to an endothermic event arising from desorption and removal of physically adsorbed water in the porous material.

Wang et al. have demonstrated a heterogeneous sol-gel system of ethyl silicate (ES40) in acidified water-ethanol cosolvent that successfully produced silica microstructure with tailored microporosity [63]. To shed further light on the structural formation of the xerogels as a function of the drying temperature, TGA was conducted by holding the sample at constant temperatures until complete dryness. It can be observed from Fig. 3.25A that, for the gels dried at 60°C, a sharp weight loss occurs after 2h. This was much faster than that for the sample dried at 25°C (Fig. 3.25B). The system keeps losing weight due to the evaporation of solvent and by-product of sol-gel reactions until equilibration is achieved after 4h at 60°C or 20h at 25°C.

6 Estimation of molar mass of metal

The molar mass of the starting materials in case of thermal treatment of metals can be calculated by TGA. In this case, metal oxide is formed by the heating of metals in air, which is heavier than the original sample and the signal therefore increases. The molar mass of the starting metal (x), which is oxidized by oxygen (molar mass of 16g/mol), can be calculated by Eq. (3.1):

$$\frac{x\left(\frac{g}{mol}\right)}{100\%}=\frac{16\left(\frac{g}{mol}\right)}{y\%} \tag{3.1}$$

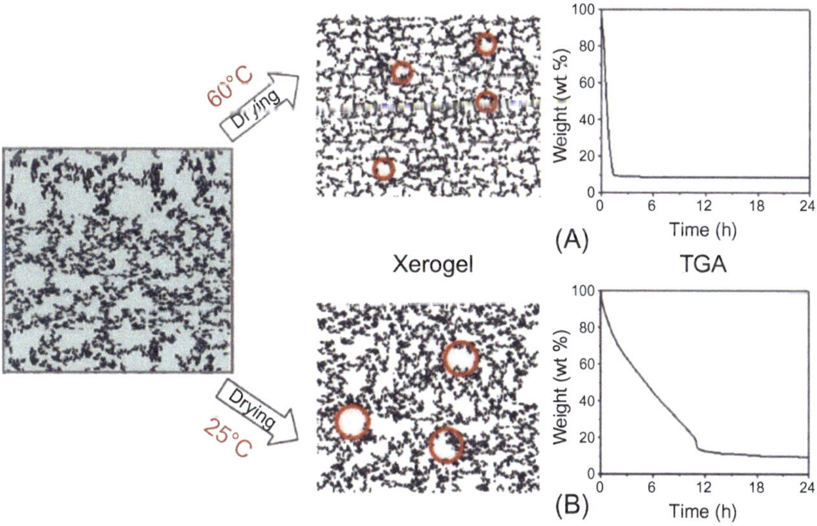

FIG. 3.25 TGA curves of silica sol-gel solutions and schematic silica matrix obtained after drying at (A) 60°C and (B) 25°C. *(Reproduced from S. Wang, D.K. Wang, S. Smart, J.C.D. da Costa, Ternary phase-separation investigation of sol-gel derived silica from ethyl silicate 40, Sci. Rep. 5 (2015) 14560. Copyrights© 2015, Springer Nature.)*

where x (g/mol) is the molar mass of the starting metal, 100% is the weight of the starting metal, 16 (g/mol) is the molecular weight or molar mass of oxygen, and y% is the weight gained by heating.

For example, let us say we are starting with 100% of metal. That corresponds to the entire mass of the original sample. Then by heating the sample up, the mass increases to 165% of the original value. In this case, an oxide is formed, which is heavier than the original sample and the signal therefore increases. What we know now is that the final product is 1.65 times heavier than the starting material (or in other words, we have gained 65% of the weight we started with on top of the starting mass). Now assuming that the mass gain is entirely caused by oxygen (molar mass of 16 g/mol), the molar mass of the starting material can be calculated with a simple ratio according to the following equation:

$$\frac{x\left(\frac{\mathrm{g}}{\mathrm{mol}}\right)}{100\%}=\frac{16\left(\frac{\mathrm{g}}{\mathrm{mol}}\right)}{65\%}=24.6\,\mathrm{g/mol} \tag{3.2}$$

which happens to be close to the tabulated value of Mg (within the error range).

7 Estimating oxygen content in metal oxides

TGA involving hydrogen reduction has been commonly used for the determination of oxygen stoichiometry [64]. For example, $YBa_2Cu_{2.82}Zn0_{.18}O_7$ sample is thermally treated in oxygen to remove all gases and moistures. Then, reduction is performed by hydrogen without exposing the sample to atmosphere. This is achieved by replacing ambient oxygen by Ar+8%H_2 gas mixture in the thermo analyzer. The use of this gas mixture minimizes errors due to buoyancy effect. In a typical run, the sample temperature was increased, under a flow (200 mL/min) of Ar+8%H_2 gas mixture, at 10°C min^{-1} up to 930°C and kept at this temperature for sufficient time till a constant weight was attained. The Cu content was reduced to metallic state at about 900°C in all these oxide compounds. However, in the case of Zn-substituted oxide compound, the sample continued to lose weight even after holding for more than 4 h at 920°C due to vaporization loss of Zn (Zn contents are also reduced to the metallic state). The reduction temperature in such cases was raised to 980°C to enhance the rate of vaporization loss. The temperature was limited to 980°C to avoid melting of the reduced product. The sample attained a constant weight at 980°C within 1.5 h (Fig. 3.26) indicating the complete removal of Zn from the charge. It is therefore possible to establish composition of the reduced product by this approach.

Saiz et al. determined the oxygen content in $YBa_2Cu_3O_{7-x}$ by applying the same method [65]. TGA analysis of this powder was performed following the weight loss as a function of temperature in a 5% H_2 in Ar reducing atmosphere from room temperature to 950°C with a heating rate of 10°C min^{-1} and a flux of 2.1 H_2/13.2 Ar 1/h, maintaining this temperature until the mass loss stopped (Fig. 3.27A). The weight loss was simultaneously recorded using 100-mg scale and 10-mg expanded scale. In order to estimate the accuracy of the method, the same analysis was performed on high-purity CuO powders (Fig. 3.27B).

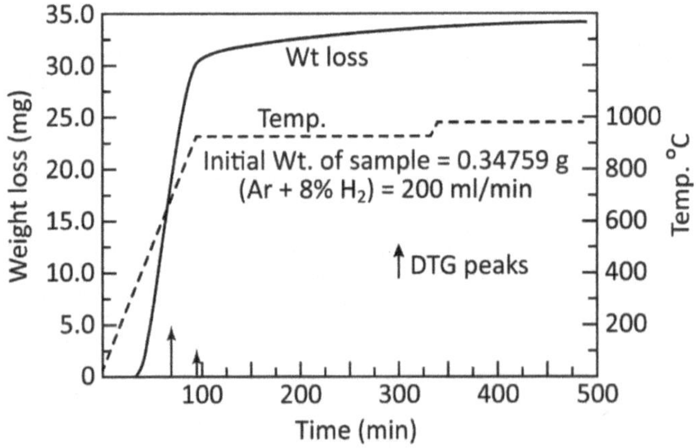

FIG. 3.26 Change in weight of oxide, $YBa_2Cu_{2.82}Zn_{0.18}O_7$ during reduction under (Ar+8%H_2) flow determined by TGA.

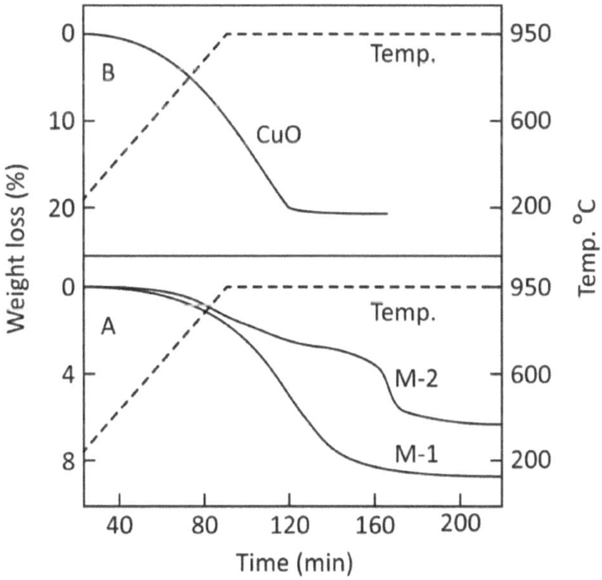

FIG. 3.27 TGA curves corresponding to (A) $YBa_2Cu_3O_{7-x}$ (samples M-1 and M-2) and (B) pure CuO (full scale 100mg). These tests were performed in 5% H, + Ar atmosphere.

Kao et al. calculated the oxygen content in bismuth superconductor ($Bi_2Sr_2Ca_2Cu_4O_{11}$) using totally different TGA approach [66]. The basic idea behind their approach was to continuously monitor the weight changes of the metal precursor mixture during heat treatment process. Their technique directly measures the oxygen content without exposing the sample to ambient environment, without assuming complete reduction at 900°C, and without making any assumption about the phases (idealized structure) present.

A common method for commercial production of aluminum nitride (AlN) powders, used for oxygen-containing precursors, is carbothermal nitridation aluminum nitride (AlN) [67]:

$$Al_2O_3 + 3C + N_2 = 2AlN + 3CO$$

Unreacted carbon can be removed by treatment in ammonia at 100°C or oxidation around 700°C. Total oxygen and nitrogen concentrations in ceramic samples can be measured by hot gas extraction. The method consists of mixing the crushed sample with graphite powder and heating it around 3000°C in an inert atmosphere. The resultant amount of carbon dioxide is proportional to the oxygen content and is measured by infrared absorption. The nitrogen, released at the same time, is measured using a gas conductivity cell. The method can be modified to discriminate between weakly bonded surface oxygen, lattice-dissolved oxygen, and oxygen contained in oxide phase. A similar method is used for carbon analysis, except that the sample is heated in air. However, as with oxygen analyses, the evolved carbon dioxide is detected.

Nicolaescu et al. reported a TG method for determining carbon, oxygen, and nitrogen content in aluminum nitride [67]. TG methods produce good results when used for quantitative analyses. However, for materials derived from a carbothermal nitridation process, multiple reactions take place, resulting in various weight losses. First, a small weight loss occurs between 100°C and 350°C due to the release of adsorbed gases. This weight loss is attributed to the removal of hydroxyl groups present on the aluminum nitride surface resulting from hydrolysis. Second, weight loss because of carbon oxidation at temperatures up to around 700°C can be determined by the following reactions:

$$C + O_2 = CO_2$$

$$2C + O_2 = 2CO$$

Third, aluminum nitride ($4AlN + 3O_2 = 2Al_2O_3 + 2N_2$) is oxidized through two stages. Initially it proceeds linearly with time, and the rate is proportional to the surface area. At a later stage, diffusion through the oxide layer becomes rate-controlling and the oxidation proceeds at a slower rate, according to a parabolic rate law. As revealed by an infrared spectroscopy study of Al-H vibrating band shift, the reaction seems to involve the formation of intermediate oxynitrides at temperatures as low as 600°C. The reaction is very slow up to 800°C. During carbothermal nitridation, in addition to the generation of aluminum nitride, unreacted α- or γ-Al_2O_3 have been frequently observed, as well as oxynitride phase.

Since the above reactions take place in different temperature ranges, there is a possibility of separating them dynamically using appropriate heating rate. Conditions for optimal reaction resolution must be established.

The TGA curve of the nitridation product is shown in Fig. 3.28. At temperatures ranging from 50°C to 400°C, a slight weight loss was observed, which was attributed to the loss of adsorbed species and moisture. At temperatures ranging from 400°C to 730°C, a larger weight loss was observed, which was attributed to carbon oxidation. At temperatures in the range 730–1400°C, weight gain was observed, which was interpreted to be due to aluminum nitride oxidation.

8 Carbonization

Energy-related applications (e.g., storage, conversion, generation, and distribution) are more important today than at any time in human history. Carbon-based nanostructures lie at the heart of fundamental advances in the development of

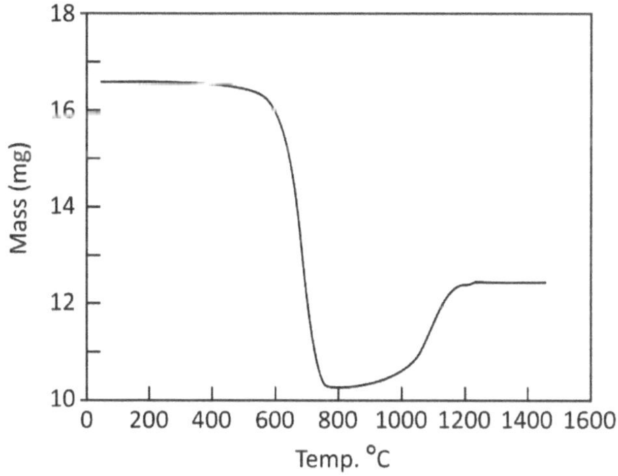

FIG. 3.28 TGA trace of aluminum nitride and C (55 mL/min, and 20°C min^{-1}).

efficient electrodes for electrical double layer capacitor (EDLC) [68], batteries [69], and catalytic applications [70]. Thus, development of carbon nanostructured materials, including novel mesoporous carbon nanoparticles, carbon fibers, carbon nanotubes, and core shell and yolk shell carbon nanoparticles with hierarchical porous structures, have attracted the interest of researchers [71]. The carbonization, which is studied by thermal analysis methods of carbon sources, is a promising way to obtain carbon nanostructures with high specific surface area for energy-related applications [72].

The carbonization process proceeds by thermo-oxidative stabilization in the heat treatment line for the continuous conversion of organic precursors (e.g., proteins, enzymes, and polymers) into carbon [73–76]. The conversion process under inert atmosphere is carried out in two stages: precarbonization and carbonization. The precarbonization process is done in an open low-temperature furnace up to 850°C under controlled thermal treatment processes. The precarbonized products are then carbonized by introducing into an open, graphite muffle furnace with temperature up to 1350°C. The carbon fibers with required level of carbon content and a combination of mechanical properties are obtained by altering the carbonization process conditions. For achieving high modulus, an optional graphitization process is carried out in a graphite furnace up to 2400°C. The carbonized/graphitized carbon fibers are then forwarded to surface treatment unit.

Cho et al. found that β-sheet-rich silk protein structure is transformed into an sp^2-hybridized carbon hexagonal structure by simply heating to 350°C [77]. The pyrolysis of spider and worm silk proteins was studied by TGA (Fig. 3.29). The pyrolysis up to 800°C with quite high residual masses of around

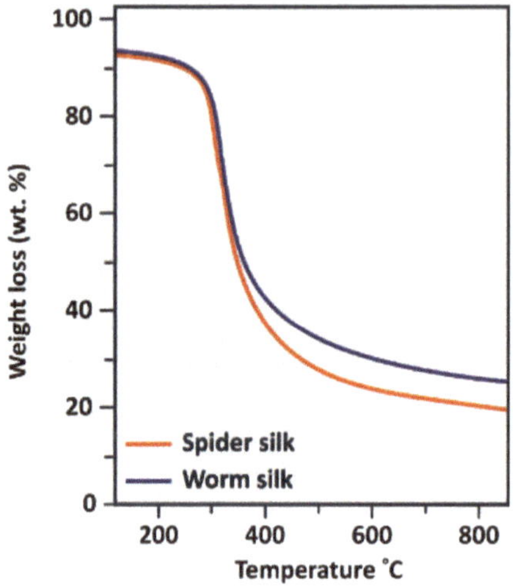

FIG. 3.29 TGA curves of spider silk and worm silk.

23.4 and 33.2 wt% was observed. These results suggest that the silk protein is affiliated with a char-type polymeric precursor without melting. This pyrolysis behavior differs from other types of natural protein-rich materials or amide-linked polymers. The β-sheet-rich protein acquires a carbon residue percentage of around 22.1 wt% following pyrolysis at 800°C. The pyrolysis behavior of β-sheet-rich proteins with no aromatic backbone or cross-linked structures indicates that the protein molecules are restructured to form a conjugated system following pyrolysis.

The thermal degradation of silk proteins initiates at temperatures around 250°C, as shown by the TGA results (Fig. 3.29). The critical chemical changes in silk proteins take place at temperatures in the range 300–350°C.

Another example is the TGA study of the thermal stability of carbons formed by carbonization of divinylbenzene (DVB) as a precursor in MCM-48 template under an O_2 atmosphere (Fig. 3.30) [78]. Several studies have reported the thermal stability of various ordered mesoporous carbon materials with graphitic pore wall (known as CMK-nG) by TGA (Fig. 3.31) [79]. The CMK-3G carbon and a reference multiwall carbon nanotube (MWCNT) exhibited almost the same weight-change profiles. In contrast, CMK-3FA and CMK-3SC showed a different weight-loss profile at a lower temperature. Significant weight losses corresponding to the oxidation of the carbons occurred between 520°C and 680°C for CMK-3G and MWCNTs, and between 400°C and 640°C for other CMK-3 carbons prepared using furfuryl alcohol (CMK-3FA) and sucrose (CMK-3SC). Thus, the oxidation temperature for CMK-3G is like that of nanotubes, and significantly higher than that of mesoporous carbons with furfuryl alcohol (CMK-3FA) and sucrose (CMK-3SC) as carbon sources.

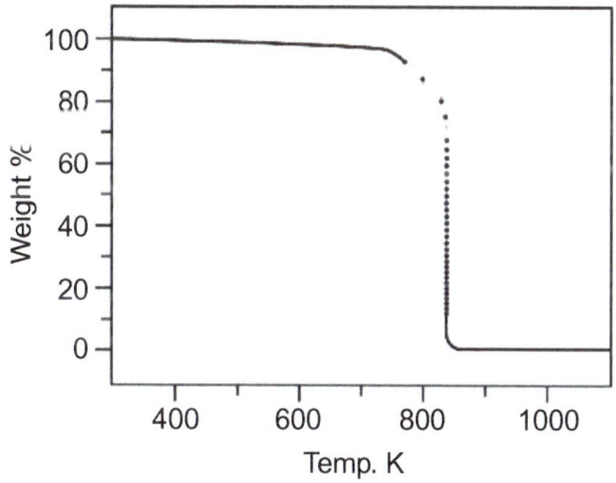

FIG. 3.30 TGA curves in O_2 atmosphere of MCM-48 templated carbons prepared by the carbonization of DVB.

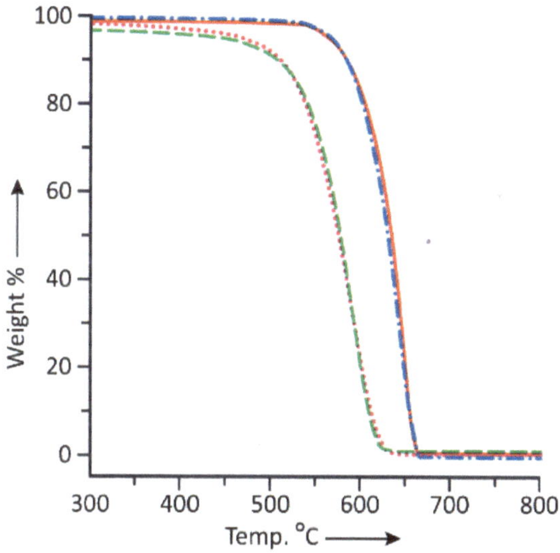

FIG. 3.31 TGA curves for the CMK-3G (—), MWCNT (- . - . - . -), CMK-3FA (……..), and CMK-3SC (- - - -) carbon materials in air. The heating rate was 5°Cmin^{-1}.

These results provide additional confirmation of the graphitized nature of the CMK-3G carbon frameworks, which is responsible for the much improved thermal stability in air.

Su et al. demonstrated the thermal decomposition behavior of mesoporous carbon synthesized using silica as sacrificial templates. The silica is removed by etching using HF. The complete thermal degradation of carbon sample confirmed the removal of all silica templates in HF [80].

Kopecká et al. demonstrated the thermal conversion of polypyrrole nanotube into nitrogen-containing carbon at 650°C nanotube [81]. TGA follows the course of carbonization reaction of globular and nanotubular polypyrroles. There is no significant difference in the thermal stability between globular and nanotubular forms of polypyrroles. A residue of around 5 wt% corresponds to iron oxides produced from the oxidant, iron (III) chloride.

The residue in both the abovementioned cases is 50 wt% at 650°C in an inert nitrogen atmosphere. Also, there is no substantial difference in the stability of globular and nanotubular polypyrroles. This is logical because the thermal stability is established by molecular rather than supramolecular structure.

MOFs served as excellent precursors for nanoporous carbon structures through carbonization [82]. For example, zeolitic imidazolate frameworks (ZIFs) have been foregrounded as structures with exceptional, intrinsic chemical and thermal stability. ZIF-8 isothermal TGA decomposition kinetics were studied by James et al. in air, argon, H_2/CO_2, and nitrogen environments by exposing ZIF-8 to each gas for 20 h at temperatures of 200, 250, and 300°C, respectively (Fig. 3.32) [82].

ZIF-8 crystallinity was preserved under experimental isothermal conditions at 200°C in each atmosphere, but crystallinity collapsed at higher temperatures. Decomposition kinetics data show that the rate of ZIF-8 carbonization significantly increases at temperatures above 200°C irrespective of environment. ZIF-8 decomposition in H_2/CO_2 reducing mixture exhibits the slowest

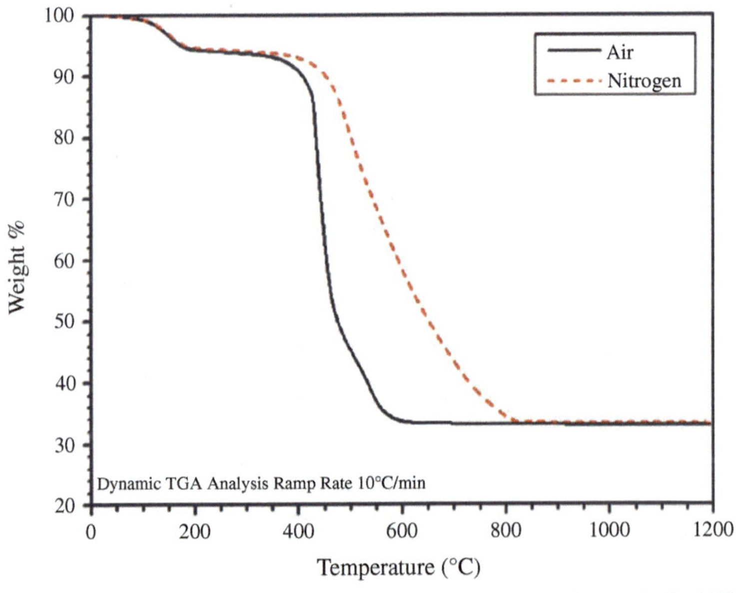

FIG. 3.32 Temperature-dependent TGA weight decomposition traces of as-synthesized ZIF-8 in air (—) and nitrogen (——) environments.

decomposition kinetics at all temperatures and the greatest morphological change. At 300°C, oxidative effects enhance ZIF-8 decomposition in air. At lower temperatures, the decomposition rate in air behaves more similarly to that of nitrogen and argon. The residual carbonized ZIF structures after decomposition at 300°C were estimated.

Hydrothermal carbonization of biomass has attracted increasing interest in the field of carbon materials and biofuels. The full potential of hydrothermally carbonized materials has only been discussed in recent years through the work of Titirici and his colleagues, in several fields such as catalysis, CO_2 sequestration, agriculture, and renewable energy products [83].

In this method, new materials based on carbon are increasingly viewed as ideal candidates. These materials are ideally produced via inexpensive, low energy consumption methods, using renewable resources as precursors (e.g., biomass). With flexible morphologies, pore structures and functionalities should be a feasible solution for the efficient storage of energy and gases. At the end of life, such materials ideally must act to improve soil quality and act as potential CO_2 storage sinks.

Briefly, in hydrothermal carbonization, pure carbohydrates as well as raw lignocellulosic biomass (e.g., agricultural wastes) are converted into valuable carbon materials using mild processing temperatures (<200°C) and water as the carbonization medium, under self-generated pressures.

For example, Cai et al. reported the hydrothermal carbonization of tobacco stalk for the production of fuel [84]. The derivation of hydrochar and combustion reaction was studied by TGA/DTG in air (Fig. 3.33). The hydrothermal carbonization reaction has passed through three distinct zones: the water evaporation phase, devolatilization and combustion phase, and char combustion phase processes, respectively [85]. The first zone was in the range 30–190°C and the weight loss was only 3.9%, indicating the samples contained relatively little water, the second zone with a weight loss of 65.1% was between 190°C and 380°C, and the zone from 380°C to 540°C was the third zone which possessed a weight loss of 26.6%.

There was an obvious decrease in weight loss in the second zone with the increment of operation temperature and this trend became more remarkable under a longer reaction time. In the third zone, in contrast, the weight loss and temperature intervals both increased with the increase in operating temperature and time. These results could be attributed to the decomposition and carbonization processes of lignocellulosic structures during hydrothermal carbonization reaction, which resulted in increasingly lower volatile matters (devolatilization at 150–350°C) and higher fixed carbon (combustion at 350–550°C).

The changes in residue levels were in accordance with the ash contents, and the high residues of hydro-chars derived under a relatively long operation time indicate that in order to obtain high-quality solid fuel from hydrothermal carbonization, the reaction time must be appropriate.

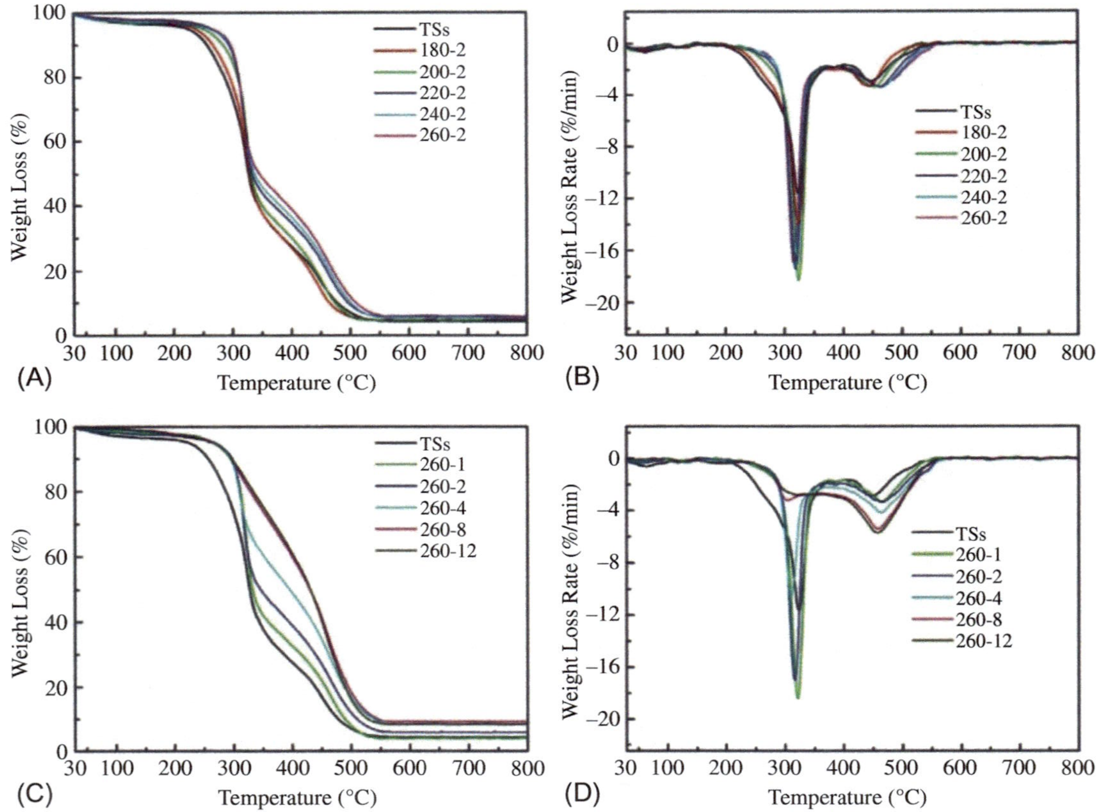

FIG. 3.33 TG (A and C) and DTG (B and D) curves of tobacco stalks (TSs) and corresponding hydro-chars. Code: 180-2, 200-2, 220-2, 240-2, and 260-2 are TS hydrothermal products at 180°C, 200°C, 220°C, 240°C, and 260°C, respectively, for reaction time of 2 h; 260-1, 260-2, 260-4, 260-8, and 260-12 are TS hydrothermal products at 260°C and holding times of 1, 2, 4, 8, and 12 h, respectively.

9 Summary

This chapter focuses on the potential of TA in materials science with applications in chemistry and engineering. It is used for studying thermal conversion of materials yielding new functional candidates for the recent advanced applications and for determining contents of elements and moisture. To the best of our knowledge, this is the first book that reports the applications of thermal analysis techniques in materials science and chemical engineering. Very recently, the organic and organic-inorganic hybrids and biofuel have been used for synthesis of functional inorganic materials, carbon, and metal and alloys through solid-state pathway for energy-related applications and electronics. The researchers track the materials thermal decomposition, conversion, crystallization, and formation of new materials using TA techniques especially TG/DTA.

References

[1] H.C. Zhou, S. Kitagawa, Metal-organic frameworks (MOFs), Chem. Soc. Rev. 43 (2014) 5415–5418. M.B. Zakaria, V. Malgras, T. Takei, C. Li, Y. Yamauchi, Layer-by-Bayer motif hybridization: nanoporous nickel oxide flakes wrapped into graphene oxide sheets toward enhanced oxygen reduction reaction, Chem. Commun. 15 (2015) 16409–16412. M. B. Zakaria, M. Hu, M. Pramanik, C. Li, J. Tang, A. Aldalbahi, S.M. Alshehri, V. Malgras, Y. Yamauchi, Synthesis of nanoporous ni-co mixed oxides by thermal decomposition of metal-cyanide coordination polymers, Chem. Asian J. 10 (2015) 1541–1545. A. Azhar, Y. Li, Z. Cai, M.B. Zakaria, M.K. Masud, S.A. Hossain, J. Kim, W. Zhang, J. Na, Y. Yamauchi, M. Hu, Nanoarchitectonics: a new materials horizon for prussian blue and its analogues, Bull. Chem. Soc. Jpn 92 (2019) 875–904.

[2] M.B. Zakaria, M.S.A. Hossain, M.J.A. Siddiky, M. Shahabuddin, E. Yanmaz, J.H. Kim, A.A. Belik, Y. Ide, M. Hu, S. Tominaka, Y. Yamauchi, Cyano-bridged trimetallic coordination polymer nanoparticles and their thermal decomposition into nanoporous spinel ferromagnetic oxides, Chem.-A Eur. J. 22 (2016) 15042–15048. M.B. Zakaria, M. Hu, R.R. Salunkhe, M. Pramanik, K. Takai, V. Malgras, S. Choi, S.X. Dou, J.H. Kim, M. Imura, S. Ishihara, Y. Yamauchi, Controlled synthesis of nanoporous nickel oxide with two-dimensional shapes through thermal decomposition of metal-cyanide hybrid coordination polymers, Chem.-A Eur. J. 21 (2015) 3605–3612.

[3] R.R. Salunkhe, Y.V. Kaneti, J. Kim, J.H. Kim, Y. Yamauchi, Nanoarchitectures for metal-organic framework-derived nanoporous carbons toward supercapacitor applications, Acc. Chem. Res. 49 (2016) 2796–2806.

[4] M.B. Zakaria, T. Chikyow, Recent advances in prussian blue and prussian blue analogues: synthesis and thermal treatments, Coord. Chem. Rev. 352 (2017) 328–345. A. Azhar, Y. Yamauchi, A. Enaiet Allah, Z.A. Alothman, A.Y. Badjah, M. Naushad, M. Habila, S. Wabaidur, J. Wang, M.B. Zakaria, Nanoporous iron oxide/carbon composites through in-situ deposition of prussian blue nanoparticles on graphene oxide nanosheets and subsequent thermal treatment for supercapacitor applications, Nanomaterials 9 (2019) 776. S. Tanaka, M. B. Zakaria, Y.V. Kaneti, Y. Jikihara, T. Nakayama, M. Zaman, Y. Bando, M.S.A. Hossain, D. Golberg, Y. Yamauchi, Gold-loaded nanoporous iron oxide cubes derived from prussian blue as carbon monoxide oxidation catalyst at room temperature, ChemistrySelect 3 (2018) 13464–13469.

[5] R.R. Salunkhe, Y.V. Kaneti, Y. Yamauchi, Metal-organic framework-derived nanoporous metal oxides toward Supercapacitor applications: progress and prospects, ACS Nano 11 (2017) 5293–5308.

[6] R. Wu, D.P. Wang, V. Kumar, K. Zhou, A.W. Law, P.S. Lee, J. Lou, Z. Chen, MOFs-derived copper sulfides embedded within porous carbon octahedra for electrochemical capacitor applications, Chem. Commun. 51 (2015) 3109–3112. M.B. Zakaria, Y. Guo, J. Na, R. Tahawy, T. Chikyow, W.A. El-Said, D.A. El-Hady, W. Alshitari, Y. Yamauchi, J. Lin, Layer-by-layer motif heteroarchitecturing of N,S-codoped reduced graphene oxide-wrapped Ni/NiS nanoparticles for the electrochemical oxidation of water, ChemSusChem 13 (2020) 3269–3276.

[7] S.H. Hsu, C.T. Li, H.T. Chien, R.R. Salunkhe, N. Suzuki, Y. Yamauchi, K.C. Ho, K.C. Wu, Platinum-free counter electrode comprised of metal-organic-framework (MOF)-derived cobalt sulfide nanoparticles for efficient dye-sensitized solar cells (DSSCs), Sci. Rep. 4 (2014) 6983. M.B. Zakaria, D. Zheng, U. Apfel, T. Nagata, E.S. Kenawy, J. Lin, Dual-heteroatom-doped reduced graphene oxide sheets conjoined CoNi-based carbide and sulfide nanoparticles for efficient oxygen evolution reaction, ACS Appl. Mater. Interfaces 12 (2020) 40186–40193.

[8] H. Zhang, J. Nai, L. Yu, X.W. Lou, Metal-organic-framework-based materials as platforms for renewable energy and environmental applications, Joule 1 (2017) 77–107.

[9] H.B. Wu, X.W.D. Lou, Metal-organic frameworks and their derived materials for electrochemical energy storage and conversion: promises and challenges, Sci. Adv. 3 (2017), eaap9252. A. Azhar, J. Earnshaw, M.B. Zakaria, P. Cheng, Y.V. Kaneti, M.S.A. Hossain, S.M. Alshehri, T. Ahamad, Y. Yamauchi, J. Na, Fabrication and characterization of prussian blue-derived iron carbide-iron oxide hybrid on reduced graphene oxide nanosheets, KONA Powder Part. J. (2020), https://doi.org/10.14356/kona.2021015.

[10] W. Xia, A. Mahmood, R. Zou, Q. Xu, Metal-organic frameworks and their derived nanostructures for electrochemical energy storage and conversion, Energy Environ. Sci. 8 (2015) 1837–1866.

[11] M.H. Yap, K.L. Fow, G.Z. Chen, Synthesis and applications of MOF-derived porous nanostructures, Green Energy Environ. 2 (2017) 218–245. S. Tanaka, R.R. Salunkhe, Y.V. Kaneti, V. Malgras, S.M. Alshehri, T. Ahamad, M.B. Zakaria, S.X. Dou, Y. Yamauchi, M.S.A. Hossain, RSC Adv. 7 (2017) 33994–33999.

[12] M.B. Zakaria, C. Li, M. Pramanik, Y. Tsujimoto, M. Hu, V. Malgras, S. Tominaka, Y. Yamauchi, Nanoporous Mn-based electrocatalysts through thermal conversion of cyano-bridged coordination polymers toward ultra-high efficiency hydrogen peroxide production, J. Mater. Chem. A 4 (2016) 9266–9274.

[13] A. Azhar, M.B. Zakaria, J. Lin, T. Chikyow, D.J. Martin, Y.G. Alghamdi, A.A. Alshehri, Y. Bando, M.S.A. Hossain, K.C.W. Wu, N.A. Kumar, Y. Yamauchi, Graphene-wrapped nanoporous nickel-cobalt oxide flakes for electrochemical supercapacitors, Chem. Select 7 (2018) 8505–8510.

[14] H. Pang, J. Deng, J. Du, S. Li, J. Li, Y. Ma, J. Zhang, J. Chen, Porous nanocubic Mn_3O_4-Co_3O_4 composites and their application as electrochemical supercapacitors, Dalton Trans. 41 (2012) 10175–10181.

[15] M.B. Zakaria, T. Chikyow, Synergetic nanoporous Mn-Ru oxides as efficient electrocatalysts for the oxygen reduction reaction, New J. Chem. 41 (2017) 8196–8202.

[16] M.B. Zakaria, A.A. Belik, C.H. Liu, H.Y. Hsieh, Y.T. Liao, V. Malgras, Y. Yamauchi, K.C. Wu, Prussian blue derived nanoporous iron oxides as anticancer drug carriers for magnetic-guided chemotherapy, Chem. Asian J. 10 (2015) 1457–1462.

[17] M.B. Zakaria, M. Hu, M. Imura, R.R. Salunkhe, N. Umezawa, H. Hamoudi, A.A. Belik, Y. Yamauchi, Single-crystal-like nanoporous spinel oxides: a strategy for synthesis of

nanoporous metal oxides utilizing metal-cyanide hybrid coordination polymers, Chem. Eur. J. 20 (2014) 17375–17384.

[18] L. Hu, Y. Huang, Q. Chen, $Fe_xCo_{3-x}O_4$ nanoporous particles stemmed from metal-organic frameworks $Fe_3[Co(CN)_6]_2$: a highly efficient material for removal of organic dyes from water, J. Alloys Compnd 559 (2013) 57–63.

[19] N. Yan, L. Hu, Y. Li, Y. Wang, H. Zhong, X. Hu, X. Kong, Q. Chen, Co_3O_4 nanocages for high-performance anode material in lithium-ion batteries, J. Phys. Chem. C 116 (2012) 7227–7235.

[20] L. Hu, N. Yan, Q. Chen, P. Zhang, H. Zhong, X. Zheng, Y. Li, X. Hu, Fabrication based on the Kirkendall effect of Co_3O_4 porous nanocages with extraordinarily high capacity for lithium storage, Chem. Eur. J. 18 (2012) 8971–8977.

[21] L. Zhang, L. Shi, L. Huang, J. Zhang, R. Gao, D. Zhang, Rational design of high-performance DeNOx catalysts based on $Mn_xCo_{3-x}O_4$ nanocages derived from metal-organic frameworks, ACS Catal. 4 (2014) 1753–1763.

[22] L. Zhang, H.B. Wu, S. Madhavi, H.H. Hng, X.W. Lou, Formation of Fe_2O_3 microboxes with hierarchical Shell structures from metal-organic frameworks and their lithium storage properties, J. Am. Chem. Soc. 134 (2012) 17388–17391.

[23] M. Hu, A.A. Belik, M. Imura, K. Mibu, Y. Tsujimoto, Y. Yamauchi, Synthesis of superparamagnetic nanoporous iron oxide particles with hollow interiors by using Prussian blue coordination polymers, Chem. Mater. 24 (2012) 2698–2707.

[24] K. Ariga, V. Malgras, Q. Ji, M.B. Zakaria, Y. Yamauchi, Coordination nanoarchitectonics at interfaces between supramolecular and materials chemistry, Coord. Chem. Rev. 320-321 (2016) 139–152.

[25] A. Azhar, M.B. Zakaria, E.M. Ebeid, T. Chikyow, Y. Bando, A.A. Alshehri, Y.G. Alghamdi, Z.X. Cai, N.A. Kumar, J. Lin, H. Kim, Y. Yamauchi, Synthesis of hollow Co-Fe Prussian blue analogue cubes by using silica spheres as a sacrificial template, ChemistryOpen 7 (2018) 599–603.

[26] Y.V. Kaneti, J. Zhang, Y.B. He, Z. Wang, S. Tanaka, M.S.A. Hossain, Z.Z. Pan, B. Xiang, Q.H. Yang, Y. Yamauchi, Fabrication of an MOF-derived heteroatom-doped co/CoO/carbon hybrid with superior sodium storage performance for sodium-ion batteries, J. Mater. Chem. A 5 (2017) 15356–15366.

[27] Y. Wang, C. Wang, Y. Wang, H. Liu, Z. Huang, Superior sodium-ion storage performance of Co_3O_4@nitrogen-doped carbon: derived from a metal-organic framework, J. Mater. Chem. A 4 (2016) 5428–5435.

[28] P.K. Gallagher, B. Prescott, Further studies of the thermal decomposition of europium hexacyanoferrate(III) and ammonium europium hexacyanoferrate(II), Inorg. Chem. 9 (1970) 2510–2512.

[29] C. Aparicio, L. Machala, Z. Marusak, Thermal decomposition of Prussian blue under inert atmosphere, J. Therm. Anal. Calorim. 110 (2011) 661–669.

[30] M.B. Zakaria, Nanostructuring of Nanoporous iron carbide spheres *via* thermal degradation of triple-shelled Prussian blue hollow spheres for oxygen reduction reaction, RSC Adv. 6 (2016) 10341–10351.

[31] M.B. Zakaria, C. Li, Q. Ji, B. Jiang, S. Tominaka, Y. Ide, J.P. Hill, K. Ariga, Y. Yamauchi, Self-construction from 2D to 3D: one-pot layer-by-layer assembly of graphene oxide sheets held together by coordination polymers, Angew. Chem. Int. Ed. 55 (2016) 8426–8430.

[32] M.B. Zakaria, H. Tan, J. Kim, A.Y. Badjah, M. Naushad, M. Habila, S. Wabaidur, Z.A. Alothman, Y. Yamauchi, J. Lin, Structurally controlled layered Ni_3C/graphene hybrids using cyano-bridged coordination polymers, Electrochem. Commun. 100 (2019) 74–80.

[33] A. Alowasheeir, M.B. Zakaria, J. Lin, A.A. Alshehri, Y.G. Alghamidi, K.A. Alzahrani, Y.V. Kaneti, Y. Yamauchi, In-situ formation of Cu–Ni cyano-bridged coordination polymer on graphene oxide nanosheets and their thermal conversion, Microporous Mesoporous Mater. 290 (2019) 109670.

[34] M.B. Zakaria, E.M. Ebeid, M.M. Abdel-Galeil, T. Chikyow, Cyanide bridged coordination polymer nanoflakes thermally derived Ni_3C and fcc-Ni nanoparticles for electrocatalysts, New J. Chem. 41 (2017) 14890–14897.

[35] Y.V. Kaneti, J. Tang, R.R. Salunkhe, X. Jiang, A. Yu, K.C. Wu, Y. Yamauchi, Nanoarchitectured design of porous materials and nanocomposites from metal-organic frameworks, Adv. Mater. 29 (2017) 1604898.

[36] P. Sennu, M. Christy, V. Aravindan, Y.G. Lee, K.S. Nahm, Y.S. Lee, Two-dimensional mesoporous cobalt sulfide nanosheets as a superior anode for a Li-ion battery and a bifunctional electrocatalyst for the Li-O_2 system, Chem. Mater. 27 (2015) 5726–5735.

[37] H. Hu, J. Zhang, B. Guan, X.W. Lou, Unusual formation of CoSe@carbon nanoboxes, which have an inhomogeneous shell, for efficient lithium storage, Angew. Chem. Int. Ed. Eng. 55 (2016) 9514–9518.

[38] C. Sun, Q. Dong, J. Yang, Z. Dai, J. Lin, P. Chen, W. Huang, X. Dong, Metal-organic framework derived $CoSe_2$ nanoparticles anchored on carbon fibers as bifunctional electrocatalysts for efficient overall water splitting, Nano Res. 9 (2016) 2234–2243.

[39] M. Liu, J. Li, Cobalt phosphide hollow polyhedron as efficient bifunctional electrocatalysts for the evolution reaction of hydrogen and oxygen, ACS Appl. Mater. Interface 8 (2016) 2158–2165.

[40] Y. Shi, B. Zhang, Recent advances in transition metal phosphide nanomaterials: synthesis and applications in hydrogen evolution reaction, Chem. Soc. Rev. 45 (2016) 1529–1541.

[41] Y. Ren, Z. Ma, P.G. Bruce, Ordered Mesoporous metal oxides: synthesis and applications, Chem. Soc. Rev. 41 (2012) 4909–4927.

[42] M.B. Zakaria, N. Suzuki, N.L. Torad, M. Matsuura, K. Maekawa, H. Tanabe, Y. Yamauchi, Preparation of mesoporous titania thin films with well-crystallized frameworks by using thermally stable triblock copolymers, Eur. J. Inorg. Chem. 2013 (2013) 2330–2335.

[43] A. Palanisamy, N.V. Salim, B.L. Fox, P. Jyotishkumar, T. Pradeep, N. Hameed, A facile method to fabricate carbon nanostructures via the self-assembly of polyacrylonitrile/poly (methyl methacrylate-b-polyacrylonitrile) AB/B′ type block copolymer/homopolymer blends, RSC Adv. 6 (2016) 55792–55799.

[44] D. Ramimoghadam, M.Z. Hussein, Y.H. Taufiq-Yap, The effect of sodium dodecyl sulfate (SDS) and cetyltrimethylammonium bromide (CTAB) on the properties of ZnO synthesized by hydrothermal method, Int. J. Mol. Sci. 13 (2012) 13275–13293.

[45] Q. Dou, A. Abdul Karim, X.J. Loh, Modification of thermal and mechanical properties of PEG-PPG-PEG copolymer (F127) with MA-POSS, Polymers 8 (2016) 341.

[46] I.M. Hung, D.T. Hung, K.Z. Fung, M.H. Hon, Synthesis and characterization of highly ordered mesoporous YSZ by tri-block copolymer, J. Porous Mater. 13 (2006) 225–230.

[47] V. Gianotti, D. Antonioli, K. Sparnacci, M. Laus, T.J. Giammaria, F.F. Lupi, G. Seguini, M. Perego, On the thermal stability of PS-*b*-PMMA block and P(S-*r*-MMA) random copolymers for nanopatterning applications, Macromolecules 46 (2013) 8224–8234.

[48] Y. Deng, T. Yu, Y. Wan, Y. Shi, Y. Meng, D. Gu, L. Zhang, Y. Huang, C. Liu, X. Wu, D. Zhao, Ordered mesoporous silicas and carbons with large accessible pores templated from amphiphilic diblock copolymer poly(ethylene oxide)-b-polystyrene, J. Am. Chem. Soc. 129 (2007) 1690–1697.

[49] Y.S. Ko, M.V. Circu, T. Geiger, S. Dünki, F.A. Nüesch, D.M. Opris, Synthesis of poly(ethylene-co-butylene)-block-poly(ethylene oxide) surfactant and its use in the synthesis of polyhydroxyethyl methacrylate nanoparticles containing Azo-dye, RSC Adv. 4 (2014) 35027–35034.

[50] C. Zhang, T. Xu, Z. Bao, L. Chen, Synthesis and characterization of fluorinated polyacrylate latex emulsified with novel surfactants, Des. Monom. Polym. 20 (2016) 118–124.
[51] C.R.E. Mansur, G. González, E.F. Lucas, Calorimetry and thermogravimetry as tools for the assessment of the thermal stability of polyoxide-based nonionic surfactants, Polym. Degrad. Stab. 80 (2003) 579–587.
[52] M.C. Burleigh, M.A. Markowitz, M.S. Spector, B.P. Gaber, Nanoporous organosilicas: periodic materials synthesized with surfactant templates in acidic media, J. Phys. Chem. B 106 (2002) 9712–9716.
[53] L. Chen, G. Zhu, D. Zhang, H. Zhao, M. Guo, W. Shi, S. Qiu, Novel mesoporous silica spheres with ultra-large pore sizes and their application in protein separation, J. Mater. Chem. 19 (2009) 2013–2017.
[54] D. Solis, E. Vigueras-Santiago, S. Hernandez-Lopez, A. Gomez-Cortes, M. Aguilar-Franco, M.A. Camacho-Lopez, Textural, structural and electrical properties of TiO_2 nanoparticles using Brij 35 and P123 as surfactants, Sci. Technol. Adv. Mater. 9 (2008), 025003.
[55] L. Zhou, D.S. Boyle, P. O'Brien, Uniform NH_4TiOF_3 mesocrystals prepared by an ambient temperature self-assembly process and their topotaxial conversion to anatase, Chem. Commun. (2007) 144–146.
[56] Y. Liu, Z. Wang, W. Teng, H. Zhu, J. Wang, A.A. Elzatahry, D. Al-Dahyan, W. Li, Y. Deng, D. Zhao, A template-catalyzed *in situ* polymerization and co-assembly strategy for rich nitrogen-doped mesoporous carbon, J. Mater. Chem. A 6 (2018) 3162–3170.
[57] C.T. Nguyen, D.P. Kim, Direct preparation of mesoporous carbon by pyrolysis of poly(acrylonitrile-b-methylmethacrylate) diblock copolymer, J. Mater. Chem. 21 (2011) 14226–14230.
[58] W.C. Chu, J.G. Li, S.W. Kuo, From flexible to mesoporous polybenzoxazine resins templated by poly(ethylene oxide-b-ε-caprolactone) copolymer through reaction induced microphase separation mechanism, RSC Adv. 3 (2013) 6485–6498.
[59] Y.S. Sun, W.H. Huang, J.Y. Liou, Y.H. Lu, K.C. Shih, C.F. Lin, S.L. Cheng, Conversion from self-assembled block copolymer nanodomains to carbon nanostructures with well-defined morphology, RSC Adv. 5 (2015) 105774–105784.
[60] Y. Deng, C. Liu, D. Gu, T. Yu, B. Tu, D. Zhao, Thick wall mesoporous carbons with a large pore structure Templated from a weakly hydrophobic PEO-PMMA diblock copolymer, J. Mater. Chem. 18 (2008) 91–97.
[61] D. Gu, F. Schuth, Synthesis of non-siliceous mesoporous oxides, Chem. Soc. Rev. 43 (2014) 313–344.
[62] J. He, P. Ma, G. Zhang, R. Li, L. Zhang, Sol-gel derived mesoporous $GaAlPO_4$ glass for heavy metal ion sequestration, RSC Adv. 6 (2016) 99149–99157.
[63] S. Wang, D.K. Wang, S. Smart, J.C.D. da Costa, Ternary phase-separation investigation of sol-gel derived silica from ethyl silicate 40, Sci. Rep. 5 (2015) 14560.
[64] S.P. Garg, R. Venkataramani, S. Mazumder, The oxygen stoichiometry of $YBa_2Cu_{2.82}Zn_{0.18}O_{7-\delta}$ between 25 and 930°C, Mater. Res. Bullet. 24 (1989) 803–807.
[65] E. Saiz, M.F. Barba, P. Ortega, J.S. Moya, Oxygen content determination in the YBa_2O_{7-x} compound, Mater. Lett. 8 (1989) 238–240.
[66] S. Kao, Y.S. Ng, Oxygen content determination of $Bi_2Sr_2Ca_2Cu_4O_{11+x}$ superconductor by thermogravimetric analysis, J. Supercond. 4 (1991) 375–384.
[67] I.V. Nicolaescu, G. Tardos, R.E. Riman, Thermogravimetric determination of carbon, nitrogen, and oxygen in aluminum nitride, J. Am. Ceram. Soc. 77 (1994) 2265–2272.
[68] Q. Li, R. Jiang, Y. Dou, Z. Wu, T. Huang, D. Feng, J. Yang, A. Yu, D. Zhao, Synthesis of mesoporous carbon spheres with a hierarchical pore structure for the electrochemical double-layer capacitor, Carbon 49 (2011) 1248–1257.

[69] X.W. Lou, C.M. Li, L.A. Archer, Designed synthesis of coaxial SnO_2@carbon hollow nanospheres for highly reversible lithium storage, Adv. Mater. 21 (2009) 2536–2539.

[70] P. Zhang, L. Wang, S. Yang, J.A. Schott, X. Liu, S.M. Mahurin, C. Huang, Y. Zhang, P.F. Fulvio, M.F. Chisholm, S. Dai, Solid-state synthesis of ordered mesoporous carbon catalysts *via* a mechanochemical assembly through coordination cross-linking, Nat. Commun. 8 (2017) 15020.

[71] C. Liang, Z. Li, S. Dai, Mesoporous carbon materials: synthesis and modification, Angew. Chem. Int. Ed. Eng. 47 (2008) 3696–3717.

[72] Y. Li, D. Cui, Y. Tong, L. Xu, Study on structure and thermal stability properties of lignin during thermostabilization and carbonization, Int. J. Biol. Macromol. 62 (2013) 663–669.

[73] I.F. Myronyuk, V.I. Mandzyuk, V.M. Sachko, V.M. Gun'ko, Structural features of carbons produced using glucose, lactose, and saccharose, Nanoscale Res. Let. 11 (2016) 508.

[74] J. Stejskal, M. Trchová, J. Hromádková, J. Kovarova, A. Kalendova, The carbonization of colloidal polyaniline nanoparticles to nitrogen-containing carbon analogues, Polym. Int. 59 (2010) 875–878.

[75] C. Chen, E.B. Kennel, A.H. Stiller, P.G. Stansberry, J.W. Zondlo, Carbon foam derived from various precursors, Carbon 44 (2006) 1535–1543.

[76] I.V. Gofman, K. Balik, M. Cerny, M. Zaloudkova, M.J. Goikhman, V.E. Yudin, Peculiarities of the initial stages of carbonization processes in polyimide-based nanocomposite films containing carbon nanoparticles, Cogent Chem. 1 (2015) 1076712.

[77] S.Y. Cho, Y.S. Yun, S. Lee, D. Jang, K.Y. Park, J.K. Kim, B.H. Kim, K. Kang, D.L. Kaplan, H. J. Jin, Carbonization of a stable β-sheet-rich silk protein into a pseudographitic pyroprotein, Nat. Commun. 6 (2015) 7145.

[78] S.B. Yoon, J.Y. Kim, J.S. Yu, Synthesis of highly ordered nanoporous carbon molecular sieves from silylated MCM-48 using divinylbenzene as precursor, Chem. Commun. 6 (2001) 559–560.

[79] T.W. Kim, I.S. Park, R. Ryoo, A synthetic route to ordered mesoporous carbon materials with graphitic pore walls, Angew. Chem. Int. Ed. Eng. 42 (2003) 4375–4379.

[80] F. Su, J. Zeng, X. Bao, Y. Yu, J.Y. Lee, X.S. Zhao, Preparation and characterization of highly ordered graphitic mesoporous carbon as a Pt catalyst support for direct methanol fuel cells, Chem. Mater. 17 (2005) 3960–3967.

[81] J. Kopecká, M. Mrlík, R. Olejník, D. Kopecký, M. Vrnata, J. Prokeš, P. Bober, Z. Moravková, M. Trchová, J. Stejskal, Polypyrrole nanotubes and their carbonized analogs: synthesis, characterization, gas sensing properties, Sensors 16 (2016) 1917.

[82] J.B. James, Y.S. Lin, Kinetics of ZIF-8 thermal decomposition in inert, oxidizing, and reducing environments, J. Phys. Chem. C 120 (2016) 14015–14026.

[83] M.M. Titirici, R.J. White, C. Falco, M. Sevilla, Black perspectives for a green future: hydrothermal carbons for environment protection and energy storage, Energy Environ. Sci. 5 (2012) 6796–6822. M. Pecchi, F. Patuzzi, D. Basso, M. Baratieri, Enthalpy change during hydrothermal carbonization of biomass: a critical review, J. Therm. Anal. Calorim. 141 (2020) 1251–1262.

[84] J. Cai, B. Li, C. Chen, J. Wang, M. Zhao, K. Zhang, Hydrothermal carbonization of tobacco stalk for fuel application, Bioresour. Technol. 220 (2016) 305–311.

[85] C. Peng, Y. Zhai, Y. Zhu, B. Xu, T. Wang, C. Li, G. Zeng, Production of char from sewage sludge employing hydrothermal carbonization: char properties, combustion behavior and thermal characteristics, Fuel 176 (2016) 110–118.

Chapter 4

Applications of TA

1 TA in food industry

Thermal analysis is used in most industrial laboratories as a quality control tool [1]. One food reaction that has been subjected to a great deal of thermal analysis is starch gelatinization. It involves the incorporation of water molecules inside the starch crystals, resulting in the swelling of the starch granules. An important nutritional consequence of this is that enzymes hydrolyze this carbohydrate material into glucose. If high starch food systems have not been sufficiently "cooked" (i.e., gelatinization is not complete), serious digestion problems occur. Gelatinization is an endothermic event, occurring near 60°C, and the enthalpy change is on the order of 8–20 J/g. A DSC peak at ca. 60°C indicates incomplete gelatinization process (Fig. 4.1).

A DSC method has been used in food chemistry to study "free" or "unbound freezable" water [2]. Starches, gums, sugars, proteins, and emulsifiers tightly bind water, thus affecting many properties in food systems. The size of the DSC endotherm observed on heating a frozen sample through 0°C is a measure of free water in the sample. Bound water does not undergo freezing. The heat of fusion of ice is 80 cal/g. Fig. 4.2 shows a DSC calorigram of a frozen corn sample.

The total water content in food products and pharmaceuticals (e.g., effervescent tables) can be accurately determined using the new module called "moisture evolution analyzer (MEA)." The sample is placed in an oven and heated so that the water is volatilized and carried to the detector by a predried carrier gas (nitrogen). In the electrolytic cell detector, the water is adsorbed by phosphorous pentoxide coated on platinum electrodes and electrolyzed to hydrogen and oxygen. The electrolysis current is integrated, scaled, and displayed as μg water. Detection levels of 1–10 ppm are now available. Excessive moisture in effervescent tablets can increase the dissolving time of the finished tablets and decrease the appearance stability, causing mottling and discoloration.

DSC is also used to profile fats used in chocolate, margarine, dairy products, etc. The Solid Fat Index (SFI) measures the amount of solid fat in the sample at specific temperatures. The SFI is an important quality control tool in the food

Thermal Analysis. https://doi.org/10.1016/B978-0-323-90191-8.00004-X

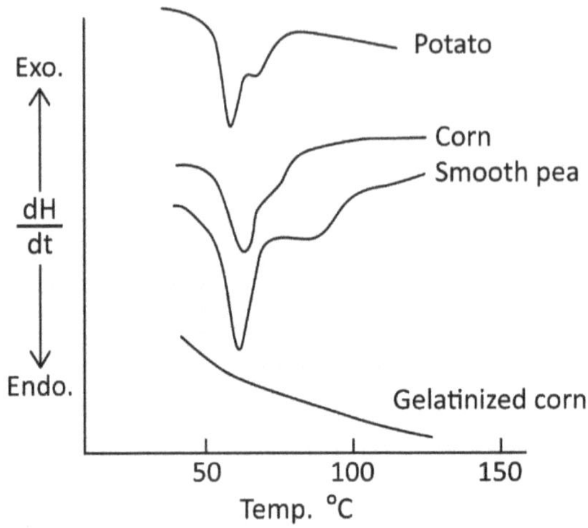

FIG. 4.1 DSC thermograms of starches in the presence of moisture.

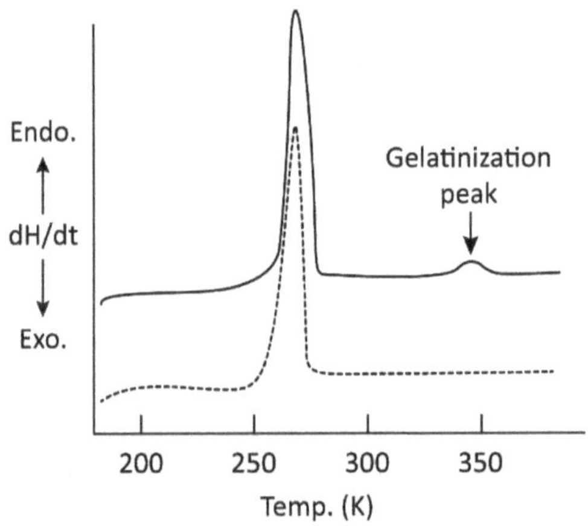

FIG. 4.2 DSC calorigrams of (——) frozen corn sample and (- - - -) an ice sample.

industry. It can be used to decide whether the raw material should be accepted or rejected. The amount of solid fat is proportional to the unswept area up to a certain temperature.

Fats often are composed of more than one crystalline phase. The percentage of each phase can be obtained from the areas of their melting DSC peaks. Fig. 4.3 shows a DSC calorigram of butterfat showing the melting of its two polymorphs.

The phases of a fat can interconvert. If this happens during production, the product may become unstable and/or unacceptable to consumer.

Food has been traditionally packed in cans. Due to some health hazards and to reduce the total weight of the products, food manufactures have begun

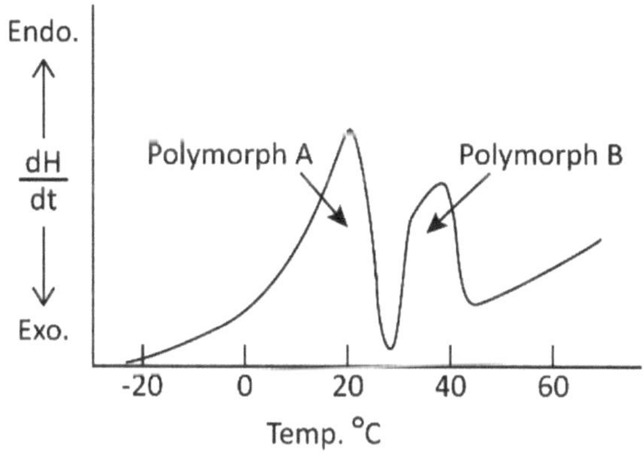

FIG. 4.3 DSC calorigram of butterfat.

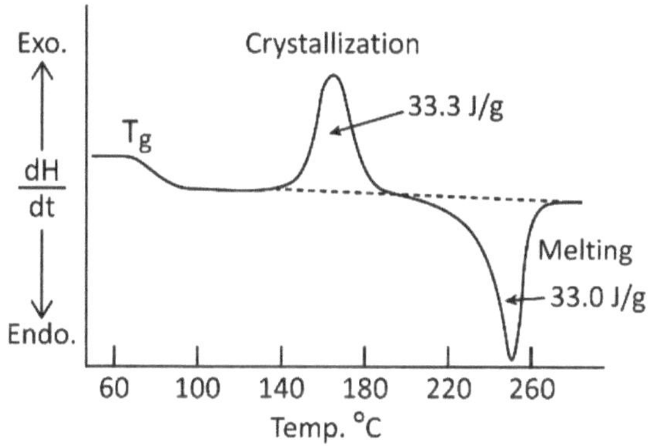

FIG. 4.4 DSC of polyethylene terephthalate (PET).

looking at light-weight packaging, including plastics. DSC has been used extensively to study the purity of these plastics and their thermal stability. Fig. 4.4 shows a DSC calorigram of polyethylene terephthalate (PET, also known as polyester).

Containers can also be made of polypropylene-EVAL sandwich. Fig. 4.5 shows a DSC calorigram of polypropylene-EVAL sandwich. The first endothermic peak is the melting of polypropylene, the second endotherm peak is the ethylene and vinyl alcohol (EVAL) melting.

DSC is also a useful tool for investigating the thermally induced conformational transitions of proteins [3]. The thermogram can be used to calculate the van't Hoff enthalpy.

Food systems that contain unsaturated oils (for their nutritional value) can go rancid. Rancidity is another term for lipid oxidation, whereby the allylic hydrogens are replaced by oxygen causing an increase in molecular weight and thus the sample gains weight.

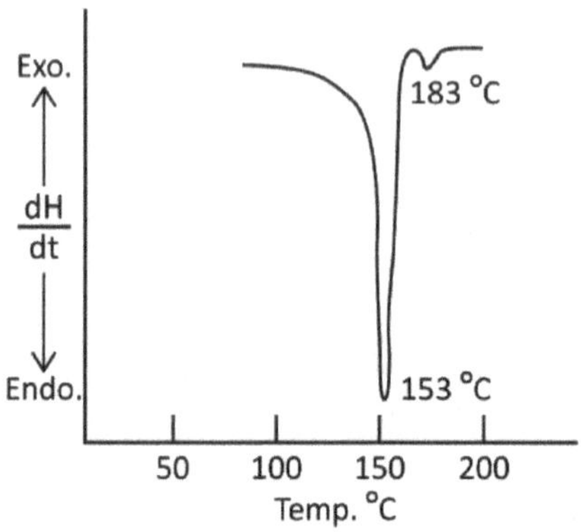

FIG. 4.5 DSC calorigram of polypropylene-EVAL sandwich.

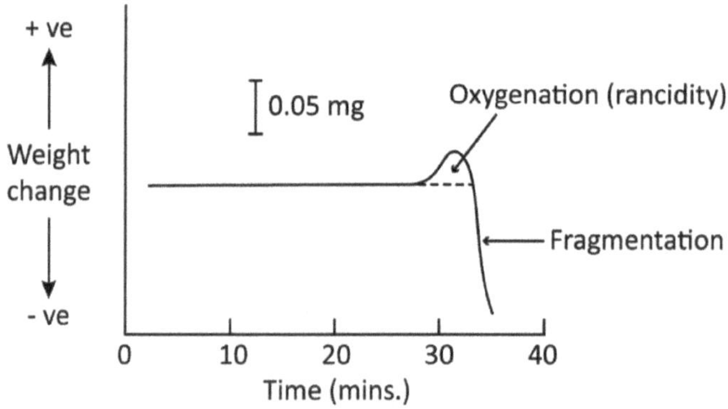

FIG. 4.6 Isothermal TGA measurement of vegetable oil oxidative stability. Sample size was 25 mg in oxygen atmosphere at 50 cc/min and isothermal run at 65°C.

In contrast, these oxygenated lipids (called hydroperoxides) undergo bond breaking to form volatile hydrocarbons, aldehydes, and ketones. This step results in a sample weight loss. TGA is sensitive to both the initial weight gain and the subsequent weight loss (Fig. 4.6).

2 TA of soap and detergents

A surfactant when mixed with water and heated will begin to dissolve at a definite characteristic temperature. This solution temperature is known as Krafft point (Fig. 4.7).

The Krafft point of reheated sample is lower due to the rapid crystallization conditions. The two additional endotherms are probably due to liquid crystal transitions. The cloud point is the temperature at which the first stage of crystallization is induced in the sample on cooling. The onset of crystallization is a

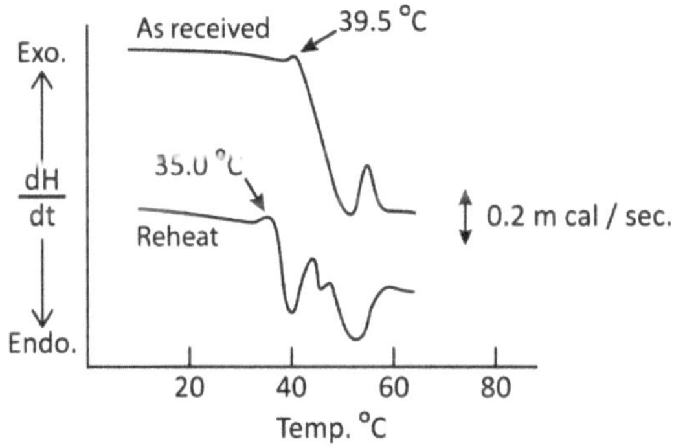

FIG. 4.7 DSC of sodium dodecyl sulfate (SDS) at a heating rate of 5°C/min. The weight of the sample was 5 mg in water.

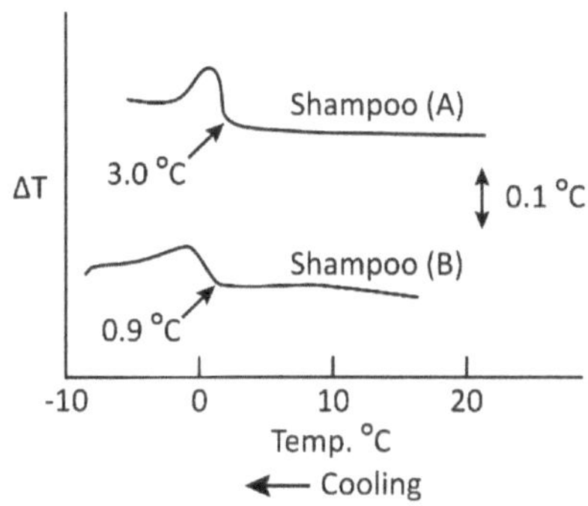

FIG. 4.8 DTA curves illustrating the cloud points of two shampoo samples.

measure of the cloud point. It is lower (about 1°C) than the optically determined cloud point due to the high cooling rate (Fig. 4.8) [4,5].

Fig. 4.9 shows the TG curves of two cleansers (A) and (B). Cleanser (A) contains 40% volatile material that was driven off in the temperature range from ambient to 150°C. The rapid weight loss starting at 300°C represents decomposition of the material. In cleanser (B), the two abrupt decomposition stages at higher temperatures indicate that the cleanser might contain two major detergents.

3 TA in electronics industry

Thermal analysis techniques have demonstrated their usefulness to the electronics industry in measuring such quality control parameters as degree of cure, thermal stability, oxidative stability, coefficient of expansion, glass transition temperatures, and heats and temperatures of transition [6]. Applications to specific materials include encapsulating compounds, solder glass, printed circuit board, molded parts, liquid crystals, and wire coating.

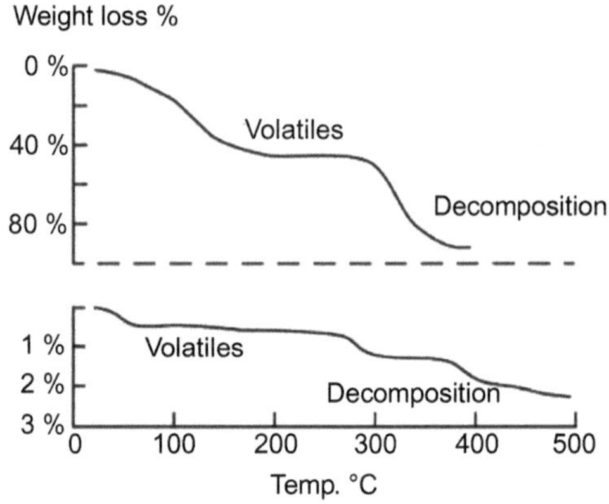

FIG. 4.9 TGA curves of sail liquid detergent *(top)* and a Dutch cleanser *(bottom)*.

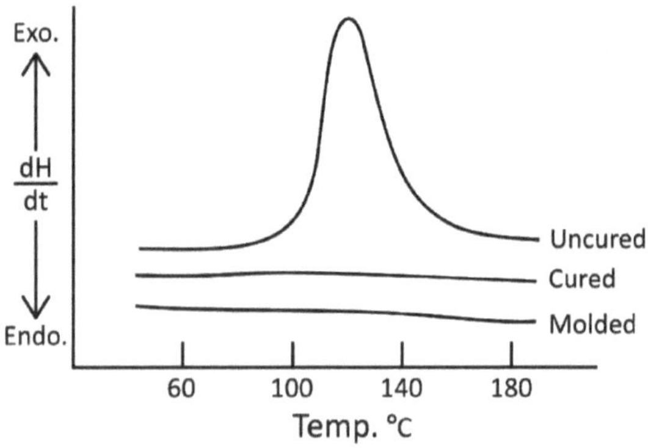

FIG. 4.10 Thermoset cure of an alkyd resin.

The uses of TA in the electronics industry are expected to expand even more as the uses of polymeric materials increase. Most of the polymer used in the electronics industries are thermosets (e.g., epoxies, phenolics, diallyl phthalates, etc.). Thermosetting resins are polymeric materials which, when heated, catalyzed, irradiated, or otherwise initiated, undergo an exothermic chemical reaction. This reaction, a cross-linking of the various portions of the polymeric molecules, produces a material that is often infusible, insoluble, hard, and quite thermally stable. Such compounds, when pressed into molds before cross-linking, become "set" to the shape of the mold after the cross-linking has occurred. The amount of cross-linking, or "degree of cure" as it is called, is an extremely important parameter of the thermosets for it is this "degree of cure" which dictates the physical properties of the finished product.

Fig. 4.10 illustrates the exothermic curing reaction as measured by DSC. The top trace is a DSC scan made on uncured alkyd resin. A broad exothermic peak is seen beginning near 100°C. The area under this peak is proportional to

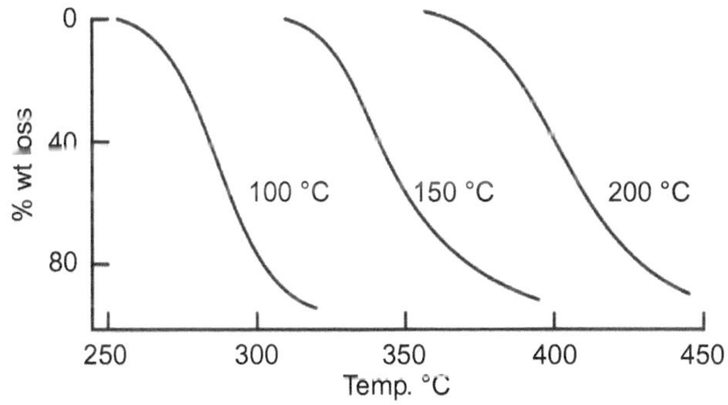

FIG. 4.11 TG thermograms of decomposition of epoxies cured at differing temperatures.

the heat of cure reaction. The middle trace is the same sample rerun under the same conditions. The lack of any peak indicates that the sample is fully cured. Of course, a partially cured material would have an exothermic peak some place between these two extremes. Finally, the bottom trace is a scan taken from a molded printed circuit card connector made from this same alkyd resin. Clearly the molded part is completely cured as well.

Fig. 4.11 shows a series of TGA scans made on epoxies cured for the same length of time at different temperatures. The sample cured for 24 h at 100°C begins to decompose at 252°C while the one cured for the same length of time at 200°C begins to decompose at 360°C.

DSC has been used for assessing sample purity. The concentration of impurity is determined by analyzing the melting endotherm curve. The higher the concentration of impurity, the lower the sample's melting point and the broader its melting range. A method for determining the purity of organic compounds using the DSC technique was presented previously [7]. The heat of fusion (ΔH_f) and temperature during melting are measured and a curve is plotted, the slope of which is the melting point depression (ΔT) due to impurities in the sample, and whose zero intercept represents the melting point of a theoretical sample with zero impurity (T_o). The total purity is computed by inserting these values into Eq. (4.1):

$$\text{mol}\,\%\,\text{impurity} = 100\left(\Delta H_f / R T_o^2\right) \times \Delta T \tag{4.1}$$

Organophosphates, amides, ureas, carbamates, heterocyclics, chlorophenoxy acids, esters, and halogenated compounds were analyzed. The method is accurate only for samples over 98% pure and does not measure impurities that are soluble in the solid phase (capable of forming solid solutions) or insoluble in the melt. It is also inapplicable to chemicals that decompose at their melting points or have inordinately high vapor pressure.

TGA is used to determine filler level in a molding compound. When the sample is heated and the organic portion of the compound is volatilized, only

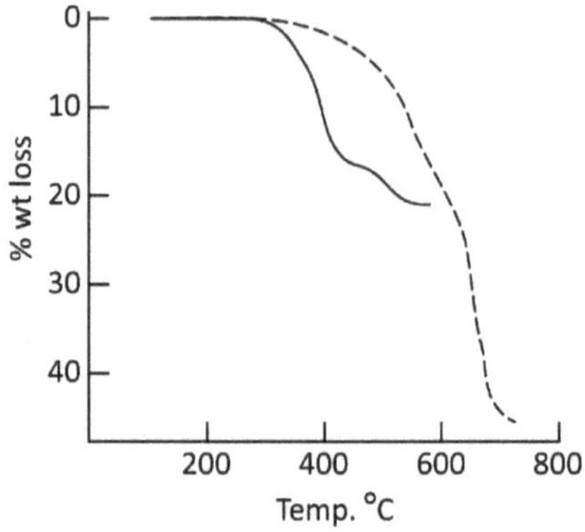

FIG. 4.12 TG thermograms illustrating the thermal stability of encapsulating thermosets (——) DAP and (- - - - - -) phenolic compound.

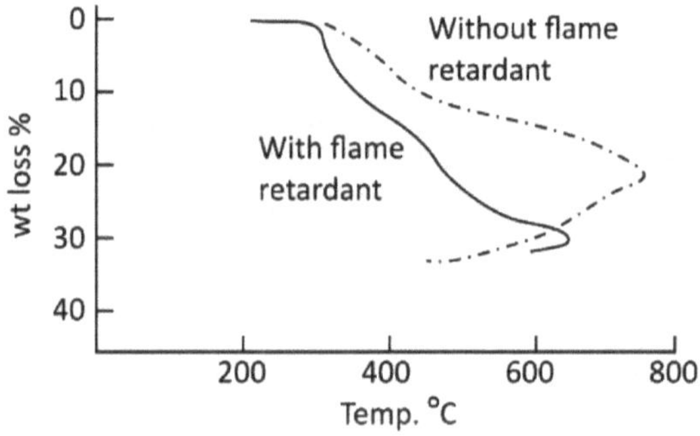

FIG. 4.13 TG curves illustrating thermal stability in O_2 of an epoxy molding compound.

the inorganic filler remains. In the sample used in Fig. 4.12, the diallyl phthalate (DAP) contained 78.9% filler while the phenolic compound contained only 54.4% filler.

To improve fire and safety performance of encapsulated materials, molding compounds often contain fire retardants. The "burning" phenomenon is displayed in Fig. 4.13 on the x-axis.

As a result of sample combustion, the sample temperature leaps ahead of the oven temperature as indicated by the return to a much lower temperature as the sample cools back down to the oven temperature following complete combustion. The effect of retardant is reflected on the "distance" traveled along the x-axis.

3.1 Solder glasses

While encapsulation using organic thermosets is acceptable for nonservice of the electronic package, more rugged materials are required for severe service.

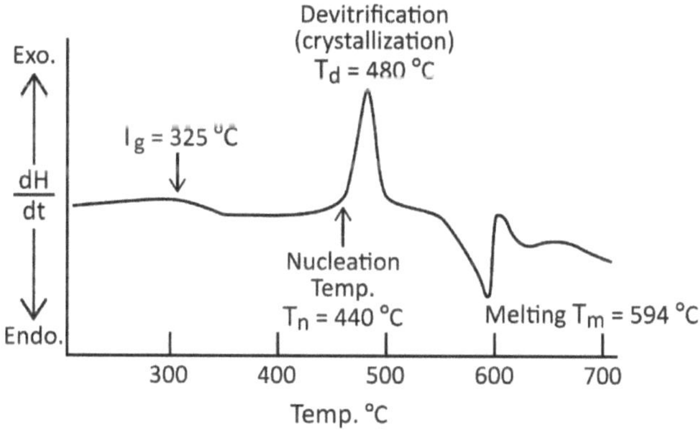

FIG. 4.14 DSC curve of CV-98 solder glass.

One such encapsulation process involves sealing the device between two pieces of ceramic using a low-melting glass called "solder glass" as the "cement" sealing the two portions of the package together [8]. The process involves a stage where the package is fired at a temperature at which the solder glass devitrifies (crystallizes) forming a very strong, hermetic seal. Fig. 4.14 shows the DSC of a solder glass illustrating a number of temperatures of interest.

Because so much useful information can be obtained from a single TA scan of solder glasses, this technique is becoming extremely widespread as a quality control tool in the manufacture of both the solder glass and the electronic packages.

3.2 Liquid crystals

Liquid crystal displays are particularly attractive because of their low power consumption [9,10]. TA techniques particularly DSC and TGA are widely used in research and quality control work on liquid crystals (e.g., determination of temperatures and heats of transitions). Fig. 4.15 shows a DSC calorigram of cholesteryl-n-propionate. The transition temperatures and heats of transitions are as follows:

$$\text{Solid} \xrightarrow[\;]{\substack{120^\circ\text{C} \\ \Delta H = 5.76 \\ \text{k cal mol}^{-1}}} \text{Liquid crystal} \xrightarrow[\;]{\substack{110^\circ\text{C} \\ \Delta H = 0.102 \\ \text{k cal mol}^{-1}}} \text{True liquid}$$

3.3 Wires and cables

One test commonly used to estimate the lifetime of polymeric wire insulators is ANSI/ASTM procedure D-2307. In this procedure, twisted pairs of insulated

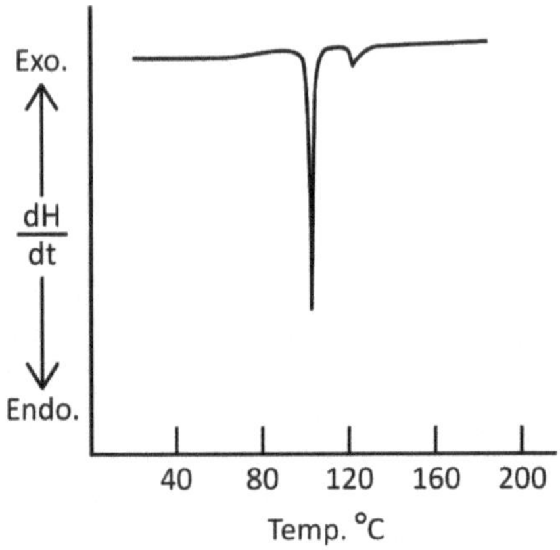

FIG. 4.15 DSC of cholesteryl-n-propionate.

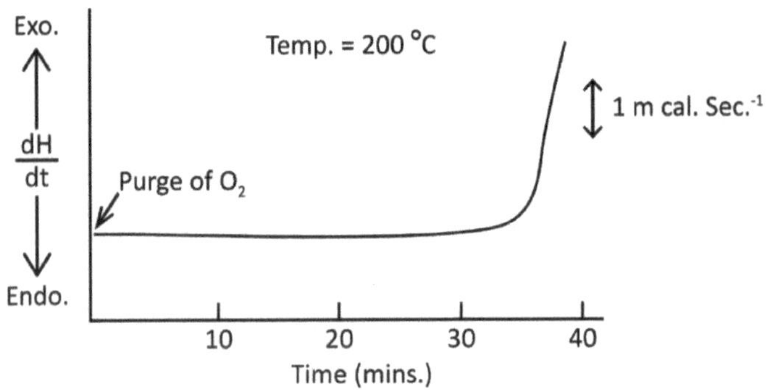

FIG. 4.16 Isothermal DSC calorigram illustrating polyethylene oxidative stability.

wire are oven aged (for up to 50 days) at elevated temperatures (up to 240°C) until voltage breakdown occurs. The method assumes first-order kinetics and uses extrapolation to estimate the long lifetimes encountered at normal use temperatures. This procedure, while useful, is very time consuming, often taking many months particularly for highly stable materials. As more and more stable polymeric electrical insulation materials are introduced, the time needed for a full series of tests becomes excessive.

TGA offers a rapid approach to estimate polymer lifetimes. From a series of perhaps four experiments, carried out in a single day, high-quality kinetic data is obtained (vide infra).

DSC has also been applied extensively to study the oxidative stability of polymeric coatings used to protect copper and aluminum wire and cable. In this method, the sample is allowed to reach isothermal equilibrium at 200°C in a nitrogen atmosphere. The purge gas is then changed from nitrogen to air or oxygen and simultaneously, time-based DSC scan is begun. As shown in Fig. 4.16, no reaction is seen for some period then an exothermic oxidation begins. The time to the onset of oxidation is taken as a measure of oxidative stability and is used to estimate the lifetime of the material. This procedure is widespread in the polymer and wire and cable manufacturing industry largely due to the Western Electric specification based on this procedure.

3.4 Oxidative stability DSC methods of polymers and oils

The team of the METTLER TOLEDO materials characterization group conducted experiments on oxidative stability DSC methods of polymers and oils [11].

Two DSC methods the oxidation induction time (OIT) and the onset oxidation temperature (OOT) that are used to measure the oxidative stability of polymers and oils were applied. These methods simulate the accelerated chemical aging of products and allow information to be obtained about their relative stability.

Different materials can be compared and analyzed to determine the influence of an additive. In practice, the method is widely used for PE (polyethylene). The application example described below also uses a sample of PE because the decomposition of PET is overlapped by melting and reesterification and cannot be clearly identified.

The OIT measurement of PE (Fig. 4.17) is often performed in crucibles made of different metals in order to determine the influence of a particular metal

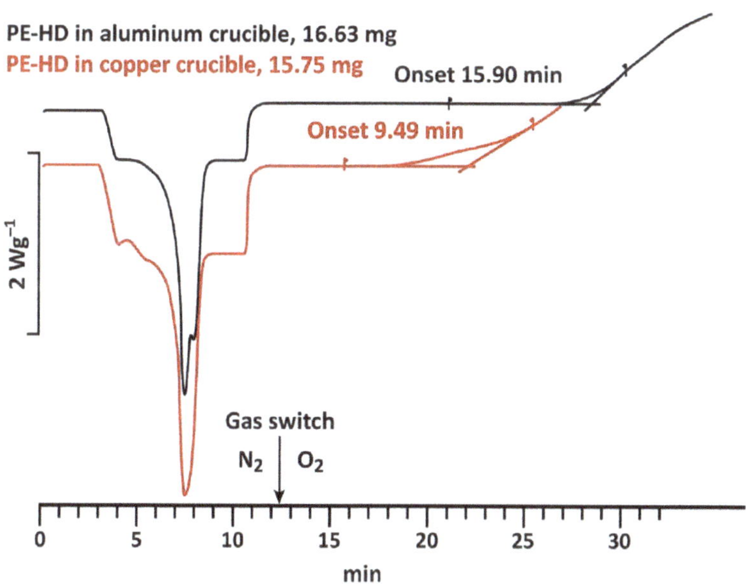

FIG. 4.17 OIT measurements of PE in Al *(top)* and Cu *(bottom)* open 40-μL crucibles.

on the stability of the PE. In this example, measurements were performed in open 40-μL aluminum and copper crucibles for comparison. The measurement was started in a nitrogen atmosphere according to the following temperature program: 3 min at 30°C. After 2 min, the gas was switched to oxygen. The measurement was stopped as soon as oxidation was observed. The OIT is the time interval from the purge gas switching to oxygen to the onset of oxidation. Oxidation clearly takes place much earlier in the copper crucible than in the aluminum crucible. The copper acts as a catalyst and accelerates the decomposition of PE.

The oxidative stability of samples can also be compared by measuring the Onset Oxidation Temperature (OOT). In this method, the sample is heated in an oxygen atmosphere and the onset temperature at which oxidation begins is evaluated. Since OIT measurements are easy to perform and do not take much time, they are often used in quality control to compare the stability of products.

4 Polymer crystallinity

The degree of crystallinity is the percentage crystalline content of a semicrystalline substance. Thermoplastics normally exhibit a degree of crystallinity of up to 80%. The degree of crystallinity of a material depends on its thermal history.

Crystallinity of a polymer affects its physical properties (e.g., modulus, stiffness, permeability, density, melting point, stability, etc.). It can be easily determined using DSC data. Given ΔH_{st} is the heat of fusion of a 100% crystalline material and ΔH_{sample} is the heat of fusion of the sample under investigation, then

$$\text{Crystallinity}\ \% = \frac{\Delta H_{sample}}{\Delta H_{st}} \times 100 \tag{4.2}$$

A major difficulty in the determination of degree of crystallinity is finding a suitable standard (100% crystalline) or locating in the literature the heat of fusion for 100% crystalline polymers (e.g., in Polymer Handbooks) [12]; 100% crystalline materials can be determined by X-ray diffraction.

5 Polymer blends

In the plastics industry, polymers are often blended to give products possessing characteristics not held by either of the parent polymers. The processor of such polymers is often faced with the task of determining the actual composition of his products. This can be easily done by analyzing the characteristic peak areas. Fig. 4.18A shows a DSC calorigram of a polyethylene/polypropylene polymer blend. Fig. 4.18B shows a calibration curve in which linearity exists between 14%–20% polyethylene and its characteristic peak area.

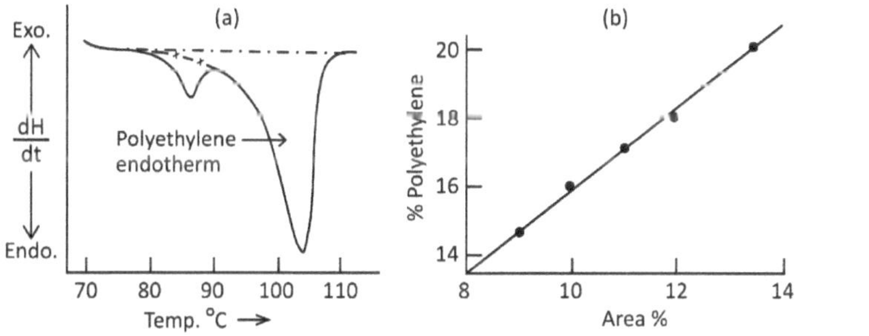

FIG. 4.18 (A) DSC melting endotherm of polyethylene/polypropylene blend and (B) calibration curve.

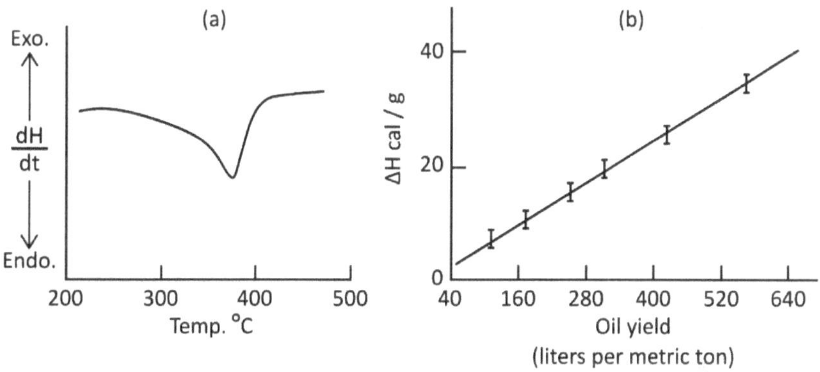

FIG. 4.19 (A) DSC calorigram of oil shale. Heating rate is 10°C/min, N_2 atmosphere, and sample mass = 13.3 mg. (oil yield: 136 L/metric ton). (B) Correlation of the enthalpy of decomposition (ΔH) with oil yields.

6 TA in fossil fuel research

Oil shales are sources of energy in the form of petroleum. As a quick test about the content of oil in oil shales, DSC has been applied to estimate ΔH of decomposition that is an estimation of the oil content (Fig. 4.19) [13]. From an energy viewpoint, the most important constituent of oil shales is the organic matter which comprises both bitumen and kerogen. The amount of indigenous organic matter in a shale deposit determines the economic value of that geological formation. The Fischer assay [14–16] has been the traditional method for estimating the potential oil yield of oil shale deposits. This method, however, is quite time consuming.

TGA was applied to study the adsorption of O_2 on coal. Fig. 4.20 shows the adsorption isotherm at 110°C. An inert gas (e.g., helium) is replaced by oxygen at a certain time while the mass of the sample is monitored as a function of time. In this case, oxygen penetrates the top layers of the coal thus producing a modification in the macromolecular structure.

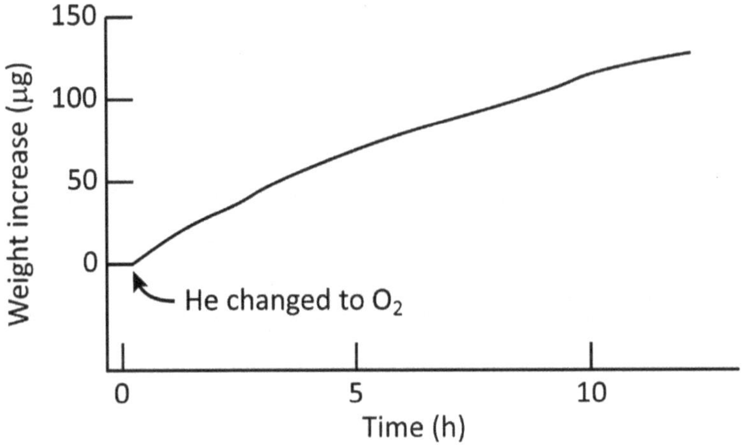

FIG. 4.20 Adsorption of O_2 on coal at 110°C as a function of time.

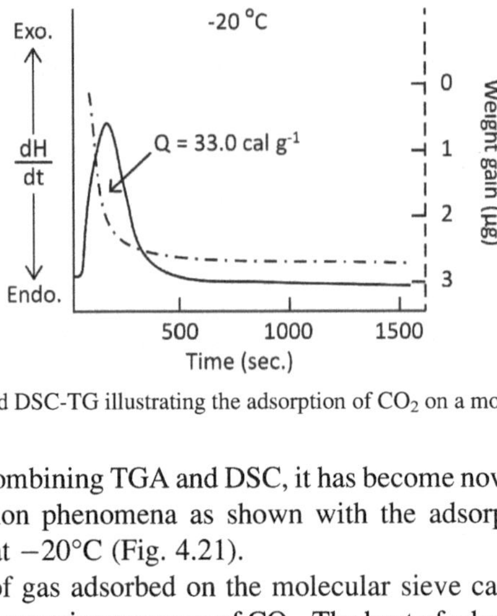

FIG. 4.21 Combined DSC-TG illustrating the adsorption of CO_2 on a molecular sieve at −20°C.

By virtue of combining TGA and DSC, it has become now possible to investigate physisorption phenomena as shown with the adsorption of CO_2 on a molecular sieve at −20°C (Fig. 4.21).

The amount of gas adsorbed on the molecular sieve can also be recorded according to the increasing pressure of CO_2. The heat of adsorption can be evaluated at the same time.

DSC has been used as a rapid and accurate technique for the determination of the specific heat capacity of petroleum fuels and solids (e.g., metals, metal hydroxides, polymers, and inorganic salts). The specific heat capacity data can then be used in the evaluation of other basic thermodynamic properties of the material [17–19].

When a sample material is subjected to a linear temperature program, the heat flow rate into the sample is proportional to its instantaneous specific heat. By recording this heat flow rate as a function of temperature and comparing it with the heat flow rate into a standard material under the same conditions, the specific heat of the sample is determined as a function of temperature [19].

7 TA in clays

TA is widely used in the identification and characterization of clays [20]. The assignment of the three predominant DTA peaks for clays goes as follows:

Peak		Assignment
25–220°C	(Endo)	The removal of the moisture and the interlayer water.
500–840°C	(Endo)	The loss of structural OH^- ions and lattice breakdown.
≃950°C	(Exo)	Recrystallization

In Egypt, the thermal study of clays is generally important since they are used to replace Nile silt in the manufacture of building bricks. Nile silt is becoming scarce because of the construction of the High-Dam. Both DTA and TGA of Nile silt have been carried out [21]. Heated Nile silt acts as a good adsorbate of cationic impurities from aprotic solvents. The DTA of Nile silt shows three broad endotherms commencing at 50°C, 375°C, and 525°C and reaching their respective maxima at 130°C, 420°C, and 575°C. The endothermic peak at 130°C is associated with about 6% weight loss and is assigned to the release of adsorbed water. The endothermic peaks at 420°C and 575°C are assigned to the thermal effect that accompanies the breakdown of the clay mineral lattice structure with subsequent expulsion of interlayer water. Slight weight loss is observed at this stage as revealed by the corresponding TGA calorigram.

The dominant minerals in Nile silt samples were identified using X-ray powder diffraction and were found as follows:

Mineral	X-ray diffraction peak(s) ($2\theta°$)
Kaolinite	12.75, 45.55
Feldspar	28.0
Calcite	30.50

The intensity of X-ray powder diffraction peaks decreases as a result of heating Nile silt at ca. 650°C. This is due to the thermal decomposition of the corresponding minerals giving metal oxides (the dominant reddish color at this heating stage is mainly due to iron oxide). The difference in the DTA peak shapes has been recognized as a means for distinguishing kaolinite (580°C) and halloysite (530°C) for some time [21].

Kinetic data were obtained using thermal desorption technique for the evolution of hydrocarbons from the decomposition of organic compounds on mineral surfaces. These measurements may be extrapolated to lower temperatures to estimate the extent to which reactions could proceed in sediments. Catalytic processes are expected to contribute to hydrocarbon formation since the surfaces of clays may strongly chemisorb organic material. After a subsequent deposition in sediments, the breakdown of surface-bonded carbonaceous material represents a source of hydrocarbons [22].

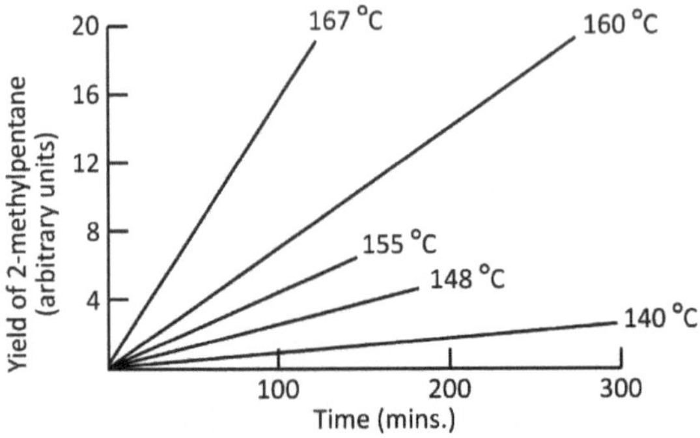

FIG. 4.22 Thermal desorption technique for the evaluation of 2-methylpentane following reaction of n-octadecanol on montmorillonite. The activation parameters can be obtained from the slopes at different temperatures.

From studies of the reactions of alcohols on montmorillonite, it was concluded that (i) Surface reactions yielded n-, iso-, and cycloalkanes. These hydrocarbons are abundant in petroleum. (ii) Yields were sufficiently large to account for the size of natural accumulations. (iii) The rates of these reactions, at realistic source-rock temperatures, were sufficiently great to indicate that surface processes could significantly participate in the formation of hydrocarbons in nature. The formation of isohexanes from their reactions of ethanol and n-dodecanol indicated the occurrence of cracking, polymerization, skeletal isomerization, and hydrogen transfer processes. Product mixtures were analyzed by gas-liquid chromatography using the argon ionization detector.

Fig. 4.22 shows the formation of 2-methylpentane during the thermal desorption reaction of n-octadecanol on montmorillonite under conditions of possible product and water vapor readsorption [22].

8 TA in cement industry

TA is currently applied in studying the thermal behavior of cement [23]. Fig. 4.23 shows the TGA and DTA curves of raw cement. Decarbonation (accompanied by weight loss of the sample) occurs between 700°C and 1000°C. The heat of decarbonation can be evaluated.

Cement can be defined as a substance that sets and hardens due to chemical reactions that occur upon mixing with water. Various types of cement have been known in use since ancient times. The Egyptians used calcined impure gypsum ($CaSO_4{\cdot}2H_2O$), the Greeks and Romans used calcined limestone ($CaCO_3$), and later added aggregate-sand and crushed-stone to from concrete.

Modern Portland cement originated in Britain in the 19th century when high temperatures were first used. Silicates, Ca_2SiO_4 and Ca_3SiO_5, were produced during the high-temperature reactions of clay and chalk (that is $CaCO_3$). Nowadays, various raw materials are used (e.g., chalk, limestone, or gypsum) as a

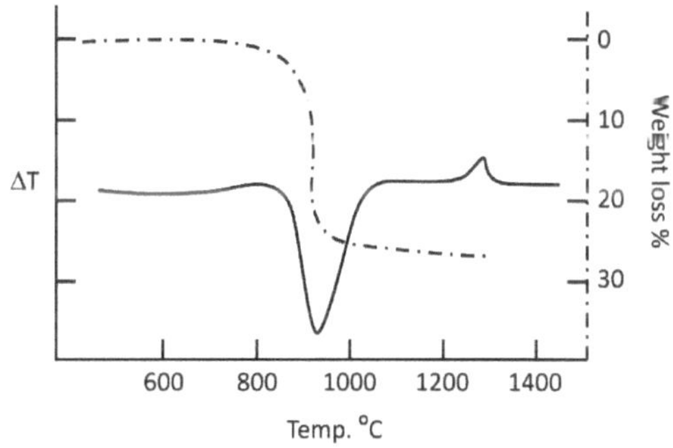

FIG. 4.23 Heating of raw cement in an argon atmosphere.

source of lime together with sand, clay, and iron oxide. The name "Portland" derives from the similarity in color and appearance to Portland stone, found in Dorset, England.

In the commercial production of Portland cement, the raw materials are first ground and intimately mixed, either dry or as a slurry with water. This feed is introduced into the top, cool end of a long rotary kiln whose axis is inclined slightly to the horizontal. The bottom of the kiln is heated to 1300–1500°C by coal, oil, or natural gas. As the feed passes down the kiln, it gradually heats up, first losing H_2O and CO_2. Further on, reactions between the solids begin to occur and these are completed in the hottest zone where partial melting also occurs. An oxidizing atmosphere in the kiln ensures that any iron present is in the Fe^{3+} state. The partially fused black lumps, known as Clinker, are discharged from the bottom of the kiln, cooled rapidly in an air blast, and crushed. Usually, gypsum is added at this stage to prevent "Flash set" from occurring (i.e., rapid setting powder is the familiar Portland cement which, when mixed with water and appropriate aggregate, hydrates to form set cement or concrete). The following oxides undergo combination because of heating:

$$2CaO + SiO_2 \rightarrow Ca_2SiO_4$$

$$3CaO + SiO_2 \rightarrow Ca_3SiO_5$$

$$3CaO + Al_2O_3 \rightarrow Ca_3Al_2O_6$$

$$4CaO + Al_2O_3 + Fe_2O_3 \rightarrow Ca_4Al_2Fe_2O_{10}$$

$$\underset{\substack{(\text{e.g.}, CaO, Al_2O_3,\\ SiO_2, Fe_2O_3, \text{etc.})}}{\text{Raw materials}} \xrightarrow[\sim 1500^\circ C]{\text{Kiln}} \underset{\substack{(\text{e.g.}, Ca_2SiO_4,\\ Ca_3Al_2O_6, \text{etc.})}}{\text{Clinker}} \xrightarrow[25^\circ C]{H_2O} \underset{\substack{(\text{Calcium silicate}\\ \text{hydrates})}}{\text{Set cement}}$$

The thermal process occurring at ca. 1500°C is called "clinkerization" and can be easily detected as an exothermic peak in DSC and DTA calorigrams. The exothermic peaks between 1200°C and 1400°C in Fig. 2.23 occur without variation in mass and are associated with the clinkerization process of cement. The following oxide ratios are used in the preparation of a typical Portland cement:

Oxide	CaO	SiO_2	Al_2O_3	Fe_2O_3	SO_3	MgO	$\begin{pmatrix} K_2O \\ Na_2O \end{pmatrix}$
Wt%	63	20	6	3	2	2	1

Other additives constitute about 3%.

By varying the composition of cement clinker or the amounts of various additives, the properties of cement can be varied. Ordinary Portland cement is the most common, general purpose cement but it is susceptible to sulfate and so is not used in contact with seawater, for example. The remedy used in sulfate-resisting cement is to reduce the Ca_3SiO_5 content. This is achieved by increasing Fe_2O_3/Al_2O_3 content in the raw materials.

The silicates and aluminates present in cement react with water to form products of hydration and, in time, these set to a hard mass. The main hydration product and the most important one for high strength is a poorly crystalline calcium silicate hydrate (C—S—H), sometimes called C—S—H gel. Besides C—S—H gel, hardened cement contains unreacted clinker, $Ca(OH)_2$, aluminate hydrates, alumino-sulfate hydrates and water.

Macro-defect-free (MDF) cement is a new cement that has far superior mechanical properties to conventional Portland cement and concrete. It has a toughness of $\simeq 1\,kJ/m^2$ compared to $0.02\,kJ/m^2$ of conventional cement. This value is about one order of magnitude less than the corresponding values for good quality steels and offer exciting prospects for future improvements and development. In the preparation of MDF cement, improved mixing and molding methods are used together with the addition of surface-active organic chemicals whose rheological properties enable the cement grains to achieve better compaction.

9 TA in explosives

The capability of handling very small samples has now led to wide acceptance of TA techniques in the study of explosives. Thermal stabilities, heats of explosion, decomposition, kinetics of reaction, and temperature of ignition have all been measured using TA [24]. For example, the formation of Al-Ni compounds by ball-milling of elemental powders has been investigated by means of in situ thermal analysis, X-ray diffraction, and scanning electron microscopy. The results show that AlNi forms in an explosive, exothermic reaction, driven by the large heat of compound formation. The mechanism discovered is suggested to explain the mechanical alloying of other systems [24].

Olsen et al. [24] reported microdifferential thermal analysis method for detecting the adsorbed explosive molecules using microfabricated bridges.

They showed that the very low thermal mass of micromechanical sensors can be used to produce unique responses that can be used for achieving chemical selectivity without losing sensitivity or reversibility. It was demonstrated that this method can differentiate explosive vapors from nonexplosives and is additionally capable of differentiating individual explosive vapors such as trinitrotoluene, pentaerythritol tetranitrate, and cyclotrimethylene trinitramine. The method is based on a microfabricated bridge with a programmable heating rate, producing unique and reproducible thermal response patterns within 50 ms that are characteristic to classes of adsorbed explosive molecules [24].

10 TA in testing the loading efficiencies of drugs

TGA was applied to test the loading efficiencies in a novel antituberculosis drug delivery system composed of mesoporous silicate nanoparticles (MCM-41 generation) loaded with selected antituberculosis drugs of different sizes, charges, and polarities (ethambutol, isoniazid, pyrazinamide, and pyrazonic acid) in addition to fluorescein as a fluorescent tagging molecule. The loaded particles were then coated with the highly cationic polymer polyethyleneimine (PEI) for the pH-stimulated release of cargo molecules. Finally, the particles have been decorated with mannose as a macrophage cells specific moiety [25].

TGA was used to determine the loading efficiencies. Figs. 4.24–4.27 describe the TGA thermograms of the drugs loaded by the rotavap method. The TGA experiments were done prior to the application of PEI/mannose coatings to exclude any interference that may occur by the PEI and/or mannose molecules in the data analysis.

There are four main weight loss steps in the TGA curves of loaded MCM-41:

1. Step I, Initial weight loss below 100°C of adsorbed moisture and solvent.
2. Step II, commencing at ca. 200°C that is assigned to the decomposition of the surface-bound functional groups such as amino and thiol groups.
3. Step III, occurring at 250–450°C that is assigned to the decomposition of the functional groups (e.g., amino groups bound to the inside pore walls).
4. Step IV occurring above 600°C that is related to the dehydration (decomposition) of the particle's silanol groups.

The amounts of drugs loaded within the particles could be estimated by the weight loss between 100°C and 600°C. The amounts of the loaded drugs are presented in Table 4.1.

According to data in Table 4.1, all four drugs and fluorescein were successfully encapsulated into the nanoparticles (NPs). The loaded amounts of the drugs used were considerably high upon using the rotavap method compared to the conventional impregnation method. Further advantage of the rotavap method is the independence of the drug class, charge, and size.

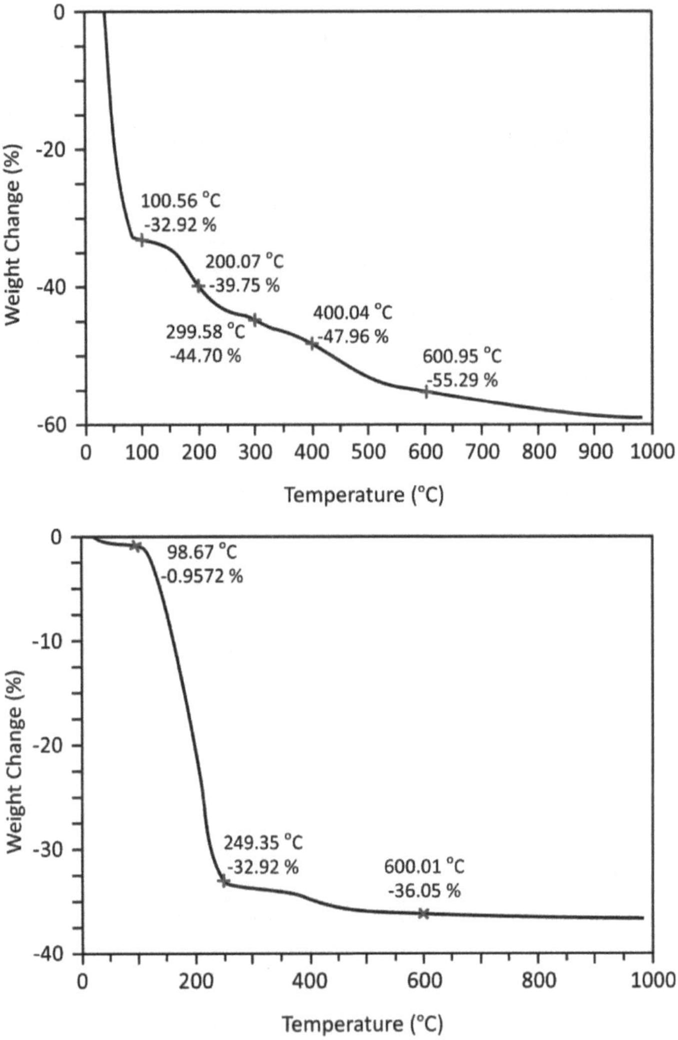

FIG. 4.24 TGA curve of nanoparticles loaded with pyrazonic acid (POA) *(top)* and pyrazinamide (PZA) *(bottom)* drugs using rotavap method.

11 TA in solid-state reaction kinetics

Any technique which offers the possibility of yielding the energy of activation of a solid-state reaction from a single run is valuable for two main reasons. First, only small quantities (mg amounts) are required. Second, the difficulty encountered with conversional isothermal methods, in which reaction may occur during the heating process, is circumvented [26a]. Very recently, Muravyev et al.

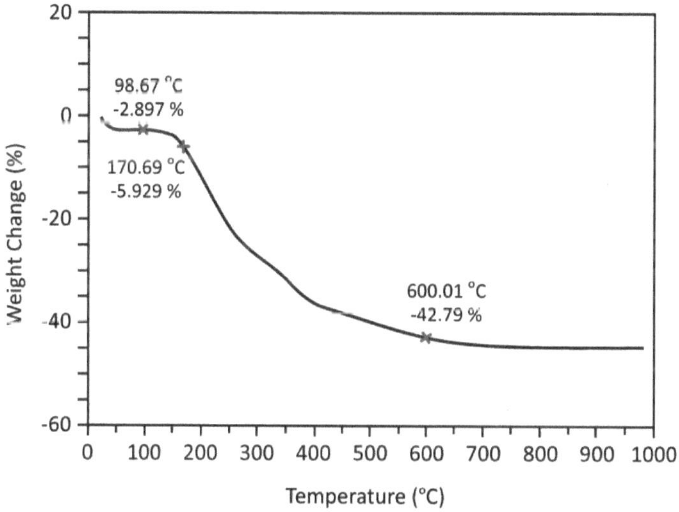

FIG. 4.25 TGA curve of nanoparticles loaded with isoniazid drug using rotavap method showing the weight loss below 100°C, and the weight loss between 100°C and 600°C. The experiment was carried out in duplicates with a standard deviation of 2.8.

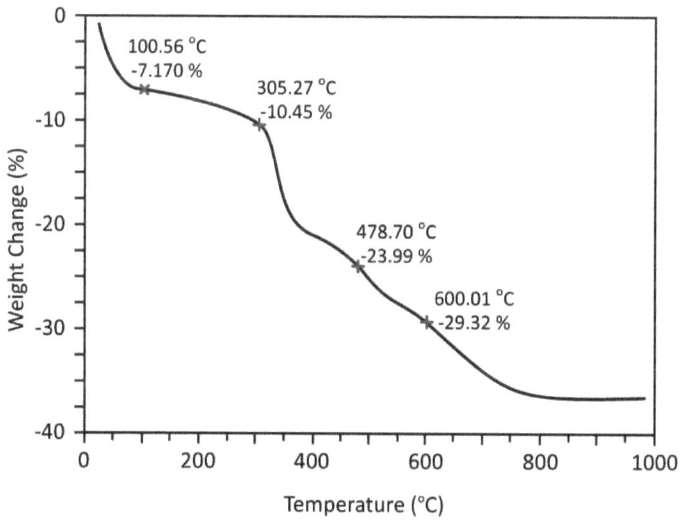

FIG. 4.26 TGA curve of nanoparticles loaded with fluorescein using rotavap method showing the weight loss below 100°C, and the weight loss between 100°C and 600°C. The experiment was carried out in duplicates with a standard deviation of 1.4.

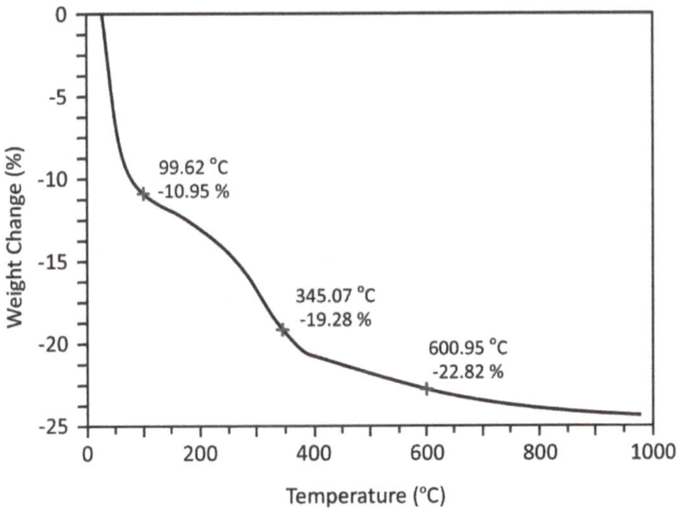

FIG. 4.27 TGA curve of nanoparticles loaded with ethambutol drug using rotavap method showing the weight loss below 100°C, and the weight loss between 100°C and 600°C. The experiment was carried out in duplicates with a standard deviation of 1.4.

reported a model-fitting approach using TG and DSC data for studying kinetics of the complex processes in solids [26b], as shown in the scheme below:

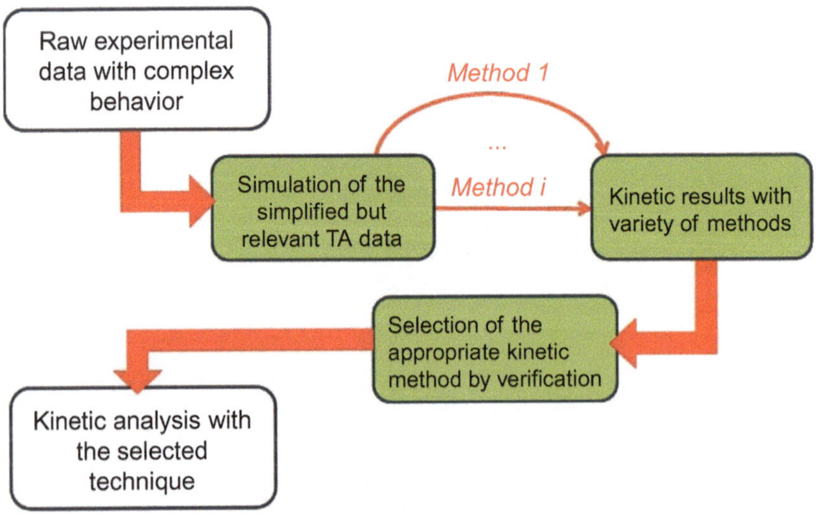

For the kinetic analysis of the single-step reaction, the peak maximum of DTG and DSC curves contain important kinetic information as formalized in the Kissinger method [26c]: activation energy (E_a) value is evaluated from the shift

TABLE 4.1 Comparison between the loading amount efficiencies of the selected drugs as estimated by TGA for the rotary-evaporator and the impregnation methods.

Loaded drug	Impregnation method			Rotavap method		
	Weight loss % (100°C to 600° C)	*Amount of loaded drug in µg/mg MSN*	*Standard deviation*	*Weight loss % (100°C to 600° C)*	*Amount of loaded drug in µg/mg MSN*	*Standard deviation*
Isoniazid	6.0	28.8	1.7	40.0	400	2.8
Pyrazinamide	8.9	106.8	2.1	35.0	350	2.1
Pyrazonic acid	6.5	90.6	2.8	22.5	225	1.7
Ethambutol	5.0	12.5	1.4	12.0	120	1.4
Fluorescein	6.0	75.0	1.4	22.2	221	1.4

The experiments were carried out in duplicate.

of the peak temperature. Therefore, the Arrhenius parameters determined by Kissinger method for the first and second peaks of DTG and DSC data fully agree with the correct kinetic parameters. On the other hand, the kinetic analysis of the TA data using the isoconversional methods is recommended by the International Confederation for Thermal Analysis and Calorimetry (ICTAC) [26d]. The isoconversional technique is strong in catching the change of E_a in course of reaction and its weakness in case of significant overlapping.

Many solid-state reactions take place in three stages as follows:

1. Induction: formation of reaction sites or nuclei.
2. Acceleration: reaction interface increasing.
3. Deceleration: reaction interface decaying.

The fractional change (α) is used as a measure of the reaction rates in solids where

$$\alpha = \frac{\text{amount reacted}}{\text{total amount}}.$$

The reaction rate is expressed as $d\alpha/dt$ and it is proportional to the area of the interface where the reaction is localized as it proceeds. The underlying mechanism in a solid-state reaction also changes at different reaction stages.

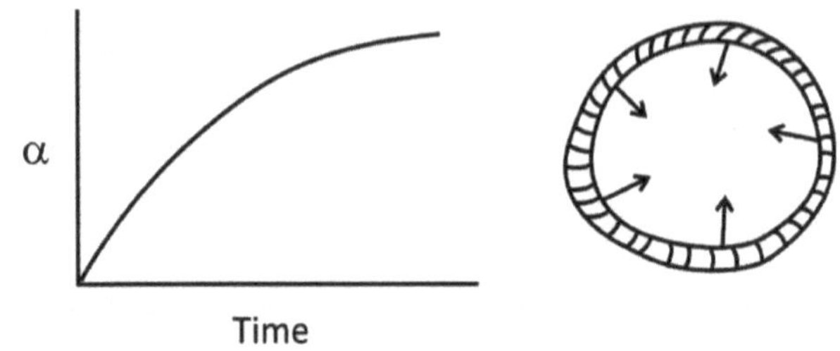

FIG. 4.28 The shape of $\alpha - t$ plot of a solid-state reaction undergoing the contracting sphere model.

Fig. 4.28 shows the $\alpha - t$ plot representing "a contracting sphere model" solid-state reaction. Since the reaction starts at the surface of the grains or microcrystallites and propagates inside, the area of the reaction front becomes smaller as the reaction proceeds and the $\alpha - t$ relation shows an initial high rate $d\alpha/dt$ that gets smaller as the reaction proceeds. This is usually the case in dehydration reactions. The rate equation for the contracting sphere model is given by the relation:

$$1 - (1 - \alpha)^{1/n} = \frac{1}{n} kt \tag{4.3}$$

where k is the rate constant and n is an integer number that is also called the "order" of the reaction.

In another solid-state reaction model, consider that we have one nucleus per crystallite. In this case, the solid will decompose according to a sigmoidal expression that comprises an initial induction period during which $d\alpha/dt$ is very small. This is then followed by an acceleratory region which is later succeeded by a decay period (Fig. 4.29).

The induction period may be prolonged, shortened, or even virtually disappears as in the solid-state decomposition of lead azide $Pb(N_3)_2$.

For a particular solid-state reaction, the appropriate kinetic equation is obtained by trial and error method. The correct rate equation is the one that gives the best fit when we substitute for kinetic experimental data.

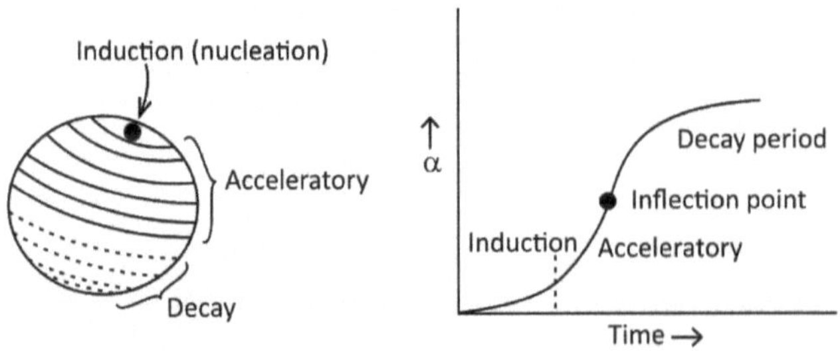

FIG. 4.29 The shape of $\alpha - t$ plot of a solid-state reaction undergoing the sigmoidal model.

No complete fit to the whole decomposition curve may be obtained but different kinetic equations may be found to fit the data over different ranges of α. The decay period in Fig. 4.29 undergoes a unimolecular decay law in many solid-state decomposition reactions where the reaction rate is proportional to the undecomposed material $(1-\alpha)$. The remaining reacting molecules become dispersed in the product matrix and these isolated molecules possess an equal probability for decomposition. The rate equation in such a case is written as

$$\text{Rate} = -\frac{d\alpha}{dt} = k\,(1-\alpha) \tag{4.4}$$

The negative sign indicates a decrease in the reaction rate as a function of time in this reaction stage, and the order of the reaction $n=1$ since the reactant molecules become dispersed in the product phase at this stage and the solid behaves as a solution:

$$-\frac{d\alpha}{(1-\alpha)} = k.dt \tag{4.5}$$

Integration of Eq. (4.5) gives

$$\ell n \frac{1}{(1-\alpha)} = kt \tag{4.6}$$

Eq. (4.6) compares with the first-order kinetic law where

$$\ell n \frac{a}{(a-x)} = kt \tag{4.7}$$

If $d\alpha/dt$ does not change with time, i.e., equals constant, then

$$\frac{d\alpha}{dt} = k\;;\;\alpha = kt \tag{4.8}$$

It is clear from Table 4.2 that the function of extent of reaction $f(\alpha)$ is often assumed as $f(\alpha)=(1-\alpha)^n$ where n is the order of the reaction, but in more general cases; $f(\alpha)=\alpha^m\,(1-\alpha)^n$, where m and n are constants.

Now suppose we have a solid-state reaction of the $\alpha-t$ characteristics as shown in Fig. 4.30A. We first locate $t_{0.5}$ (the value of time corresponding to $\frac{1}{2}\alpha$). From the different α values at the corresponding time intervals t, we tabulate the data as follows:

t
$t/t_{0.5}$
A
$f(A)=A^2$

A plot of the reduced time $t/t_{0.5}$ on the x-axis vs α^2 on the y-axis gives a straight line (Fig. 4.30B) whose slope = A. The value of A in this case should be 0.2500 which is consistent with $f(\alpha)=\alpha^2$ that represents a one-dimensional

TABLE 4.2 The different kinetic expressions are given different notations to facilitate their recognition (note: $f(\alpha) = kt$).

Reaction type	A-value	Description of $(\alpha - t)$ curves	$f(\alpha)$
		Acceleratory	
		(i) Power law	$\alpha^{1/n}$
		(ii) Exponential law	$\ell n\,\alpha$
		(iii) Contracting envelope	$n[1-(1-\alpha)^{1/n}]$
		Sigmoidal	
		(i) Avrami-Erofeev	$-\ell n\,(1-\alpha)]^{1/n}$
A_n	0.8326		$n=2$
	0.8850		$n=3$
		(ii) Prout-Tompkins	$\ell n[\alpha/(1-\alpha)]$
		Deceleratory (decay)	
		(A) Based on geometry	
R_n	0.2929	(i) Contracting area	$1-(1-\alpha)^{1/2}$
	0.2063	(ii) Contracting volume	$1-(1-\alpha)^{1/3}$
		(B) Based on order	
F_1	−0.6931	(i) First order	$-\ell n\,(1-\alpha)$
		(ii) Second order	$1/(1-\alpha)$
		(iii) Third order	$[1/(1-\alpha)]^2$
		(C) Based on diffusion	
D_1	0.2500	(i) One-dimensional	α^2
D_2	0.1534	(ii) Two-dimensional	$[(1-\alpha)\ell n(1-\alpha)] + \alpha$
D_3	0.0426	(iii) Three-dimensional (Jander Function)	$[1-(1-\alpha)^{1/3}]^2$
D_4	0.0367	(iv) Three-dimensional (contracting sphere model; Ginstling-Broustein function)	$1-\frac{2\alpha}{3}-(1-\alpha)^{2/3}$

diffusion model. It is thus clear that the "reduced time method" is also based on trial and error yet the comparison of A values offers a higher precision in choosing the proper mechanism.

In this case, we have a zero-order reaction, the rate constant (k) represents the specific rate of advance of an interface.

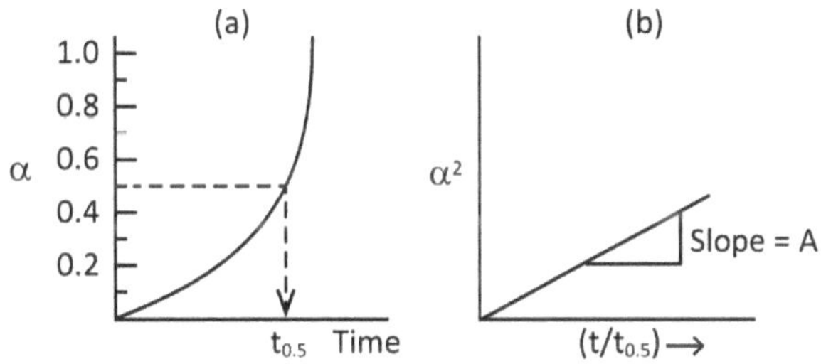

FIG. 4.30 (A) $\alpha - t$ plot of a solid-state reaction and (B) a plot of the reduced time $(t/t_{0.5})$ vs the kinetic expression $f(\alpha) = \alpha^2$.

Knowing k values at different temperatures, the activation parameters can be determined from Arrhenius and/or Eyring equation.

For more complex solid-state reactions, the simple practice of guessing the reaction order is not applicable and a more general method must be used to find the reaction mechanism. Many authors have derived expressions which reflect the nature of various reaction types as reported in Table 4.2.

The suitable expression can be recognized by the trial and error procedure as has been mentioned earlier. A convenient method, however, in recognizing the suitable expression is the so-called "reduced time plot" method by Sharp et al. [27]. In this method all forms of kinetic expressions are written in the form

$$f(\alpha) = A\frac{t}{t_{0.5}} \tag{4.9}$$

where $f(\alpha)$ is the kinetic expression under test, $t_{0.5}$ is the reaction's half-life (i.e., time to 50% conversion), and A is a constant that is characteristic to this kinetic expression. The values of A for various expressions are mentioned in Table 4.2.

In Table 4.2, only nine values of A are mentioned, and these values are useful in choosing the corresponding kinetic expression.

Some experimental techniques, including DTG and DSC operated in the isothermal mode, produce an output proportional to the reaction rate $d\alpha/dt$ as a function of time (vide infra).

11.1 DSC in solid-state reaction kinetics

Reaction kinetics of solid-state reactions can be studied using DSC calorigrams taken isothermally (the sample is heated isothermally at a constant temperature and dH/dt is monitored as a function of time) or dynamically (dH/dt is monitored as a function of the sample temperature).

The advantages of evaluating reaction kinetics by the dynamic mode are that considerably less experimental data are required than in the isothermal method. Moreover, the kinetics can be probed over an entire temperature range in a continuous manner without any gaps. In addition, a sample undergoes considerable reaction during raising the temperature of interest. The results obtained by an isothermal method of investigation are often questionable [28].

DSC dynamic mode

Thomas and Clarke [26] derived the following general equation for DSC (4.9) (Fig. 4.31):

$$\frac{d\alpha}{dt}=\frac{1}{A}\cdot\frac{dH}{dt} \tag{4.9}$$

where dH/dt is the deviation from the base line and A is the total area of the peak. The suitable substitution for $(d\alpha/dt)$ must be made. For a zero-order reaction, for example, $(d\alpha/dt)=k$, then

$$k=\frac{1}{A}\cdot\frac{dH}{dt} \tag{4.10}$$

for a first-order rate equation $(d\alpha/dt)=k\ (1-\alpha)$:

$$k\,(1-\alpha)=\frac{1}{A}\cdot\frac{dH}{dt}$$

$$k=\frac{1}{A\,(1-\alpha)}\cdot\frac{dH}{dt}=\frac{1}{A-\alpha A}\cdot\frac{dH}{dt}$$

$$=\frac{(dH/dt)}{(A-\alpha)} \tag{4.11}$$

where a is the area swept until a temperature T. Knowing k at different temperatures, the activation parameters can be obtained.

Both dynamic DSC and DTA techniques have been applied to calculate kinetic values from the peak temperature-heating rate relationship. In this technique, temperatures at which the reaction peak maxima occur are plotted as a function of their respective heating rates (heating rates from 1 to 20K/min) [ANSI/ASTM E 698-79].

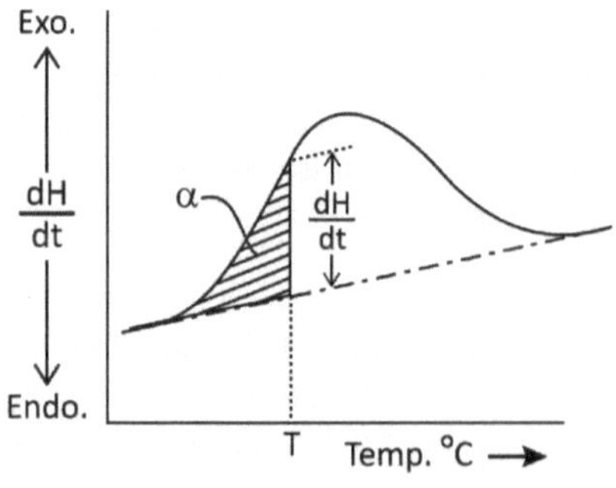

FIG. 4.31 Relation between α and temperature in a DSC dynamic run.

Assuming a heating rate β K/min, a plot of $\ell n\ \beta$ vs $1/T$, where T is the peak temperature in Kelvin, gives a straight line whose slope $= -E_a/R$, thus $E_a = (8.32 \times \text{slope})\ \text{kJ mol}^{-1}$ (Fig. 4.32). The values of β and T can be tabulated as in Table 4.3 before graphic representation.

This method, however, has the following limitations:

- Not applicable in cases where $\ell n \beta - \frac{1}{T}$ plots are not linear.
- Not applicable to isothermal modes.
- Not applicable in simultaneous or consecutive reactions.
- Not applicable in materials undergoing phase transitions if the reaction rate is significant at the transition temperature.

The Arrhenius preexponential factor can be calculated as follows:

$$Z = \beta E_a e^{Ea/RT} / RT^2 \left(\text{min}^{-1}\right) \quad (4.12)$$

where β is the heating rate from the middle of the range and T is the corresponding peak temperature.

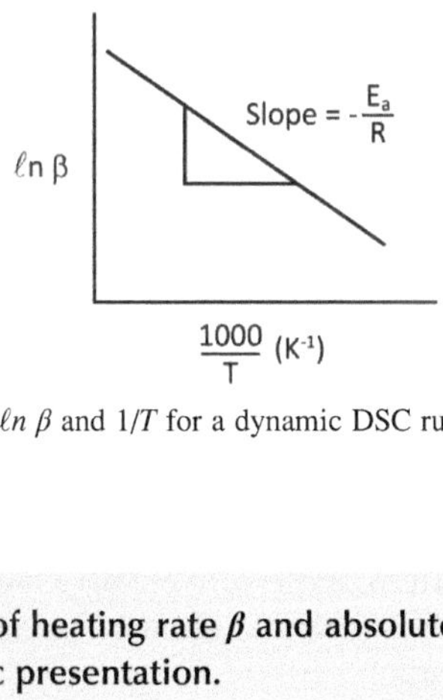

FIG. 4.32 Relation between $\ell n\ \beta$ and $1/T$ for a dynamic DSC run.

TABLE 4.3 Tabulation of heating rate β and absolute temperature T values can be prior to graphic presentation.

Heating rate, β, K/min	1	3	5	7	10	15	20
$\ell n\ \beta$	0	1.1	1.61	1.95	2.3	2.71	3.0
Peak temp. (K)	131.0	155.7	166.6	178.5	184.6	196.6	205
1000/ T (K^{-1})	2.474	2.332	2.275	2.214	2.185	2.129	2.09

The data can be verified using an isothermal test that is carried out as follows:

- Knowing E_a and Z values, calculate the rate constants k at different temperatures using the relation.

$$k = Z e^{-E_a/RT}$$

- Calculate the half-life times of reaction for different k values from Eq. (4.13):

$$t_{1/2} = 0.693/k \tag{4.13}$$

For example

T (K)	350	360	370	380
k (min^{-1})	0.00289	0.00503	0.00847	0.01390
$t_{1/2}$	240	138	82	50

- Select a temperature requiring at least 1-h aging time, and age the sample isothermally for the calculated half-life. Quench immediately to some temperature at least 50 K below the aging temperature so that no significant reaction occurs during subsequent holding time.
- Run the aged sample dynamically and record its reaction peak.
- Run similar but unaged sample in the same way and record its reaction peak.
- On an equal weight basis, the peak area or displacement from baseline of the aged sample should be one-half that of the unaged sample. If so, the reaction kinetics are confirmed for the temperature range explored.

The precision and accuracy of this method depend on reaction complexity, instrument capability, and the amount and range of data taken.

Under optimum conditions, a set of five data points over at least a fivefold range of heating rates has resulted in a standard deviation within ±2% for the calculated activation energy.

Regarding the isothermal aging test, a difference of less than 10% relative between predicted and observed results can be considered a reasonable test of accuracy.

DSC isothermal mode

So far, the polymerization kinetics was investigated using DSC [29a]. For example, the kinetics of the thermal polymerization in the solid state of 2,4-hexadiyne-1,6-diol bis(p-toluene sulfonate) have been studied using isothermal DSC [29b]. A common method for studying 1,4-addition polymerization involves extracting residual monomer with a suitable solvent at different stages of the reaction and weighing the insoluble polymer fraction. However, the method is limited by the fact that it is difficult to extract monomer at high degrees of conversion, resulting in reduced accuracy. The kinetics of the titled compound was studied using isothermal DSC as a quick and reliable technique. The thermal polymerization process obeys first-order kinetics in the range 20%–90% conversion.

From Fig. 4.33, the amount reacted up to time (t) is the shaded area (x), the total amount of material is proportional to the total area of the polymerization peak (a). A plot of $\ell n\ a/(a-x)$ vs time (t) gives straight line of slope k. Other isothermal calorigrams are obtained at different temperatures (T). From T and the corresponding k values, the activation parameters are evaluated.

The underlying kinetic equation is $kt = \ell n \frac{a}{(a-x)}$.

For this polymerization reaction:

$$k = 3.6 \times 10^{12} \exp.\frac{-22.3 \pm 0.9}{RT} \min^{-1};$$

$$E_a = 22.3 \pm 0.9 \text{ k cal mol}^{-1}$$

Practically, the isothermal DSC peaks are obtained after a certain time known as the induction period t_0.

Fig. 4.34A shows typical isothermal DSC calorigrams for the titled polymerization reaction and Fig. 4.34B shows the $\ell n \frac{a}{(a-x)}$ vs time plots at different temperatures giving the rate constant (k).

Fig. 4.35 shows the Arrhenius plot. From the induction periods, one can calculate the energy of activation of the nucleation stage. For an isothermal run at temperature T (K), the rate constant is assumed to be $t_0^{-1}(\min^{-1})$. A plot of $\ell n\ (t_0^{-1})$ vs $1000/T$ gives the Arrhenius plot from which the energy of activation E_a can be obtained, as shown in Fig. 4.35.

The activation energy is in excellent agreement with similar systems studied by other techniques (e.g., diffuse reflectance spectroscopy) [30].

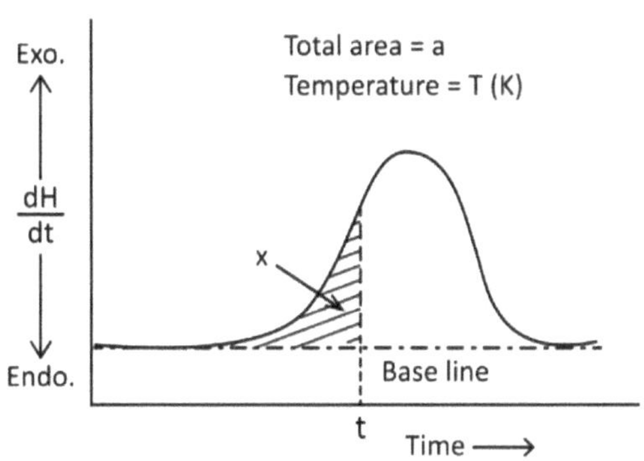

FIG. 4.33 Isothermal DSC polymerization calorigram.

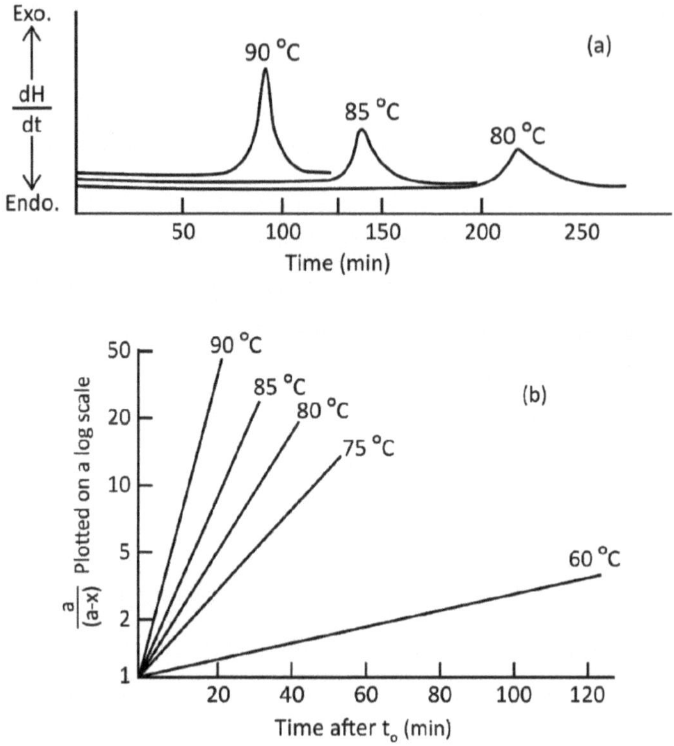

FIG. 4.34 (A) Isothermal DSC polymerization and (B) first-order kinetic plots of log $\frac{a}{(a-x)}$ vs t; the time after the induction period is t_0.

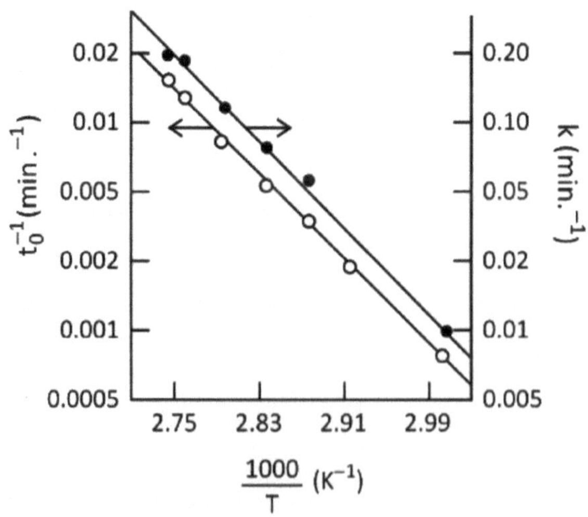

FIG. 4.35 Arrhenius plots of a thermal polymerization reaction.

The heat of polymerization was calculated from the area of the exothermic peak yielding a value $\Delta H_P = -30.2 \pm 0.7\,\text{kcal mol}^{-1}$ from 60°C to 90°C. The induction period accounts for additional contribution of ca. 1–2 kcal mol^{-1} which results only in a small base line shift.

Note: Several materials are used as references in the determination of ΔH (e.g., indium for which the heat of fusion $\Delta H_f = 6.80\,\text{cal/g}$).

Isothermal DSC has been used to study the kinetics of solid-state phase transformation reactions. Fig. 4.36A shows the isothermal DSC runs of 9-cyanoanthracene photodimer (9 CNAD). Considering the rate constant for nucleation of the new phase (k) as the reciprocal of the induction time ($1/t_0$) min^{-1} (Fig. 4.36B), the activation parameters were obtained as follows [31]:

$$E_a = 143.0 \pm 2.4\,\text{kJ mol}^{-1} \tag{4.14}$$

$$\Delta H_a = 141.0 \pm 2.4\,\text{kJ mol}^{-1} \tag{4.15}$$

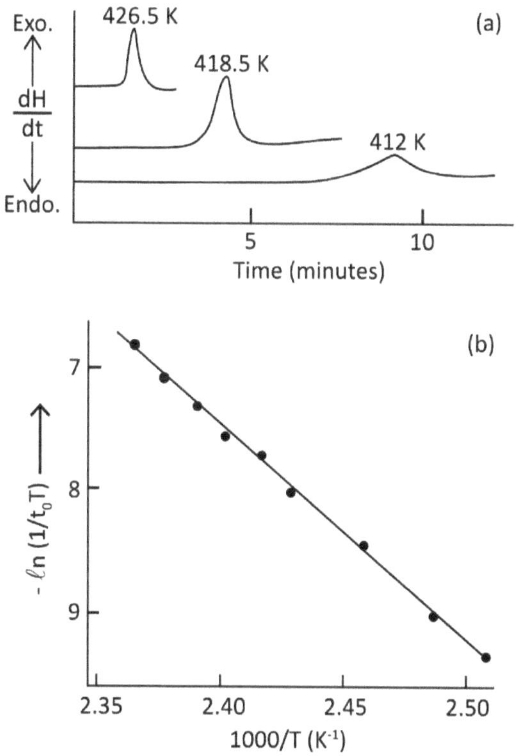

FIG. 4.36 (A) Isothermal DSC runs for the phase transformation in 9 CNAD and (B) Eyring plot constructed from (A).

The evaluation of kinetic parameters from DTA

Most approaches to the determination of energy of activation from DTA stem are attributed to the theoretical treatment of Borchardt and Daniels [32], in which the rate constant (k) is given by the following equation:

$$k=\left(\frac{KAV}{m_0}\right)^{n-1}\left\{\frac{C_P\left(\frac{d\,\Delta T}{dt}\right)+K\,\Delta T}{\left[K\,(A-a)+C_P\,\Delta T\right]^n}\right\} \tag{4.16}$$

where k is the constant characteristic of the apparatus, A is the total area of the peak, C_P is the total heat capacity of the reactant or reference material, ΔT is the differential temperature, t is time, a is the area of the peak up to time t, m_0/V is the initial concentration of reactant (Fig. 4.37).

To a first approximation, the heat capacity (C_P) is quite small [33]. For a first-order reaction ($n=1$), Eq. (4.16) is reduced to Eq. (4.17):

$$k=\frac{\Delta T}{(A-a)} \tag{4.17}$$

Fig. 4.38 shows the Arrhenius plot for the de-aquation (loss of water molecules) of $[Cu(en)\,(H_2O)_2]SO_4$ complex according to the reaction:

$$\left[Cu(en)\,(H_2O)_2\right]SO_4\,(s)\rightarrow[Cu(en)]SO_4\,(s)+2H_2O\,(g)$$

Another approximate method for determining E_a from a single DTA curve recorded at an arbitrary rate of heating was described by Piloyan et al. [34]. The deviation from the baseline (Fig. 4.39) in the initial stages of the reaction was described by

$$\Delta T=A\frac{d\alpha}{dt} \tag{4.18}$$

where ΔT is the deviation from the baseline (°C), A is the area of the thermal effect (°C × sec), $d\alpha/dt$ is the rate of the reaction, and α is the fractional change

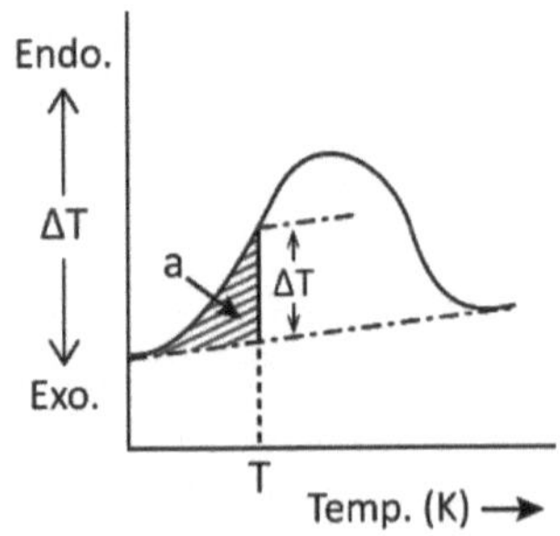

FIG. 4.37 DTA curve for kinetic parameters extraction.

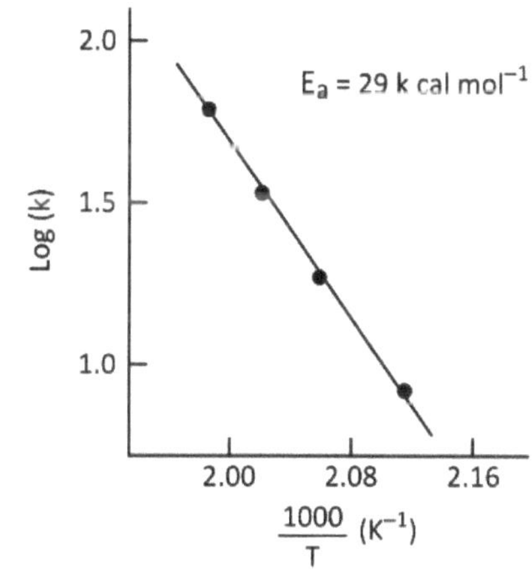

FIG. 4.38 The Arrhenius plot for the deaquation $[Cu(en)(H_2O)_2]SO_4$ complex.

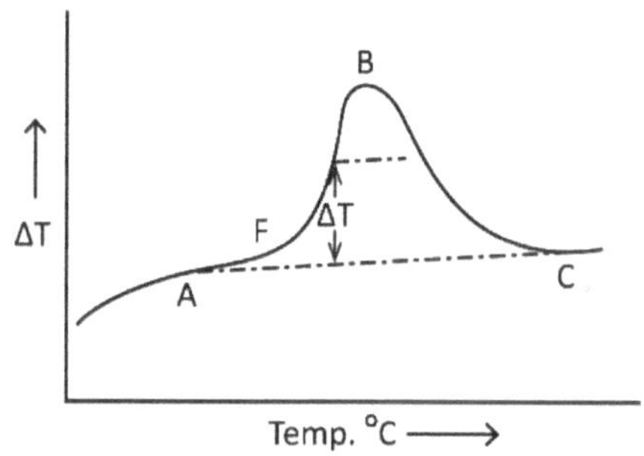

FIG. 4.39 The deviation from the baseline for a dynamic DTA run.

(extent of the reaction). The rate of a chemical reaction is usually described by the relation:

$$\frac{d\alpha}{dt} = C' f(\alpha)\, e^{-E_a/RT} \tag{4.19}$$

where C' is a constant and $f(\alpha)$ is a function of the general form $f(\alpha) = \alpha^m (1-\alpha)^n$, where m and n are constants.

Substituting Eq. (4.18) into Eq. (4.19) and taking logarithms, the following equation can be obtained:

$$\ell n\, \Delta T = C'' - \ell n f(\alpha) - \frac{E_a}{RT} \tag{4.20}$$

where C'' is a constant which combines all the constants of Eqs. (4.18), (4.19) and T is the temperature in Kelvin corresponding to a given value of ΔT on the y-axis.

It was assumed, to a first approximation, that the change in temperature (T) has a greater effect on ΔT than the change in α at normal heating rates. The term $\ell n\, f(\alpha)$ can be neglected and Eq. (4.20) can be written as

$$\ell n\, \Delta T = C - \frac{E_a}{RT} \tag{4.21}$$

The value of ΔT (at different temperatures) are taken directly from the DTA curve in units of length (cm or mm) as is shown in Fig. 4.39. The errors associated with this method are estimated as $\pm 15\%–20\%$ (note the high uncertainty in this technique).

The interval measured is approximately represented by the portion FB of the curve in Fig. 4.39. The measurements are taken not from the point at which the thermal effect begins (point A; Fig. 4.39), but at some distance from it (approximately from the point of maximum curvature, F).

Table 4.4 presents values of E_a determined for the decomposition reactions of various substances.

TGA methods for studying solid-state reaction kinetics

Methods exploiting dynamic thermogravimetry

Kinetic information obtained from nonisothermal (dynamic) thermogravimetry methods have been reviewed [35–37]. The advantages of evaluating reaction kinetics by dynamic modes can be summarized as follows:

- Less experimental data are required.
- Kinetics can be probed over an entire temperature range in a continuous manner without any gaps.
- The reaction can be probed during the heating process.

Methods based on dynamic TGA can be classified as follows:

(i) Analysis of TGA-curves

The rate of change in sample weight (dW/dt) is obtained by choosing a horizontal distance corresponding to 1 min (this is done by taking the rate of heating

TABLE 4.4 Activation energies for some reactions.

		E_a (k cal mol^{-1})	
Substance	**Type of reaction**	**This method**	**Literature**
$CuSO_4{\cdot}5H_2O$	Dehydration	17.9	18.0
$CaCO_3$	Decomposition	35.4	35–48
Mg CO_3	Decomposition	32.0	32.4
$CaC_2O_4{\cdot}H_2O$	Dehydration	21.0	20–22

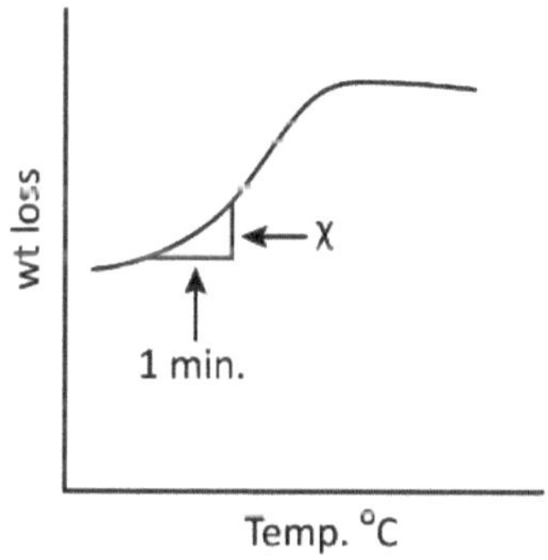

FIG. 4.40 Measuring the vertical distance (x) on a TG thermogram after 1-min interval.

(°C/min into account) and measuring the vertical distance (x) (Fig. 4.40). The following relation was derived earlier [38].

$$\frac{\frac{-E_a}{2.3\,R}\,\Delta\left(\frac{1}{T}\right)}{\Delta\log W_r} = -n + \frac{\Delta\log\left(\frac{dW}{dt}\right)}{\Delta\log W_r} \tag{4.22}$$

A plot of $\frac{\Delta\log\,(dW/dt)}{\Delta\log W_r}$ vs $\frac{\Delta(1/T)}{\Delta\log W_r}$ gives a straight line with intercept of $-n$ and slope of + or $-E_a/2.3$ R for any unique physical or chemical reaction where dW/dt is the rate of weight loss [either by tangents or (change in weight) (change in time) at a certain temperature], W_r is the remaining weight $=(W_c-W)$ where W_c is the weight loss at completion of the reaction and W is the total weight loss up to time t.

Fig. 4.41 shows the TG curve of $CaC_2O_4{\cdot}H_2O$. Three distinct stages in the temperature range 25–1000°C are observed.

(1) $CaC_2O_4{\cdot}H_2O \longrightarrow CaC_2O_4 + H_2O\uparrow$
(2) $CaC_2O_4 \longrightarrow CaCO_3 + CO\uparrow$
(3) $CaCO_3 \longrightarrow CaO + CO_2\uparrow$

The orders and E_a values of the three stages were calculated from plots in Fig. 4.42 as

Reaction	Order	E_a (k cal mol^{-1})
1	1.0	22
2	0.7	74
3	0.4	39

Clark and Thomas applied another method to get activation parameters from TG-thermograms [38]. The dehydration of Mn(II) formate dihydrate was

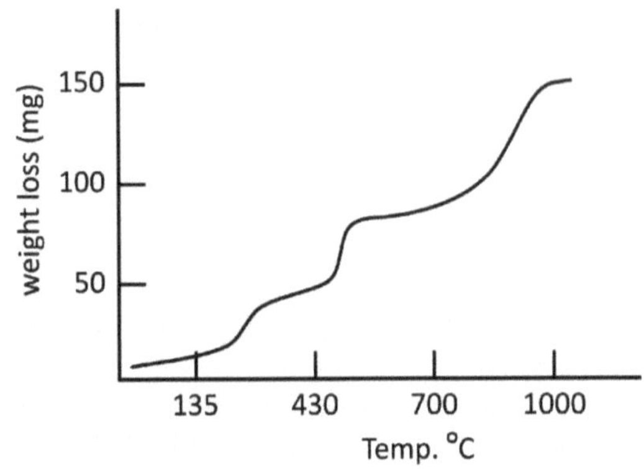

FIG. 4.41 TG thermogram of $CaC_2O_4{\cdot}H_2O$ decomposition.

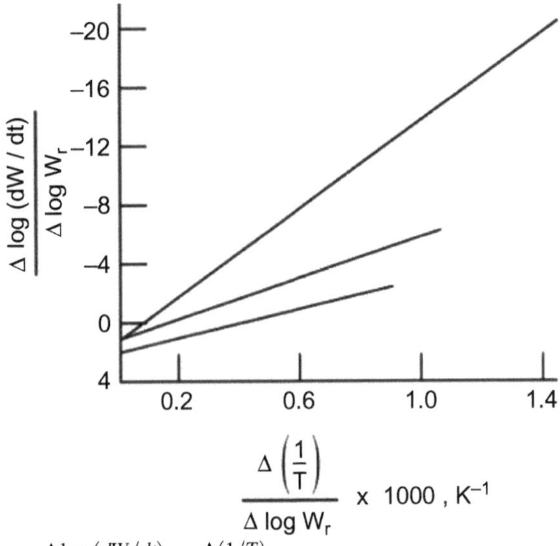

FIG. 4.42 Plots of $\frac{\Delta \log (dW/dt)}{\Delta \log W_r}$ vs $\frac{\Delta(1/T)}{\Delta \log W_r}$ for the three stages of $CaC_2O_4{\cdot}H_2O$ decomposition in the temperature range 25–1000°C.

studied as a model compound. The mechanism of dehydration was based on direct microscopic observations.

The nucleation step occurs over the fractional change range $\alpha = 0.0$–0.15. Nucleation starts at the <011> interfaces as shown in Fig. 4.43A and B. This is followed by anisotropic penetration of the reaction interfaces into the hydrate matrix in the range $\alpha = 0.15$–1.0. The $\alpha - t$ plot shows a linear region in the range $\alpha = 0.15$–0.5 during which $(d\alpha/dt)$ is constant and the rate constant k (cm s^{-1}) is defined as the rate advance of the <011> interfaces.

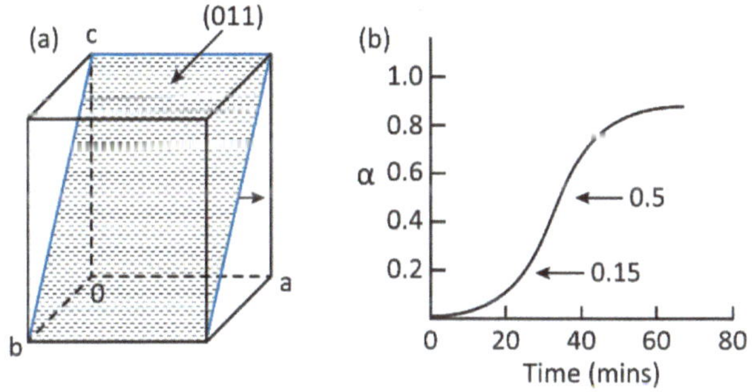

FIG. 4.43 The nucleation step occurring over the fractional change range $\alpha = 0.0$–0.15. Nucleation starts at the <011> interfaces.

From isothermal dehydration in the range 0.15–0.5:

$$\left(\frac{d\alpha}{dt}\right)_T = K'' = -\frac{1}{m_0}\left(\frac{dm}{dt}\right)_T \cdot - \left(\frac{dm}{dt}\right)_T = K''m_0 \tag{4.23}$$

where m_0 is the initial weight of the sample that is constant.

This means that the slope of a TGA curve is proportional to K'' so that a plot of log (dm/dt) vs $1/T$ should be also linear in the range $\alpha = 0.15$–0.5; further, the activation energy for the interfacial growth, as defined here, is obtained from the slope of the plot.

Because the slope of the TGA curve is proportional only to k (where k is the true rate constant), no exact significance can be attached to the value obtained for the preexponential factor, which will generally be a function of particle size distribution.

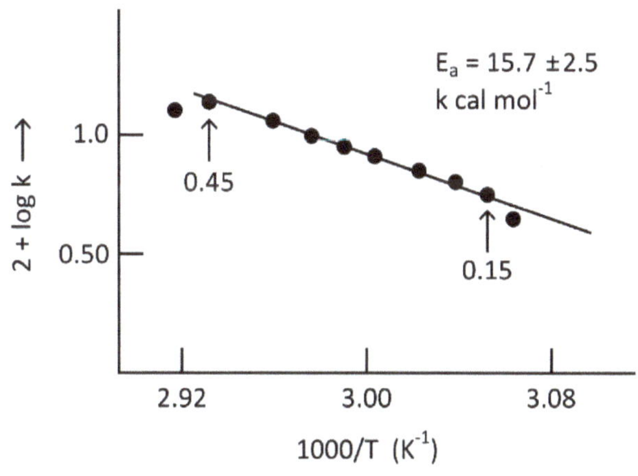

FIG. 4.44 Arrhenius plot for Mn(II) formate dihydrate dehydration.

Fig. 4.44 represents Arrhenius plot for Mn(II) formate dihydrate dehydration. The energies of activation of the dehydration reaction of Mn(II) formate dihydrate obtained by different thermo-analytical techniques are mentioned below [39]:

E_a (k cal mol^{-1})	Method	Reference
17.6±0.6	Isothermal TGA	[26a]
17.8±0.6	Isothermal TGA	[38]
16.5±0.5	DSC	[38]
15.7±2.5	Dynamic TGA	[39]

(ii) Analysis of derivative TGA curves (DTGA) Bowden et al. [40] applied DTGA to study the kinetics of thermal degradation of poly(1-butene sulfone) (PBS). The rate of degradation (R_d) may be written as follows:

$$R_d = -\frac{dW}{dt} = kW^n = \frac{Z}{\beta} e^{-E_a/RT} \cdot W^n \quad (4.24)$$

where k is the rate constant, Z is the preexponential factor, β is the heating rate, W is the weight remaining, E_a is the energy of activation, n is the order of reaction, R is the gas constant, and T is the absolute temperature.

When Eq. (4.24) is applied at two different temperatures and the resultant expression subtracted from one another, Eq. (4.25) is obtained:

$$\Delta \log (R_d) = n[\Delta(\log W)] - \left[\frac{E_a}{2.303R}\right]\Delta\left(\frac{1}{T}\right) \quad (4.25)$$

From the TGA thermogram as well as the DTGA curve, we can plot both W and dW/dt against $1/T$ as shown in Fig. 4.45.

For several $1/T$ values, we can determine from Fig. 4.45 the corresponding W and R_d (i.e., dW/dt) values. We make the necessary subtraction to get values of $\Delta\left(\frac{1}{T}\right)$, Δ (log W), and Δ log $\left(\frac{dW}{dt}\right)$. The values $\Delta\left(\frac{1}{T}\right)$ should be equal. The data can be tabulated as follows:

$\frac{1}{T}$	W	R_d	Log W	Log R_d	$\Delta\left(\frac{1}{T}\right)$	Δ log W	Δ log R_d
					constant		

As shown in Fig. 4.46, a plot of Δ (log R_d) vs Δ (log W) at constant $\Delta\left(\frac{1}{T}\right)$ is linear with a slope n of 1.05 for this particular reaction. The activation energy is calculated from the intercept as $E_a = 48.5$ k cal mol^{-1} that is in good agreement with the value determined from the isothermal study.

(iii) Heating rate (b) method in TGA decomposition kinetics In this method, a series of four or five test specimens taken from the original sample is temperature programmed each at a different heating rate, through its decomposition region. The specimen weight losses are recorded as a function of temperatures for a constant decomposition level. The kinetic activation energy is

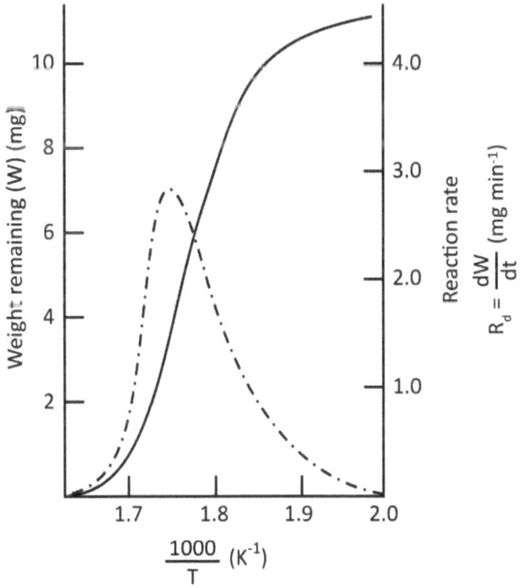

FIG. 4.45 Plots of weight remaining and rate of weight loss as a function of $1/T$. The plots are constructed from simple TGA and DTGA curves.

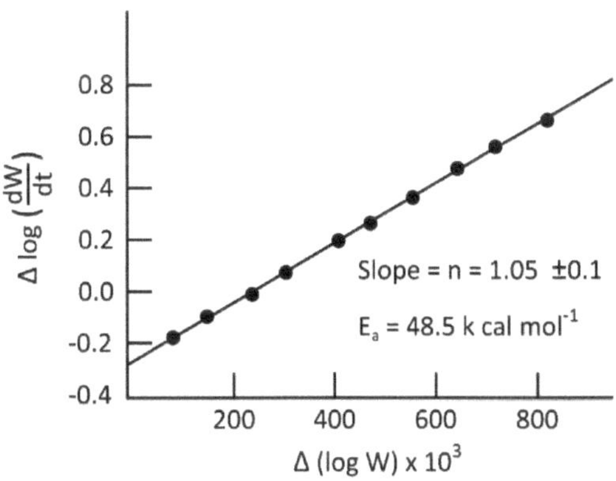

FIG. 4.46 Kinetics of thermal decomposition of poly(1-butene sulfone) plotted according to the method of Freeman and Carrol [41].

then determined from a plot of the logarithm of the heating rate vs the reciprocal of the temperature of constant decomposition level.

This activation energy may then be used to calculate estimated lifetime at a given temperature or the maximum operating temperature for a given estimated lifetime.

The specific experimental conditions used (such as temperature range and specimen atmosphere) depend on the material being tested. Experimental design and data reduction are similar for each material. In the analysis illustrated here, a high-temperature flour-polymer wire insulation material was examined under the following experimental conditions: specimen weight: 40–60 mg, program rates: 1, 2, 5, 10, or 20°C/min, temperature range 200–500°C, and recording: expanded scale to display 75%–100% weight.

The specimen temperature is programmed through the decomposition profile while recording the expanded scale weight loss vs specimen temperature (i.e., dynamic TGA mode). The first 25% of the weight loss is plotted. This experiment is repeated, each with a fresh specimen, with at least four heating rates covering at least an order of magnitude. A series of thermal curves like those shown in Fig. 4.47 should be produced.

Fig. 4.47 shows a series of lines created from the four curves by plotting log (β) vs ($1/T$) at different conversion levels. If the specimen decomposition mechanisms were the same at all conversion levels, the lines would have the same slope.

The slope of the line (or the group of lines if they have nearly the same slope) gives the derivative term ($d \log \beta$)/[d ($1/T$)]. This value is then substituted in the following relation [38]:

$$E_a = \frac{-R}{b}\left[\frac{d \log \beta}{d(1/T)}\right] \tag{4.26}$$

where E_a is the activation energy (Jmol^{-1}), R is the gas constant (8.314 J mol^{-1} K^{-1}), T is the absolute temperature at constant conversion (K), β is the heating rate (°C/min), and b is the constant ($b = 0.457$ in this particular reaction; vide infra).

The value for the constant b provided in Table 4.5 varies depending on the value of E_a/RT [42]. Thus, for a certain conversion (i), the slope:

$$\begin{aligned}&\left[d \log \beta / d\left(\frac{1}{T}\right)\right] is(-S_i), \text{then}\\ &E_a = -\frac{R}{b}(-S_i);\\ &\frac{E_a}{R} = \frac{S_i}{b}\end{aligned} \tag{4.27}$$

Given T_i is a middle temperature in the conversion range (i), then one makes an iterative process in which E_a values are calculated for different proposed b values from the relation:

$$\frac{E_a}{RT_i} = \frac{S_i}{b} \tag{4.28}$$

This process is continued until E_a no longer changes with successive iterations. Table 4.5 provides values for the activation energy and the corresponding values of E_a/RT calculated for the five conversion cases shown in Fig. 4.47.

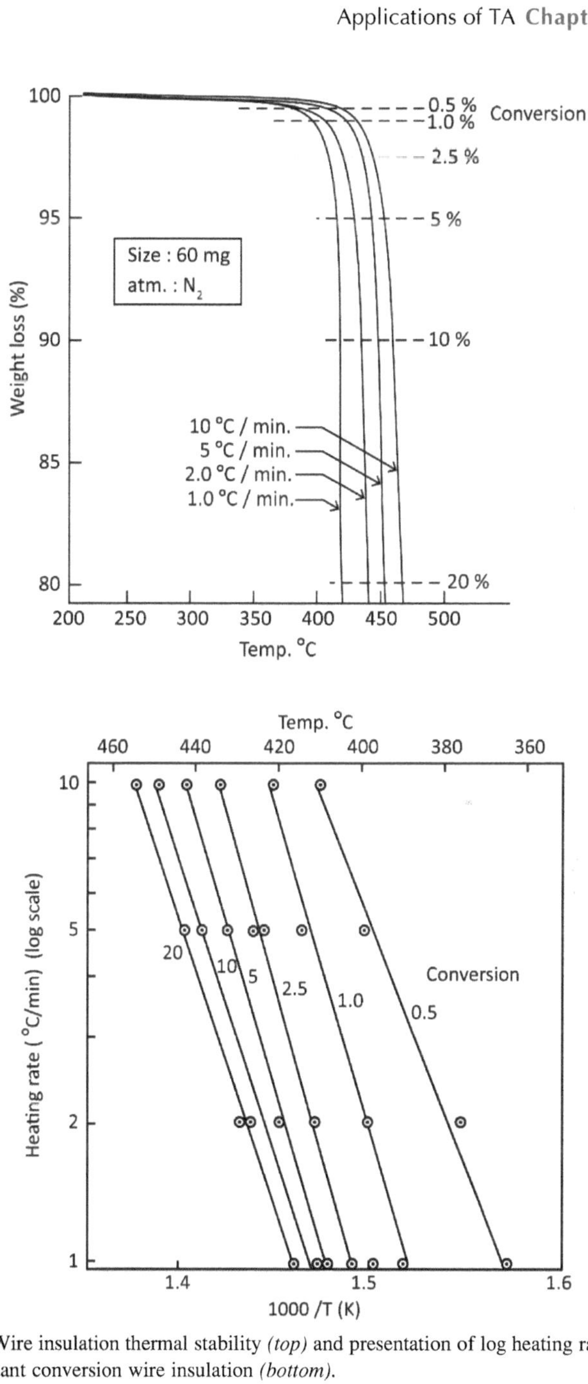

FIG. 4.47 Wire insulation thermal stability *(top)* and presentation of log heating rate vs temperature of constant conversion wire insulation *(bottom)*.

TABLE 4.5 Activation energies for wire insulation decomposition.

Conversion %	E_a/RT	E_a (kJ mol^{-1})
0.5	31	170
1.0	44	245
2.5	44	246
5.0	42	244
10	36	211
20	36	212

Note: $b = 0.457 \pm 0.004$ for $29 < \frac{E_a}{RT} < 46$.

This also means that the decomposition mechanism is not the same throughout the whole decomposition range. Toop has postulated a relationship between E_a and the estimated lifetime of some wire insulation [43].

$$\ell n\, t_f = \frac{E_a}{RT_f} + \ell n\left[\frac{E}{\beta R}\cdot P\left(\frac{E}{RT_c}\right)\right] \tag{4.29}$$

where t_f is the estimated time of failure (min) and T_f is the = failure temperature (K), $P\left(\frac{E}{RT_c}\right)$ is the value from a numerical integration table in the paper by Toop, T_c is the temperature for 5% loss at β (K), and β is the heating rate (°C/min).

Kinetic parameters can be obtained more easily from dynamic TGA in terms of the modified Coats and Redfern equation [44–46]. In this procedure, the heating rate (β) is considered.

$$\ell n\frac{F(\alpha)}{T^2} = \ell n\frac{ZR}{\beta E_a}\left[1 - \frac{2RT}{E_a}\right] - \frac{E_a}{RT} \tag{4.30}$$

where Z is the frequency factor, R is the gas constant, β is the linear heating rate, and E_a is the activation energy. A plot of $[\ell n\, F(\alpha) - \ell n\, T^2]$ vs $1/T$ gives a straight line of slope $(-E_a/R)$ as shown in Fig. 4.48 [44–46].

Isothermal gravimetry in solid-state kinetics

Isothermal gravimetry can be easily manipulated to get kinetic data for many solid-state changes such as decomposition, drying, sorption, etc.

The y-axis of the plotter is set such that a reasonable sample weight is expanded full scale. Fig. 4.49 shows isothermal TGA thermograms illustrating the drying of black powder. In the case of black powder, an overall moisture

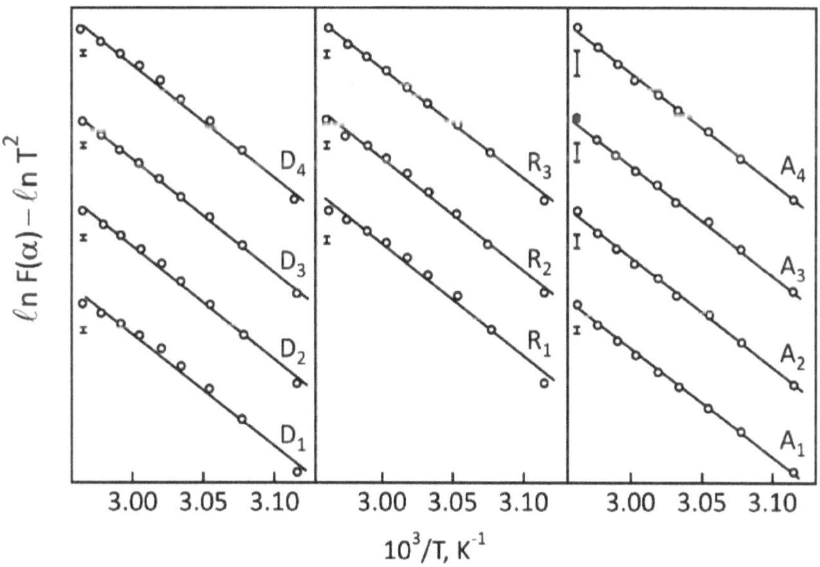

FIG. 4.48 Typical Coats and Redfern's plots for the dynamic dehydration from $CuSO_4{\cdot}5H_2O$ to $CuSO_4{\cdot}3H_2O$. The vertical segments refer to an ordination scale of 0.1.

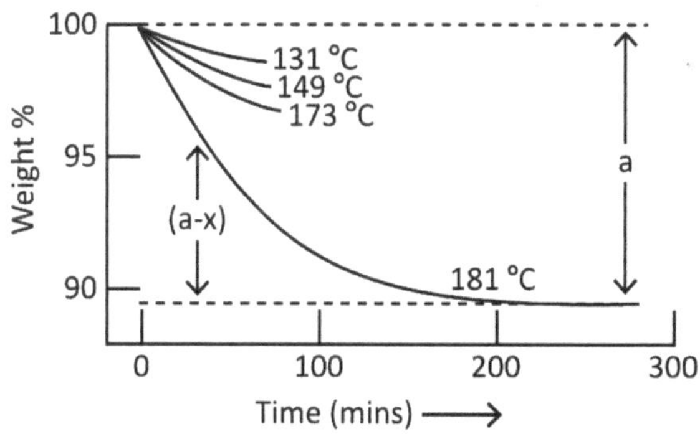

FIG. 4.49 Drying of black powder using isothermal thermogravimetry.

content of about 10% could be expected, and the material could not be safely heated above 200°C.

The y-axis of the plotter is set such that 15% sample weight is expanded full scale, with the y-axis scale running from 100% to 85% weight. Small weight losses can therefore be read directly in percent on the y-axis scale.

Many polymers drying, and degradation reactions follow first-order kinetics ($n=1$). In this case:

$$kt = \ell n \frac{a}{(a-x)} \tag{4.31}$$

knowing (a) and ($a-x$) from the graph, a linear relation between t and $ln(a/(a-x))$ indicates that the reaction undergoes first-order kinetics from which k is determined at a certain temperature. The energy of activation can be calculated from the following relation:

$$\ell n\, k = \ell n\, Z - \frac{E_a}{RT} \tag{4.32}$$

A plot of $\ell n\, k$ against $1/T$ gives a straight line, the gradient of which is E_a/R:

$$E_a = 63.2\,\text{kJ}\,\text{mol}^{-1} \tag{4.33}$$

From the intercept, a value of $187{,}000\,\text{min}^{-1}$ for the frequency factor Z was obtained.

It is now possible to predict the approximate drying time at any temperature. The experiment shows that at 181°C, the drying time would be about 4.5 h and the rate constant at this temperature is $0.0113\,\text{min}^{-1}$. The approximate drying time, t, at any other temperature can be estimated using the proportionality equation and the rate constant at the temperature of interest (k_t):

$$t(\text{drying time}) = \frac{4.5 \times 0.0113}{k_t} \tag{4.34}$$

The rate constant at the temperature of interest, k_t, can be calculated from the Arrhenius equation since the values of S_a and Z are known.

Such data could be used in the manufacturing processes to help choose the optimum drying times of a material and balancing this against the cost of long drying periods.

For more complex solid-state reactions, the proper kinetic expression can be estimated as mentioned earlier by plotting the kinetic model function $f(\alpha)$ vs the time (t) assuming the following kinetic equation:

$$f(\alpha) = kt \tag{4.35}$$

where k is the rate constant.

Fig. 4.50A shows plots of $f(\alpha)$ vs t for the isothermal dehydrations from $CuSO_4 \cdot 5H_2O$ to $CuSO_4 \cdot 3H_2O$ at 50.6°C (the symbols A, A_2, ….., D_1, D_2, .. and R_1, R_2, refer to the appropriate kinetic expression as mentioned earlier).

Fig. 4.50B shows a similar plot for $SrCl_2.3H_2O$ dehydration to $SrCl_2 \cdot H_2O$ at 67.2°C in the α-range 0.1–0.9. From the linearity of plots, both A_m (m ca. 3) and R_n (n ca. 1) models were estimated as appropriate. The best values of m and n were determined by "scanning" the values using a computer [47,48].

12 Summary

This chapter demonstrates in detail the various applications of TA in food, soap and detergents, electronic industries, solid-state reactions, fossil fuel research, clays and cement industries, chemistry of polymers and large directing agents,

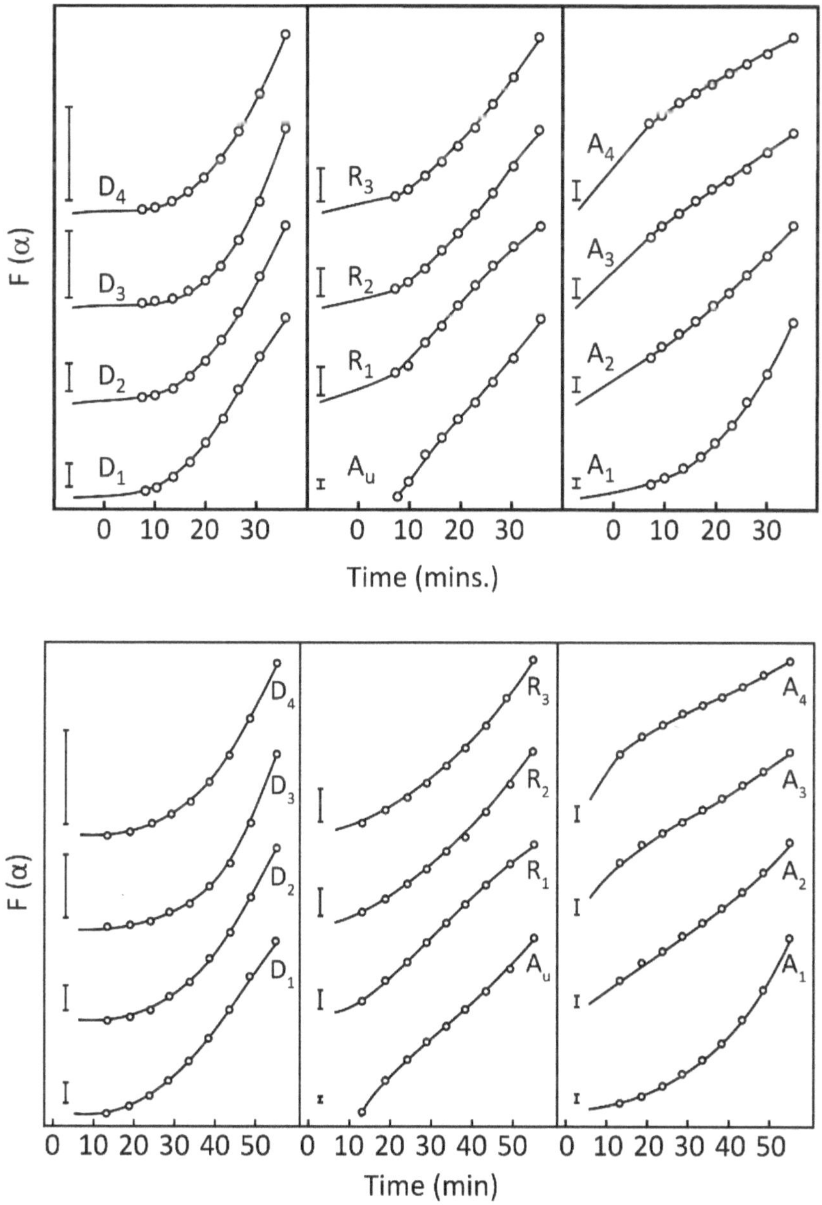

FIG. 4.50 Plots of $f(\alpha)$ vs t for the isothermal dehydration from $CuSO_4{\cdot}5H_2O$ to $CuSO_4{\cdot}3H_2O$ at 50.6°C *(top)*. The vertical segment in the lower left hand of each plot refers to an ordinate scale of 0.1 and $f(\alpha)$ vs plots for the thermal dehydration from $SrCl_2{\cdot}2H_2O$ to $SrCl_2{\cdot}H_2O$ *(bottom)* at a temperature of 67.2°C in the range $\alpha = 0.1$–0.9. Each vertical segment refers to an ordinate scale of 0.1.

explosives, drug delivery, and solid-state kinetics, etc. Part of this chapter was devoted to the application of TA in studying solid-state reaction kinetics. This includes the recently reported model-fitting approach using TG and DSC data for studying kinetics of the complex processes in solids. An introduction about solid-state reaction models together with the underlying kinetic expressions is given. Both dynamic and isothermal modes were discussed for the evaluation of kinetic parameters using DSC, DTA, TGA, and DTGA methods.

References

[1] Mettler Toledo Webinar of Thermal Analysis in the Food Industry, Properties of food products can be investigated by thermal analysis, 2017.

[2] P. Giri, C. Pal, An overview on the thermodynamic techniques used in food chemistry, Mod. Chem. Appl. 2 (2014) 1000142.

[3] Y. Fujita, Y. Noda, The effect of organic solvents on the thermal denaturation of lysozyme as measured by differential scanning calorimetry, Bull. Chem. Soc. Jpn. 56 (1983) 233–237.

[4] E. Summerton, G. Zimbitas, M. Britton, S. Bakalis, Crystallisation of sodium dodecyl sulfate and the corresponding effect of 1-dodecanol addition, J. Cryst. Growth 455 (2016) 111–116.

[5] R.M. Miller, A.S. Poulos, E.S.J. Robles, N.J. Brooks, O. Ces, J.T. Cabral, Isothermal crystallization kinetics of sodium dodecyl sulfate-water micellar solutions, Cryst. Growth Des. 16 (2016) 3379–3388.

[6] W.P. Brennan, C.M. Earnest, M.P. DiVito, R.L. Hutchinson, EIC 6th Electrical/Electronical Insulation Conference, IEEE, Chicago, IL, USA, 1983, https://doi.org/10.1109/EEIC.1983.7465108.

[7] C. Plato, A.R. Glasgow, Differential scanning calorimetry as a general method for determining the purity and heat of fusion of high-purity organic chemicals. Application to 95 compounds, Anal. Chem. 41 (1969) 330–336.

[8] J.S. Lee, C.K. Hsu, L.K. Lin, C.L. Chang, S. Borjinn, C.W. Huang, A study on the thermal behavior of solder glass, J. Therm. Anal. Calorim. 56 (1999) 131–136.

[9] G.J. Davis, Thermal, Purity, and Solubility Properties of Cholesteryl Esters and their Thermal Behavior in Lipid-water Systems, Doctoral Dissertation 1896, University of Massachusetts Amherst, 2014, p. 581. February.

[10] V.P. Arora, A. Kumar, V.K. Agarwal, Differential scanning calorimetry studies of a cholesteric liquid crystal, Natl. Acad. Sci. Let. 26 (2003) 76–78.

[11] N. Fedelich, S. Giani, E. Hempel, N. Jing, M. Nijman, R. Riesen, J. Schawe, M. Schubnell, Mettler Toledo Selected Applications Handbook: Thermal Analysis of Polymers, Mettler-Toledo AG, Global MarCom Switzerland, 2013. Oxidative stability of petroleum oil fractions, UserCom 10, 7–8; A. Hammer, The characterization of olive oils by DSC, UserCom 28, 6–8.

[12] A.P. Gray, Polymer crystallinity determinations by DSC, Thermochim. Acta 1 (1970) 563–579. R. L. Blaine, n.d. Determination of Polymer Crystallinity by DSC, TA Instruments, New Castle, DE, USA. A. Hammer, Application Handbook, Selected Applications: Thermal Analysis of Polymers, Mettler Toledo, Greifensee, Switzerland, 2013, p. 11.

[13] K. Rajeshwar, D.B. Jones, J.B. DuBow, Characterization of oil shales by differential scanning calorimetry, Anal. Chem. 53 (1981) 121–122.

[14] E.W. Cook, Green River shale-oil yields: correlation with elemental analysis, Fuel 53 (1974) 16–20.

[15] C.M. Earnest, Thermogravimetry of selected American and Australian oil shales in inert dynamic atmospheres, Therm. Acta 58 (1982) 271–288.

[16] S. Guven, S. Akin, B. Hascakir, Comprehensive spectral and thermal characterization of oil shales, in: SPE Middle East Unconventional Resources Conference and Exhibition 26 28 January, Muscat, Oman, 2015.

[17] A. Zanier, H.W. Jäckle, Heat capacity measurements of petroleum fuels by modulated DSC, Therm. Acta 287 (1996) 203–212.

[18] D. Lager, W. Hohenauer, C. Knoll, P. Weinberger, A. Werner, Methodology to determine the apparent specific heat capacity of metal hydroxides for thermochemical energy storage, J. Therm. Anal. Calor. 133 (2017) 207–215.

[19] M.J. O'Neill, Measurement of specific heat functions by differential scanning calorimetry, Anal. Chem. 38 (1966) 1331–1336.

[20] W. Smykatz-Kloss, Differential Thermal Analysis: Application and Result in Mineralogy, Springer-Verlag, Berlin, Heidelberg, New York, 1974.

[21] E.M. Ebeid, M.A. Salem, A.F.M. Habib, Some sorption properties of Nile silt, Colloids Surf. 27 (1987) 341–344.

[22] H.E. Kissinger, Reaction kinetics in differential thermal analysis, Anal. Chem. 29 (1957) 1702–1706.

[23] A. Galwey, Some chemical and mechanistic aspects of petroleum formation, J. Trans. Roy. Soc. S. Afr. 70 (2015) 9–24. A.K. Galwey, The rate of hydrocarbon desorption from mineral surfaces and the contribution of heterogeneous catalytic-type processes to petroleum genesis, Geochim. Cosmochim. Acta 36 (1972) 1115–1130. A.R. West, Solid State Chemistry and its Applications, second ed., John Wiley & Sons Ltd, United Kingdom, 2014. J. Shi, J. Wang, T.R. Muzenda, Y. Qu, P. Hou, X. Cheng, M. Liu, Effects of nanosilica on the hydration and hardening properties of blended cement-based materials under heat curing, J. Therm. Anal. Calorim. 141 (2020) 1317–1330.

[24] J. Olsen, L. Senesac, T. Thundat, A. Boisen, Trace explosives detection by micro differential thermal analysis, in: IEEE 24th International Conference on Micro Electro Mechanical Systems, 2011, pp. 984–987. M. Atzmon, In situ thermal observation of explosive compound-formation reaction during mechanical alloying, Phys. Rev. Let. 64 (1990) 487. L.R. Senesac, D. Yi, A. Greve, J.H. Hales, Z.J. Davis, D.M. Nicholson, A. Boisen, T. Thundat, Rev. Sci. Instrum. 80 (2009), 035102.

[25] S.M. Shawky, A.A. Abo-AlHassan, H. Lill, D. Bald, S.F. EL-Khamisy, E.M. Ebeid, Efficient loading and encapsulation of anti-tuberculosis drugs using multifunctional mesoporous silicate nanoparticles running title: mesoporous silicate nanoparticles as smart drug delivery system, J. Nanosci. Curr. Res. 1 (2016), https://doi.org/10.4172/JNCR.1000103, 1000103.

[26] (a) J.M. Thomas, T.A. Clarke, The evaluation of kinetic parameters from thermoanalytical techniques. dehydration of manganous formate dihydrate, J. Chem. Soc. A: Inorg. Phys. Theoret. (1968) 457–460. (b) N.V. Muravyev, A.N. Pivkina, N. Koga, Critical appraisal of kinetic calculation methods applied to overlapping multistep reactions, Molecules 24 (2019) 2298. (c) H.E. Kissinger, Reaction kinetics in differential thermal analysis, Anal. Chem. 29 (1957) 1702–1706. S. Mandal, D.E. Lee, T. Park, Crystallization kinetics of $Cu_{60}Zr_{25}Ti_{15}$ and $(Cu_{60}Zr_{25}Ti_{15})_{95}Ni_5$ bulk metallic glasses by differential scanning calorimetry (DSC), J. Therm. Anal. Calorim. (2020) doi:10.1007/s10973-020-09735-3. (d) S. Vyazovkin, A.K. Burnham, J.M. Criado, L.A. Pérez-Maqueda, C. Popescu, N. Sbirrazzuoli, ICTAC kinetics committee recommendations for performing kinetic computations on thermal analysis data, Thermochim. Acta 520 (2011) 1–19.

[27] J.H. Sharp, G.W. Brindley, B.N.N. Achar, Numerical data for some commonly used solid state reaction equations, J. Am. Ceram. Soc. 49 (1966) 379–382.
[28] E.S. Freeman, B. Carroll, The application of thermoanalytical techniques to reaction kinetics: the thermogravimetric evaluation of the kinetics of the decomposition of calcium oxalate monohydrate, J. Phys. Chem. 62 (1958) 394–397.
[29] (a) D. Florez, S. Hoppe, G.H. Hu, D. Meimaroglou, Radical bulk polymerization of styrene in the presence of rubber particles from recycled tires: a kinetic study using DSC, J. Therm. Anal. Calorim. (2020), https://doi.org/10.1007/s10973-020-09701-z. (b) A.R. McGhie, P.S. Kalyanaraman, A.F. Garito, Kinetics of thermal polymerization in the solid state: 2,4-hexadiyne-1,6 diol Bis(p-toluenesulfonate), J. Polym. Sci. C: Polym. Lett. 16 (1978) 335–338.
[30] R.R. Chance, J.M. Sowa, An examination of the thermal polymerization of a crystalline diacetylene using diffuse reflectance spectroscopy, J. Am. Chem. Soc. 99 (1977) 6703–6708.
[31] E.M. Ebeid, Some Solid-State Aspects of Organic Crystals, Doctoral Dissertation, Tanta University, Egypt, 1980.
[32] H.J. Borchardt, F. Daniels, The application of differential thermal analysis to the study of reaction kinetics, J. Am. Chem. Soc. 79 (1957) 41–46.
[33] W.W. Wendlandt, The thermal decomposition of metal complexes-IV: some amine complexes of copper (II) sulphate, J. Inorg. Nucl. Chem. 25 (1963) 833–842.
[34] G.O. Piloyan, I.D. Ryabchikov, O.S. Novikova, Determination of activation energies of chemical reactions by differential thermal analysis, Nature 212 (1966) 1229.
[35] C.J. Keattch, D. Dollimore, An Introduction to Thermogravimetry, second ed., Heyden, London; New York, 1975.
[36] W.W. Wendlandt, Thermal Methods of Analysis, second ed., Wiley, New York, 1974.
[37] L.J. Jenkins, P.E. Slade, Techniques and Methods of Polymer Evaluation, Thermal Analysis, Vol. 1, Marcel Dekker, New York, 1966.
[38] T.A. Clarke, J.M. Thomas, Extraction of meaningful kinetic parameters from thermogravimetric analysis, Nature 219 (1968) 1149–1151.
[39] R.C. Eckhardt, T.B. Flanagan, Anisotropic solid-state reaction. Dehydration of manganous formate dihydrate, Trans. Faraday Soc. 60 (1964) 1289–1298.
[40] M.J. Bowden, L.F. Thompson, W. Robinson, M. Biolsi, Thermal degradation of poly(1-butene sulfone), Macromolecules 15 (1982) 1417–1422.
[41] D.A. Anderson, E.S. Freeman, The kinetics of the thermal degradation of polystyrene and polyethylene, J. Polym. Sci. 54 (1961) 253–260.
[42] J.H. Flynn, L.A. Wall, Direct method for the determination of activation energy from thermogravimetric data, Polym. Lett. 4 (1966) 323–328.
[43] D.J. Toop, Theory of life testing and use of thermo-gravimetric analysis to predict the thermal life of wire enamels, IEEE Trans. Electr. Insul. EI-6 (1971) 2–14.
[44] T.A. Florez, G.M. Aparicio, Thermal characterization and lifetime estimation of the humus lombricospt, Am. J. Analyt. Chem. 5 (2014) 45–49.
[45] R. Ebrahimi-Kahrizsangi, M.H. Abbasi, Evaluation of reliability of Coats-Redfern method for kinetic analysis of non-isothermal TGA, Trans. Nonferrous Met. Soc. China 18 (2008) 217–221.
[46] A. Mianowski, M. Tomaszewicz, T. Siudyga, T. Radko, Estimation of kinetic parameters based on finite time of reaction/process: thermogravimetric studies at isothermal and dynamic conditions, React. Kinet. Mech. Catal. 111 (2013) 45–69.
[47] H. Tanaka, S. Ohshima, H. Negita, The problem of discerning kinetic models for solid decompositions from isothermal analyses, Thermochim. Acta 53 (1982) 387–390.
[48] H. Tanaka, S. Shimada, H. Negita, The kinetic study of thermal dehydration stages of $SrCl_2.2H_2O$ and its deuterium analog, Int. J. Chem. Kinet. 17 (1985) 149–156.

Chapter 5

Thermal analysis in studying the phase problem

1 Crystal structure transformation

A controlled thermal treatment of solid nanostructures in air or in an inert atmosphere yields nanostructured inorganic materials with multiple functions for the recent applications to date, such as catalysis, energy storage and conversion, drug vehicles, etc. A large fraction of metal centers can be utilized as a metal source, and the removable organic components, by simple calcination, provide nanopores, thereby increasing the specific surface area of the final products. The original morphology is almost retained, even after the thermal treatments, whereas in some cases, the crystal structures are tuned by heating leading to new crystalline materials with new functions. This strategy has proven to be a promising solid-state chemical synthetic method for the preparation of new materials with fine crystal structures. This method enables researchers to get various crystal structures and has a great potential to overcome difficulties in the preparation through traditional approaches, such as supramolecular templating (e.g., soft- and hard-templating methods).

This method has been used for the synthesis of various materials and studying the crystal structure changes before and after thermal treatments. For example, the ideal crystal structures of $Co_3[Fe(CN)_6]_2$ and $Fe_3[Co(CN)_6]_2$ cyano-bridged coordination polymers are shown in Fig. 5.1 [1]. The metal framework is bridged by cyanide groups in a cubic shape with face-centered cubic unit cell [2]. To prepare nanoporous FeCo-spinel oxides, the as-prepared $Co_3[Fe(CN)_6]_2$ and $Fe_3[Co(CN)_6]_2$ nanocubes were calcined in air for 4 h. During this stage, organic units/bridges (—C≡N—) were completely removed. The thermal decomposition process was studied by TGA/DTA/mass spectrometry (MS) under He/O_2 flow and the results indicate several exothermic peaks at approximately 260–450°C, accompanied by the production of CO_2 and/or N_2O ($m/z = 44$) as well as N_2 and/or CO ($m/z = 28$), which has the same behavior for the combustion reaction of a CN-containing material [1].

Even after the removal of the organic units, the products retained the original cubic shape. Meanwhile, the wide-angle XRD patterns proved that the calcined powder possessed a highly crystallized spinel structure (*Fd-3m*) without any

Thermal Analysis. https://doi.org/10.1016/B978-0-323-90191-8.00006-3

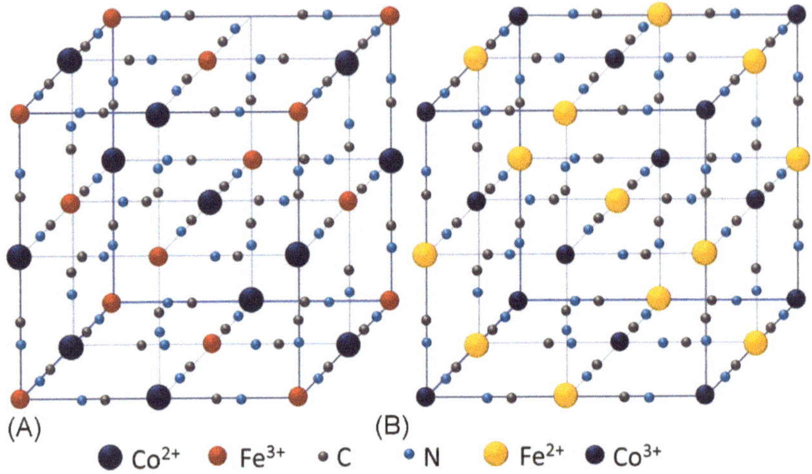

FIG. 5.1 Crystal structure models of (A) $Co_3[Fe(CN)_6]_2$ and (B) $Fe_3[Co(CN)_6]_2$ cyano-bridged coordination polymers.

formation of other crystalline phases (Fig. 5.2). The crystalline nature of the as-prepared precursors, intermediate after heating for 1 h, and the final products after heating for 4 h were examined by high-resolution TEM (Fig. 5.3). The high crystalline nature of the as-prepared $Fe_3[Co(CN)_6]_2$ cyano-bridged coordination polymer is obvious from the highly ordered crystalline domains in two directions of the nanocubes, which is a typical behavior of cyano-bridged coordination polymers. After calcination for 1 h, a single particle was composed of many crystalline domains, which were randomly oriented inside the nanocube. By further annealing, the crystalline domains were oriented in the same direction, although a few amorphous parts still existed. It is well known that a multistep transformation often happens, and it is driven by the need to decrease the surface energy. In general, amorphous phases are formed at the first step. Then the amorphous phases aggregate and form small crystalline domains with random orientations. Finally, the crystalline domains gradually rotate and attach to each other to form highly oriented crystalline phases of nanoporous metal oxide, as shown in Fig. 5.3. Nanoporous products are obtained from a recrystallization event [i.e., a transformation from the amorphous state (less dense phase) to a crystalline state (denser phase)].

Fig. 5.4 shows the (100) surface model of spinel structure of Fe_2CoO_4. The surface energy affects the crystal growth where the crystal surface with lower surface energy tends to be exposed more easily, meaning minute nanocrystals are stabilized without further crystal growth. However, the crystal surface with higher surface energy is not generally exposed, and larger crystals are formed in this case to mineralize the total surface energy by rotating the crystalline domains [1].

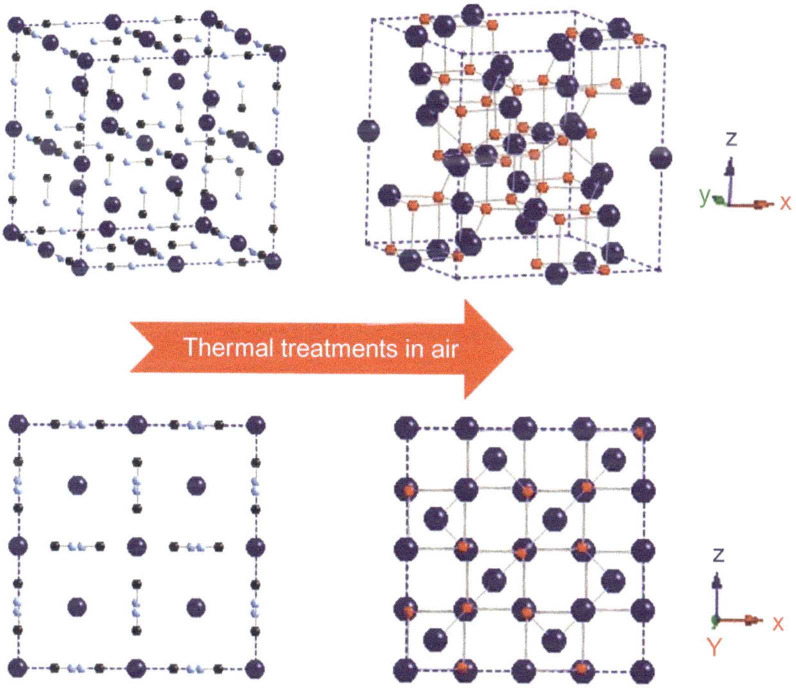

FIG. 5.2 Crystal structure models of thermal conversion of cubic $Fe_3[Fe(CN)_6]_2$ cyano-bridged coordination polymer into spinel $FeCo_2O_4$ in air.

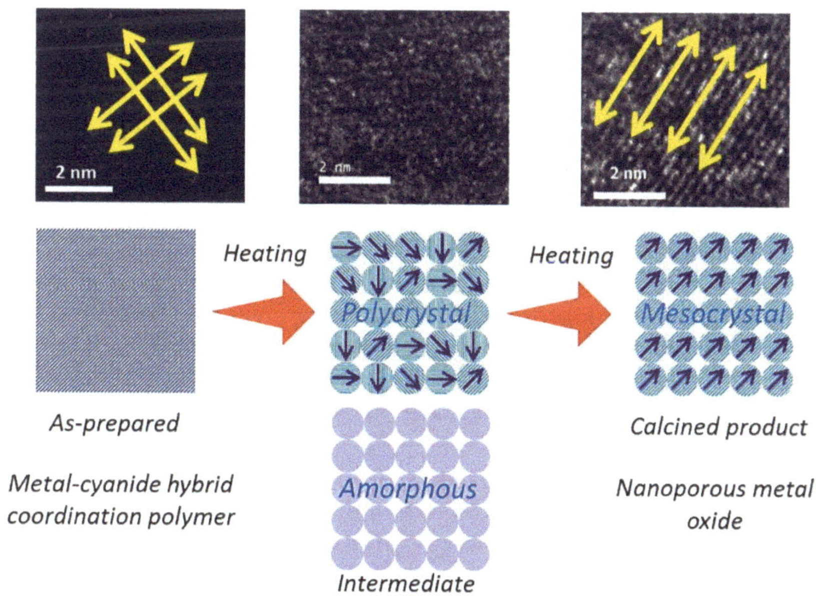

FIG. 5.3 Schematic illustration and high-resolution TEM images of the thermal conversion process of cubic $Fe_3[Fe(CN)_6]_2$ cyano-bridged coordination polymer into $FeCo_2O_4$ in air.

FIG. 5.4 Crystal model of (100) surface in Fe_2CoO_4 spinel structure.

The phase transformation in another system, an orthorhombic $Ni(H_2O_2)_2[Ni(CN)_4]\cdot 4H_2O$ nanoflakes, has been studied by controlled thermal treatment process in nitrogen environment [3]. The nanoflakes have an orthorhombic crystal structure that matches Hofmann-type (Ni-$(H_2O)_2[Ni(CN)_4]\cdot xH_2O$) with 2D layered structure without any impurities indicating the high purity of the samples. The crystal structure model illustrated in Fig. 5.5A shows the presence of interstitial water molecules, which reveals defects in the 2D crystal structure. The oxygen atoms of the water molecules may occupy the regular lattice positions instead of C or N in order to complete the crystal structure causing random defects [2]. The removal of lattice water may lead to a collapse of the $Ni(H_2O)_2[Ni(CN)_4]\cdot xH_2O$ framework. From Inductively Coupled Plasma Optical Emission Spectroscopy (ICP-OES) analysis, no alkali elements (e.g., Na^+ and K^+) could be detected, hence no defects could be attributed to them.

The transmission electron microscope (TEM) image of one particle confirms the formation of $Ni(H_2O)_2[Ni(CN)_4]\cdot xH_2O$ nanoflakes with a lateral size of ~150nm (Fig. 5.5B). The ordered lattice fringes of the same nanoflake obtained from high-resolution TEM (Fig. 5.5C) combined with the clear periodic spots of the electron diffraction (ED) pattern (inset; Fig. 5.5C) revealed that the nanoflakes are highly crystallized.

The thermal decomposition of $Ni(H_2O)_2[Ni(CN)_4]\cdot xH_2O$ nanoflakes under nitrogen was further studied by TGA [1]. In the first stage, water molecules are removed (weight loss of ~32.75%) by heating from room temperature up to ~160°C. There are $6H_2O$ molecules calculated from TG curve, thus the final structure would be $Ni(H_2O)_2[Ni(CN)_4].4H_2O$, which matches Hoffman-type. Thus, the obtained product is thermally stable until ~300°C and the second weight loss started from 300°C until ~420°C (~16.46%). This weight loss was anticipated to correspond to the loss of CN-groups, which are released as CN_2 and N_2 gases. The third weight loss starting from ~520°C may correspond to the organic residue (e.g., graphitic carbon) and in the final stage fcc-Ni metal is obtained. The theoretical calculation of Ni contribution in $Ni(H_2O)_2[Ni(CN)_4]\cdot xH_2O$ is around 31.44%, but the weight residue calculated experimentally after

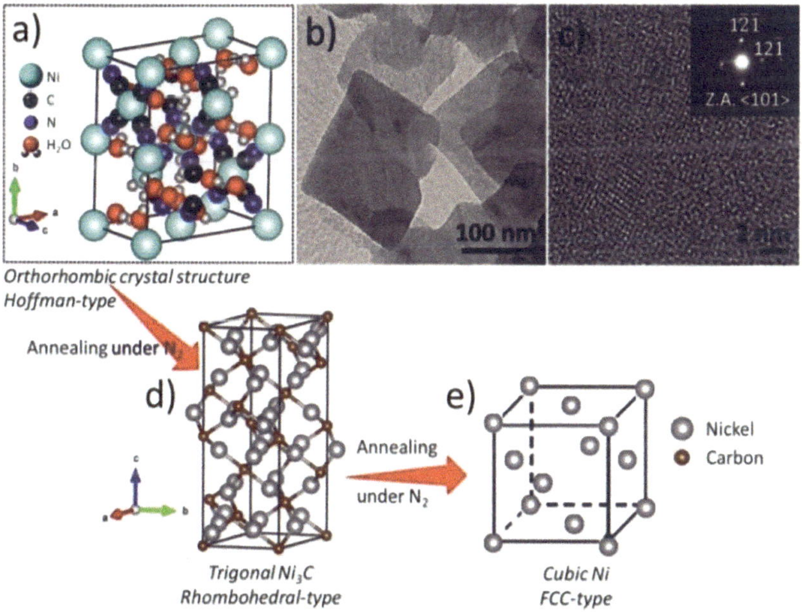

FIG. 5.5 (A) Crystal structure model, (B) TEM, and (C) high-resolution TEM (inset; electron diffraction) images of an orthorhombic $Ni(H_2O_2)_2[Ni(CN)_4]\cdot 4H_2O$ nanoflakes, and crystal structure models of (D) trigonal Ni_3C and (E) cubic Ni to illustrate the thermal conversion mechanism in an inert atmosphere.

thermal decomposition process is around 43.1%. So, Ni metal was expected to be contaminated by carbon. This result is in accordance with the reports indicating that nickel metal together with solid amorphous carbon is obtained by thermal decomposition of Ni_3C in inert atmosphere [3].

The product obtained at 650°C is a fcc-Ni nanoparticle like the product obtained at 550°C. These results reveal the thermal stability of fcc-Ni nanoparticles during heating at elevated temperatures. The crystal structure models of trigonal Ni_3C and fcc-Ni are shown in Fig. 5.5D and E, which indicate the phase transformation by annealing as well. These results reflect the transition between crystal structures by heating. In the first stage, the authors started with an orthorhombic $Ni(H_2O)_2[Ni(CN)_4]\cdot xH_2O$ phase that was transformed into trigonal Ni_3C around 420 °C in the second stage. At the final stage, cubic Ni metal with amorphous structure are obtained by further heating at 550°C.

Recently, a novel family of cyano-bridged trimetallic coordination polymers (CPs) with various compositions and shapes has been synthesized by changing the compositional ratios of Fe, Co, and Ni species in the reaction system [4]. In order to efficiently control the nucleation rate and the crystal growth, trisodium citrate dihydrate plays an important role as a chelating agent. After the obtained cyano-bridged trimetallic CPs undergo thermal treatment in air at three different

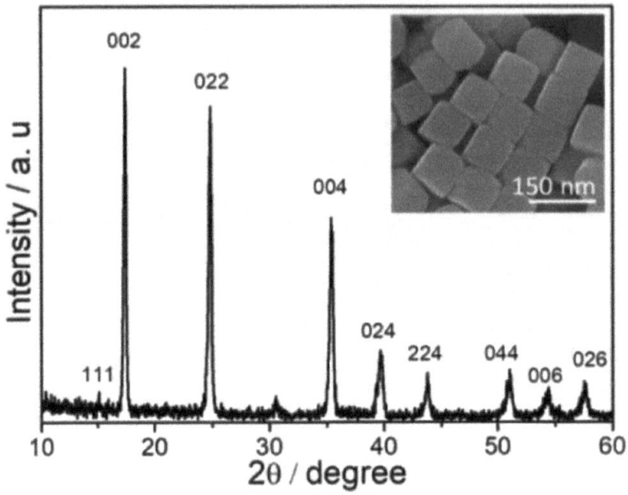

FIG. 5.6 Wide-angle XRD patterns of trimetallic $K_{0.85}Co_{0.91}Ni_{0.21}[Fe(CN)_6]{\cdot}xH_2O$ cyano-bridged coordination polymer (inset; SEM image).

temperatures (250°C, 350°C, and 450°C), nanoporous spinel metal oxides are successfully obtained. Interestingly, the obtained nanoporous metal oxides are composed of small crystalline grains, and the grains are oriented in the same direction, realizing pseudo-single crystals with nanopores. A typical sample, $K_{0.85}Co_{0.91}Ni_{0.21}[Fe(CN)_6]{\cdot}xH_2O$ shows the crystal structure transformation by heating.

$K_{0.85}Co_{0.91}Ni_{0.21}[Fe(CN)_6]{\cdot}xH_2O$ has a cubic morphology, as shown by SEM images (inset; Fig. 5.6). The phase purity and the crystal structures of the obtained $K_{0.85}Co_{0.91}Ni_{0.21}[Fe(CN)_6]{\cdot}xH_2O$ sample was examined by wide-angle XRD measurements (Fig. 5.6). The sample shows exactly the same diffraction patterns of a cubic crystal structure with a fcc unit cell as compared to Prussian blue (PB) crystals (PDF Card: 01-073-6123) [4]. The as-prepared $K_{0.85}Co_{0.91}Ni_{0.21}[Fe(CN)_6]{\cdot}xH_2O$ sample was thermally converted to the corresponding mesoporous metal oxides. Wide-angle XRD patterns revealed that the calcined sample possesses a spinel structure without any formation of other phases (Fig. 5.7). The obtained XRD pattern is in good agreement with the simulated XRD patterns of a typical spinel oxide sample.

2 Topotactic thermal conversion

Topotactic thermal conversion is known as a process that converts the chemical composition of the materials thermally from one to another without changes in the crystal structures [6]. It may be a desirable and easy method of such changes. For example, Deng et al. [6] reported a facile wet chemical route under mild

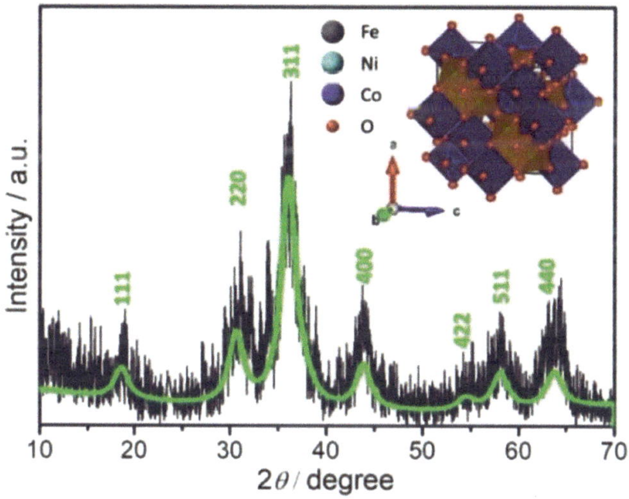

FIG. 5.7 Wide-angle XRD patterns of spinel CoNiFe-oxide (inset; crystal structure model). The simulated one was obtained by Rietveld refinement using the GSASII program [5].

experimental conditions for the large-scale fabrication of hollow, nonhollow, and porous nanospheres of hydroxyl titanium oxalate ($Ti(OH)_2C_2O_4$). Simple thermal treatment converted the spheres into anatase TiO_2 while retaining morphological integrity, as shown in Fig. 5.8. According to X-ray diffraction (XRD) patterns, both hydroxyl titanium oxalate nanospheres obtained by wet chemistry and nanoporous anatase titanium dioxide obtained by calcination have same crystal structure. On the other hand, energy dispersive spectroscopy (EDS) spectrum shows the changes in composition caused by thermal treatments. High-order assembly of subunit of anatase TiO_2 nanocrystals into spherical aggregates of about 200 nm in diameter was achieved.

An easily generalizable method for converting metal precursors nanowires topotactically by thermal treatments to their stoichiometric metal oxides was illustrated by Kolmakov and colleagues [7]. They used tin nanowires fabricated electrochemically in porous anodic oxide templates and preserved the shape of individual nanowires as well as the morphology of an entire array of parallel nanowires upon oxidation into SnO_2. The oxidation of the Sn nanowire proceeds by first forming the suboxide SnO and then the dioxide SnO_2. SnO is a p-type semiconductor and its metastability at certain temperatures makes that material interesting on its own. A crucial aspect of the structure-preserving oxidation process is the formation of a thin oxide skin through a low-temperature process before the full oxidation is carried out in temperature steps ending at a temperature high enough to form stoichiometric, thermodynamically stable SnO_2. The oxidation process was shown to proceed through a kinetically

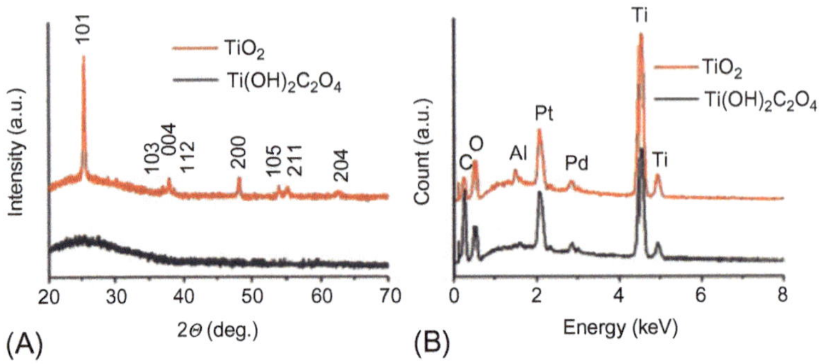

FIG. 5.8 (A) X-ray diffraction pattern (XRD) and (B) energy dispersive spectroscopy (EDS) spectrum of the hydroxyl titanium oxalate nanospheres obtained by wet chemistry *(black patterns)* and nanoporous anatase titanium dioxide obtained by calcination *(red patterns).*

controlled process, in which several coaxial core-shell metastable structures are formed, of which, initially, the inner core is molten tin (Fig. 5.9).

A report by Trávníček et al. demonstrated the topotactic dehydration process in $[Co(en)_3][Fe(CN)_6]{\bullet}2H_2O$ by thermal treatments [8]. The course of thermal decomposition in an air atmosphere up to 600°C was monitored by TGA/DSC techniques. First, the topotactic dehydration of $Co(en)_3][Fe(CN)_6]{\bullet}2H_2O$ to the hemihydrate $[Co(en)_3][Fe(CN)_6]{\bullet}1/2H_2O$ occurred preserving the single-crystal character as confirmed by X-ray diffraction analysis. The first weight loss occurs in the temperature range 50–80°C and is accompanied by endothermic effect with the minimum at 80°C, as can be seen on DSC curve (Fig. 5.10).

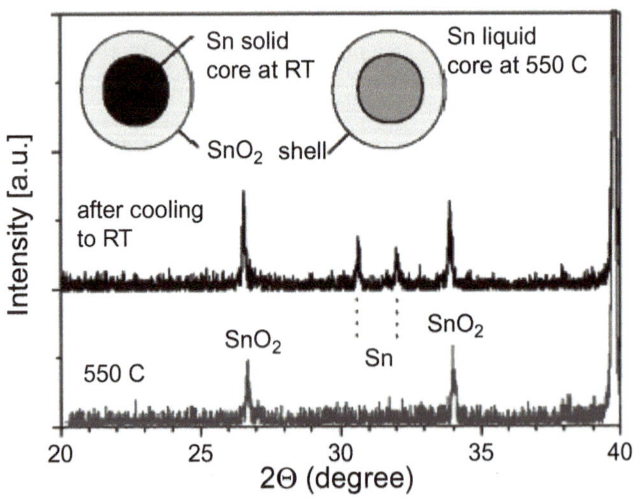

FIG. 5.9 X-ray diffraction patterns of Sn/SnO_2 with Sn in solid (top) and molten (bottom) states.

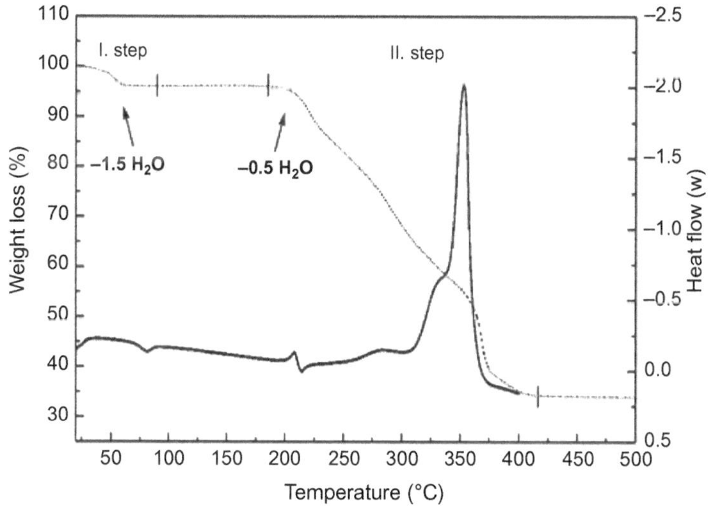

FIG. 5.10 TG (•••••••) and DSC (——) curves of complex $[Co(en)_3][Fe(CN)_6]{\cdot}2H_2O$. Data were recorded in the temperature interval of 20–500°C, and 20–400°C, respectively, with a heating rate of $5°C\,min^{-1}$.

Experimental weight loss (5.6%) is in good agreement with the calculated one (5.5%) corresponding to the release of 1.5 crystal water molecules from the formula unit, thus yielding the complex $[Co(en)_3][Fe(CN)_6]{\bullet}1/2H_2O$. The hemihydrate is thermally stable within the temperature interval of 72–187°C (Fig. 5.10).

Highly porous inorganic materials with remarkable catalytic properties have been prepared by topotactic thermal transformation and degradation of metal-organic frameworks (MOFs) and porous coordination polymers (PCPs) [9]. For example, Li and his colleagues have successfully synthesized a series of graphene-encapsulated transition metal nitrides ($Fe_xMn_{6\text{-}x}Co_4$-N@C; $0<x<6$) with well-controlled morphology by topotactic transformation of $Mn_yFe_{1-y}[Co(CN)_6]_{0.67}{\bullet}nH_2O$ (Mn_yFe_{1-y}-Co PBAs; $0<y<1$) in N_2 atmosphere. The morphology, textural property, and crystalline structure of the prepared $Fe_xMn_{6\text{-}x}Co_4$-N@C were thoroughly characterized by TG-DSC, SEM, XRD techniques (Fig. 5.11) [10].

The TGA-DSC profile of $Mn_{0.4}Fe_{0.6}$-Co *Prussian blue* analogue (PBA) under nitrogen atmosphere revealed two decomposition stages (Fig. 5.12). At the first stage, the weight loss below 215°C is ascribed to the loss of water molecules from the MOF structure. The subsequent weight loss at the second stage around 455°C might be ascribed to the conversion of MOF structure to the final products. Through thermal decomposition of Mn_yFe_{1-y}-Co PBAs at 650°C in N_2, the CN groups of PBAs serve as nitrogen and carbon sources to form nitrogen-doped graphene layers. Co and Fe atoms form FeCo alloy, while

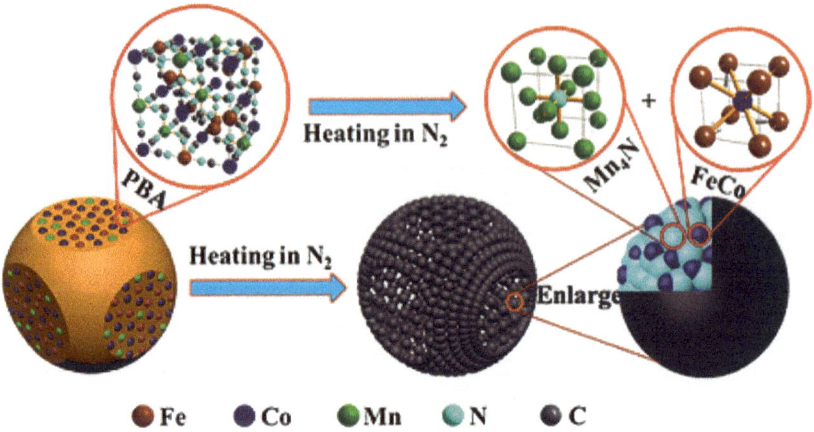

FIG. 5.11 Preparation protocol of the graphene encapsulated transition metal nitrides ($Fe_xMn_{6-x}Co_4$-N@C) topotactically with well-controlled morphology. *(Reproduced from X. Li, Z. Ao, J. Liu, H. Sun, A. I. Rykov, J. Wang, Topotactic transformation of metal-organic frameworks to graphene-encapsulated transition-metal nitrides as efficient Fenton-like catalysts, ACS Nano 10 (2016) 11532–11540. Copyright© 2016, American Chemical Society.)*

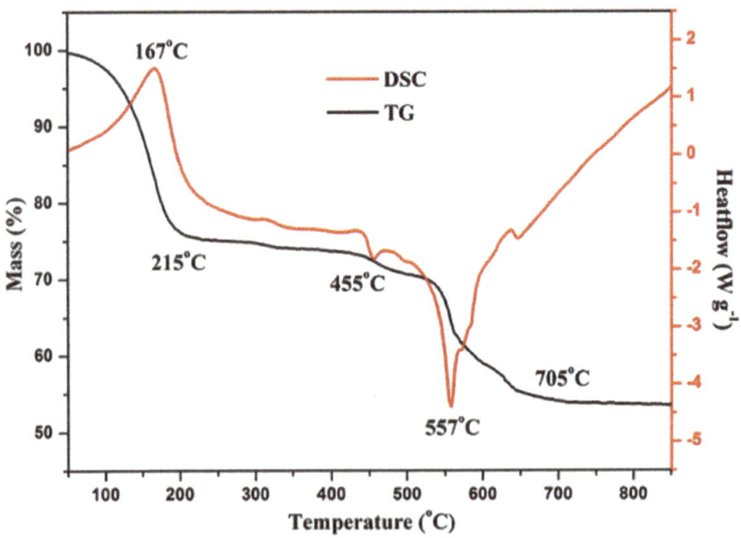

FIG. 5.12 TGA-DSC curve of $Mn_{0.4}Fe_{0.6}$-Co PBA under nitrogen atmosphere.

Mn atoms form Mn_4N nanocrystals inside, most probably due to the same crystal system (cubic) and space group (*Pm-3m*) of FeCo and Mn_4N. The final obtained $Fe_xMn_{6-x}Co_4$-N@C particles, constructed from the small FeCo and Mn_4N nanocrystals encapsulated in graphene layers, inherit the morphology of Mn_yFe_{1-y}-Co PBAs (Fig. 5.11).

3 Construction of phase diagrams using TA techniques

Thermal analysis is an experimental method to determine phase diagrams, where particular attention is paid to research carried out using thermoanalytical techniques and their application to the determination of phase diagrams [11]. A phase diagram shows conditions at which thermodynamically distinct phases can occur at equilibrium. It is determined experimentally by recording cooling rates over a range of compositions (Fig. 5.13A) and the phase transitions occur along lines of equilibrium. In contrast, thermal analysis provides information about changes in material properties as function of temperature (Fig. 5.13B).

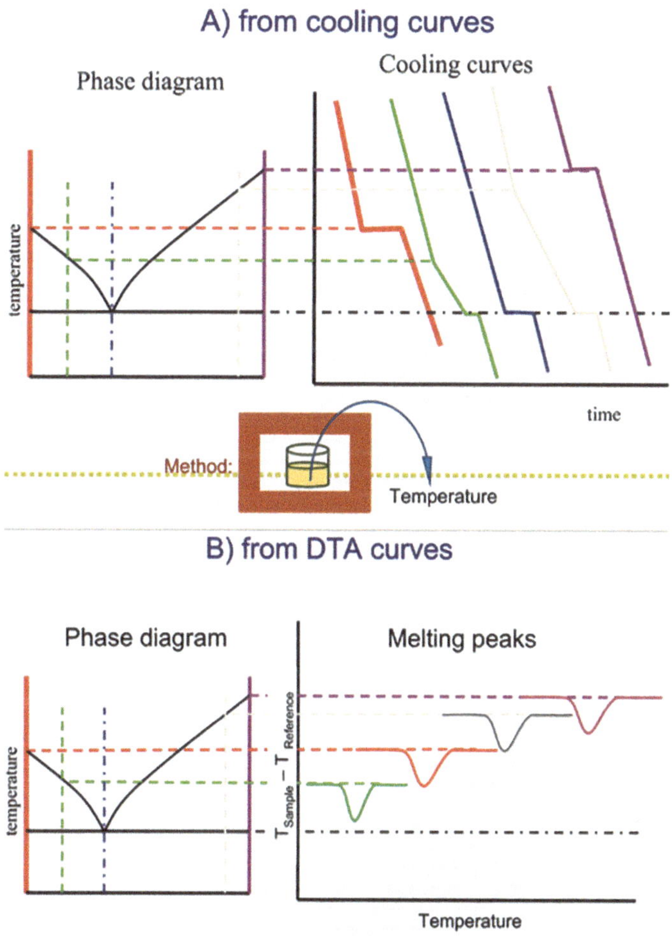

FIG. 5.13 Construction of phase diagrams by experimental methods: (A) cooling curves and (B) DTA curves.

Drug-polymer solid dispersion can markedly improve the dissolution rate of drugs and lead to higher bioavailability [12–14]. However, due to the lack of miscibility between most of the insoluble drugs and commonly used hydrophilic polymers, drugs tend to crystallize out of the initially homogeneous drug-polymer solution during storage, especially at high drug loading and/or when exposed to moisture [15]. Besides drug crystallization, amorphous phase separation may also occur [16,17] generating drug-rich and polymer-rich amorphous domains, which are still thermodynamically unstable in comparison to drug crystals. Hence, crystallization may start from these amorphous domains instead of the homogeneous drug-polymer solution [18,19]. In addition, both crystallization and amorphous phase separation may be interrupted by the glass transition [20,21]. Therefore, for an apparently simple drug-polymer two-component system, it undergoes multiple phase separation pathways and is kinetically trapped in different metastable points in the temperature-composition phase diagram. Consequently, the drug dispersion structures may vary depending on both the composition and the thermal treatment history of the solid dispersion systems. Therefore, it is highly desirable to construct a complete phase diagram as a map where the possible phases of a solid dispersion system can be located, including the liquid-solid phase transition curve (i.e., the crystallization curve, along which drug crystals coexist with a drug-polymer glass solution in equilibrium), the amorphous phase separation curves, and the glass transition curve.

Based on the polymer solution theory, a typical phase diagram of a small molecule-polymer system looks like Fig. 5.14, including the liquid-solid phase transition curve, the bimodal and spinodal amorphous phase separation curves, and the glass transition temperature (T_g) curve [20]. The liquid-solid transition

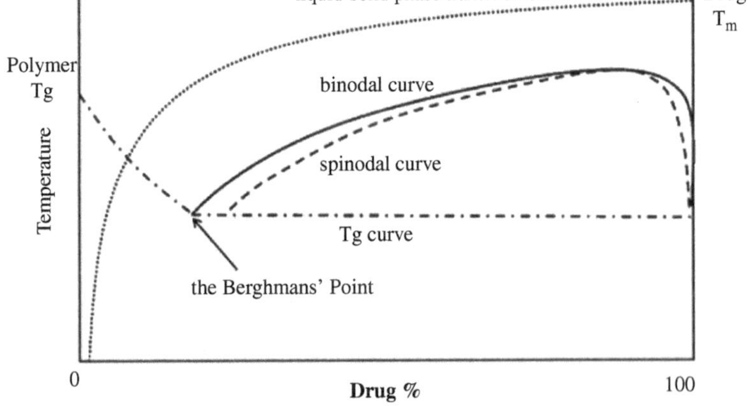

FIG. 5.14 Theoretical phase diagram of drug-polymer solid dispersions based on the polymer solution theory [22].

curve represents the equilibrium solubility of drug crystals in the polymer matrix at different temperatures. The area above this curve implies that drugs are dissolved in polymer and form an unsaturated solution. The area below the curve implies that the drug loading in the solid dispersion is above its equilibrium solubility. The drug molecules tend to precipitate out as drug crystals which exist in equilibrium with a drug-polymer solution whose composition is determined by the liquid-solid phase transition curve.

As first proposed by Mohan et al. [23] and later applied to solid dispersion systems by Taylor and Yu's groups [24], this liquid-solid transition curve can be determined by DSC measurement through the end points of the melting peaks. However, as pointed out previously [21], only part of the curve at temperatures above T_g could be obtained experimentally because of the high viscosity at lower temperatures and the difficulty for the system to reach equilibrium. When the system is cooled to temperature below the liquid-solid phase transition curve, amorphous (liquid-liquid) phase separation may take place prior to drug crystallization, and amorphous polymer-rich and drug-rich phases are formed [20]. Here, drug crystallization represents the thermodynamically stable state, while amorphous phase separation represents the metastable states.

As shown in Fig. 5.14, the amorphous phase separation is defined by two curves. The binodal curve represents the metastable amorphous phase separation line (polymer-rich and drug-rich phases coexist with each other), while the spinodal curve represents a change in the mechanism of amorphous phase separation [25]: for systems within the area between the binodal and spinodal curves, amorphous phase separation goes through a nucleation and growth pathway, where the drug-rich domains first form as small droplets and grow in size. In contrast, for systems below the spinodal curve, amorphous phase separation forms bicontinuous polymer-rich and drug-rich domains at first and then goes through a coarsening process. In principle, these transient phase structures could be fixed by cooling down the sample below the glass transition temperature, and this provides the motivation to study the phase diagram as a guiding map. In this context, Lin et al. used the felodipine-poly(acrylic acid) (PAA) system to exemplify how a complete phase diagram can be predicted from data obtained through thermal analysis [26].

Fig. 5.15 shows the predicted complete liquid-solid phase transition curve. The extent of melting point depression for felodipine in PAA is small for almost the whole composition range. At room temperature (25°C), the predicted equilibrium solubility of felodipine in PAA is less than 0.1 wt% (3.17×10^{-4} wt%). DSC, first proposed by Mohan et al. [23], has been used to measure the equilibrium solubility of drugs in small molecule solvents and polymers. This method involves heating a crystal/solvent system of a certain composition (x). The key point [23] is to realize that the solubility of the crystal in the solvent is exactly x at the final temperature of crystal dissolution, which is calculated as the intersection of the tangents on the heat flow peak (T_{end}). This method of

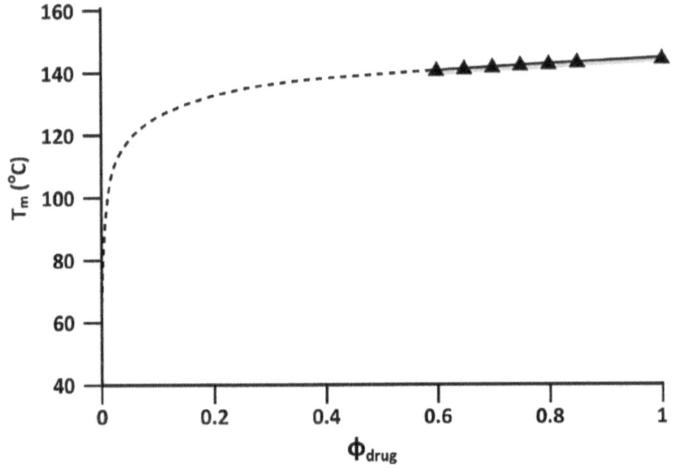

FIG. 5.15 Complete liquid-solid phase transition curve for the felodipine-PAA system.

measuring the solubility of felodipine in PAA and the results obtained are shown in Fig. 5.16.

Fig. 5.17A shows the typical results from DSC measurement to determine the glass transition temperatures. The small endothermic peak prior to the major melting peak was probably due to the melting of drug crystals of smaller size. During the first heating scan, melting of felodipine crystals was observed, while no crystallization peak was observed in the subsequent cooling scan. Instead, two glass transitions were detected in the cooling scan as well as in the second heating

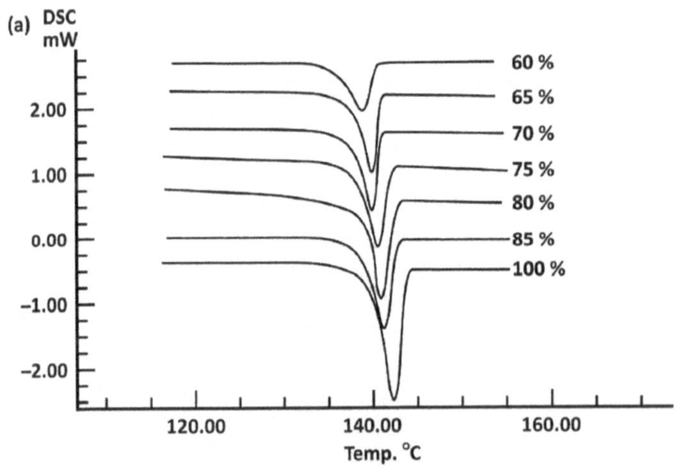

FIG. 5.16 DSC thermograms of felodipine-PAA solid dispersion measured at a heating rate of 1°C min^{-1}. The numbers on each curve indicate the drug loading in weight percentage.

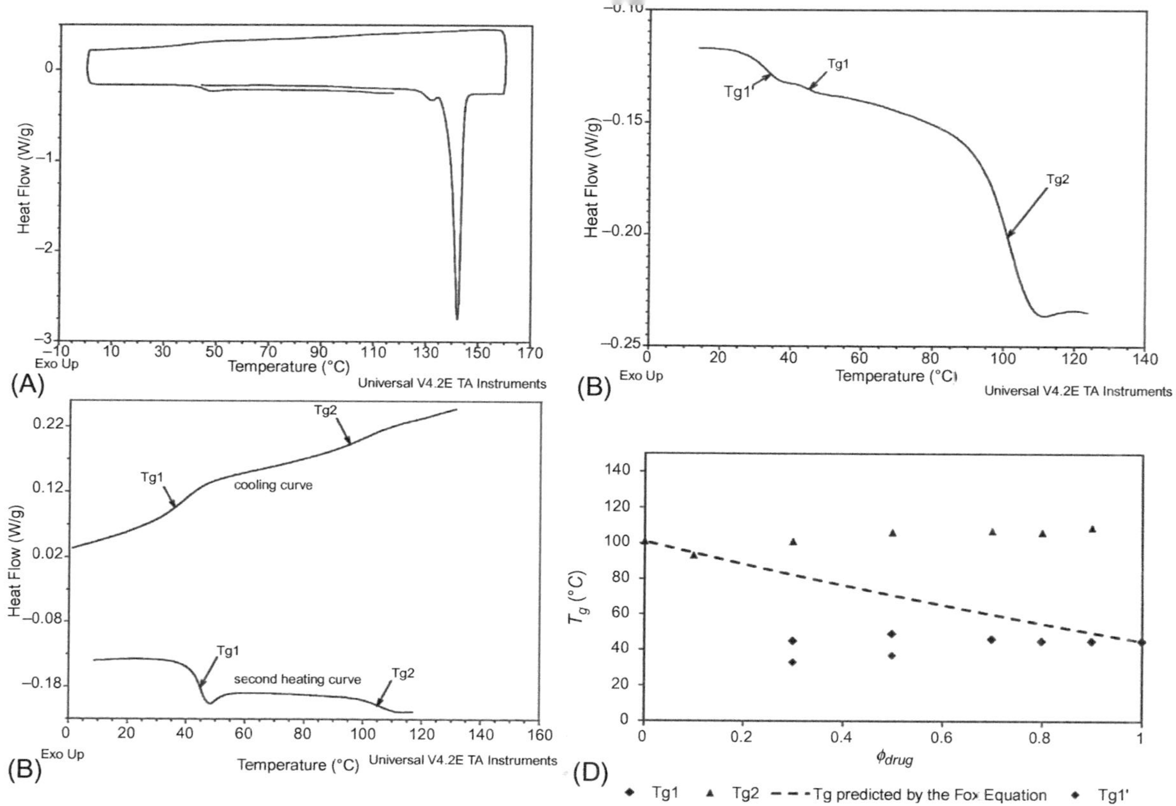

FIG. 5.17 (A) DSC thermograms of solid dispersion sample with 80 wt% felodipine, including the first heating, the subsequent cooling, and the second heating scan, all measured at a heating rate of 10°C min^{-1}. (B) The magnified cooling and reheating curves from (A), $T_{g1}=45$°C and $T_{g2}=106$°C in the second heating measurement. (C) DSC thermograms of solid dispersion sample with 30 wt% felodipine measured at a heating rate of 10°C/min (the reheating curve), $T'_{g1}=33$°C and $T_{g1}=45$°C. (D) T_{gs} of felodipine-PAA amorphous mixture at different compositional ratios of felodipine-PAA against the heating temperature.

scan (Fig. 5.17B), indicating the occurrence of amorphous phase separation during cooling instead of drug crystallization. Similar results were obtained for samples with drug contents from 30 to 90 wt%, as plotted in Fig. 5.17D, together with the predicted T_g curve described by the following equation:

$$\frac{1}{T_g} = \frac{w_1}{T_{g1}} + \frac{w_2}{T_{g2}}. \tag{5.1}$$

However, for the samples containing 30 and 50 wt% felodipine, one more T_g below the glass transition temperature of pure felodipine was observed (Fig. 5.17C). This lower T_g of felodipine was also observed by Karavas et al. [27]. In their experiments for felodipine-PVP system, T_g of drug-rich phase for 25 wt% felodipine and 50 wt% felodipine samples were at 37.6°C and 30.1°C, respectively.

Nakamura et al. have demonstrated the thermal analysis data and phase diagram of the LiF-BiF_3 system [28]. DTA up to complete melting was performed for the complete pseudo-binary system LiF-BiF_3. Melts with high bismuth fluoride concentration show severe evaporation, nevertheless even pure BiF_3 could be molten at 655°C. The system contains one intermediate compound $BiLiF_4$ which melts by peritectic decomposition due to the formation of LiF at 415°C. The eutectic between $BiLiF_4$ and BiF_3 melts at 415°C.

The DSC measurement of pure bismuth fluoride was performed at a rate of 10°C min^{-1}. The resulting curve is presented in Fig. 5.18, together with the TGA curve. Significant mass loss occurred above 575°C, which reflects the known high vapor pressure at higher temperatures which is estimated to c. 20 mbar at this temperature. Several minor endothermic peaks are present in the curve in that range, likely related to heat loss due to sublimation. The sharp peak around 655°C corresponds to melting, which became obvious by visual inspection of the sample after the measurement. It is remarkable that this

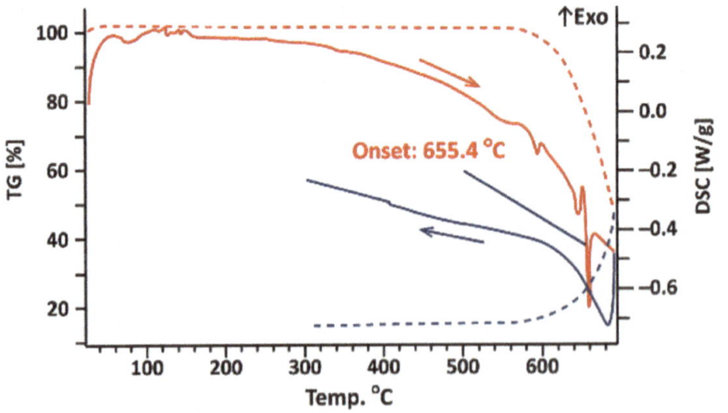

FIG. 5.18 DSC and TGA curves (heating and cooling with 10°C min^{-1}) of pure BiF_3.

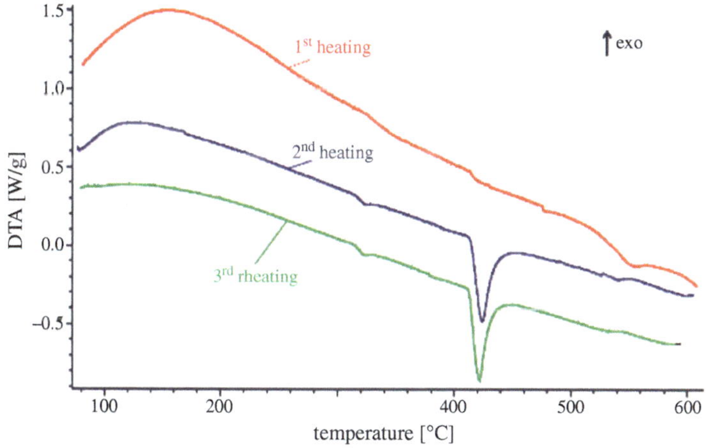

FIG. 5.19 DTA of first, second, and third heating runs (10°C min^{-1}) of a LiF-BiF_3 sample with 65% BiF_3.

value is almost identical to α-to-δ transition of Bi_2O_3 that has a high transformation heat of 29.8 kJ/mol. It seems realistic to assume that very high melting temperature has its origin in partial hydrolysis of BiF_3 to the oxide prior to, or during the measurements. The results of TA measurements with LiF were entirely compatible with the literature [28].

Fig. 5.19 shows first DTA mixing run together with the subsequent heating cycles for one sample. Three thermal events of invariant temperatures were observed for the different compositions within the LiF-BiF_3 system. An intense transition was observed at approximately 415°C in almost all compositions (peak II). By visual inspection of samples that were cooled after passing peak II, it became obvious that it is due to melting of a major part of the sample. Another transition (peak III) was detected at ca. 450°C. The separation of both peaks II and III was only possible at slower heating rates of 2 K/min; at 10 K/min both peaks overlapped. Peak III was clearly observed only in samples with bismuth content lower than 60%. For higher concentrations of BiF_3, it was either not observed or only minor. The smallest "peak I" was observed for almost all measurements over the whole concentration range between LiF and BiF_3 at a constant temperature of approximately 315°C but did never occur for pure components.

The equilibrium phase diagram in Fig. 5.20 explains all experimental results. Solid lines result from a thermodynamic assessment of the system with FactSage thermodynamic program and database system [29]. The invariant point with lowest temperature (peak I) is tentatively interpreted as a metastable eutectic between LiF and BiF_3. The formation of $BiLiF_4$ either by the peritectic reaction or from the solid components is a diffusion-controlled, and hence is a slow process. Moreover, segregation of LiF (top) and BiF_3 (bottom) resulting

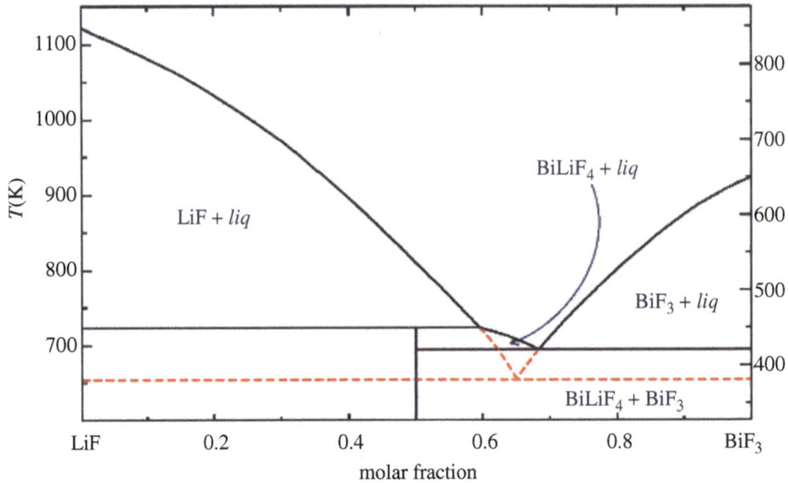

FIG. 5.20 *Full lines*: The equilibrium phase diagram LiF-BiF_3. *Dashed lines*: Nonequilibrium formation of a LiF/BiF_3 eutectic if $BiLiF_4$ formation is incomplete due to segregation.

from large density difference shows that both components are available also if the system resides within the bottom phase fields of Fig. 5.20, where $BiLiF_4$ is an equilibrium phase. Consequently, minor amounts of LiF and BiF_3 can be found inside the crucible at the same time as nonequilibrium phases and form an eutectic. The position is estimated by the intersection of the LiF and BiF_3 liquidus prolongation, which is shown in Fig. 5.20 by dashed lines. The temperature there (~380°C) is higher than peak I (~315°C), but it should be noted that this temperature is not the subject of the current (equilibrium) assessment. Besides, no liquidus data for BiF_3-rich melts were available, hence the liquidus slope there relies on the FactStage data for its heat of fusion (ΔH_f=21.6kJ/mol) [29]. A smaller ΔH_f would shift the eutectic to lower T.

4 Thermoelectric materials

Thermoelectric materials are those materials that have low thermal conductivity with large electrical conductivity. Low-dimensional materials are high-efficiency thermoelectrics because they can become thermal insulators with high electrical conductivities. Thermal insulation is widely present in chemical plants (e.g., pipes, furnaces, etc.), so thermoelectric materials could replace current insulations to harvest thermal energy and transform it into electricity. Thermoelectric devices can play a very important role in clean energy generation, conversion, and recovery. Thermoelectrics are "fuel-free" purely solid-state devices with no moving parts and therefore are extremely reliable. Thermoelectrics can harvest residual low-grade energy which otherwise is wasted [30].

High-performance thermoelectric materials in a wide range of temperatures are essential to broaden the application spectrum of thermoelectric devices. In recent years, significant progress has been made on improving thermoelectric materials [31–36]. In addition to the importance of the properties of thermoelectric materials for the device performance, establishing a temperature difference across the device is key for the application of thermoelectric generators.

For the improvement of thermoelectric materials as well for efficient heat-to-electric power conversion, device optimization is a crucial factor and one of the key challenges is how to create a large temperature across a thermoelectric generator especially in the case of a dilute incident heat flux. A solar-to-electric power conversion efficiency of ~5% could be demonstrated. A solar thermoelectric generator with a large thermal concentration, which minimizes the amount of thermoelectric nanostructure material, presents great potential to enable cost-effective electrical power generation from the sun. The Seebeck effect can be exploited to generate power from a temperature difference. The efficiency of thermoelectric power generation systems depends on the material-specific thermoelectric figure of merit, the heat source and sink temperatures, and the heat transfer into and out of the devices. The efficiency of ideal thermoelectric device (η_{te}) is determined by operating temperatures and dimensionless figure of merit (ZT) of the material, defined as $ZT = (S^2\sigma/k)T$, where S, σ, k, and T are the Seebeck coefficient, electrical conductivity, thermal conductivity, and absolute temperature, respectively [37,38]. The device efficiency can be expressed by Eq. (5.2):

$$\eta_{te} = \frac{T_h - T_c}{T_h} \frac{\sqrt{1 + (ZT)_M} - 1}{\sqrt{1 + (ZT)_M} + \frac{T_c}{T_h}} \tag{5.2}$$

where T_c is the cold-side temperature, T_h the hot-side temperature, and $(ZT)_M$ is the effective ZT of the thermoelectric materials between T_c and T_h. According to Eq. (5.2), an efficiency of approximately 8.6% can be reached by imposing a temperature difference of 200°C across an ideal thermoelectric device with $(ZT)_M = 1$ and $T_c = 20$°C.

Nithiyanantham et al. have reported a low-temperature formation process of rectangular PbTe nanocrystals and their thermoelectric properties [39]. The crystallization process of the as-prepared PbTe nanocrystals was determined by TGA and DTA using a solid as-prepared PbTe sample of 2.66 mg (Fig. 5.21). The sample was heated in the air up to 1000°C at a heating rate of 10°C min. In Fig. 5.21, curve A is the TGA curve, while curve B is the DTA curve. From the TGA curve, it is obvious that the first weight loss occurred near 109°C, which might have been because of the removal of moisture from the sample. The second weight loss occurred at 327°C because of the removal of excess organic moieties and excess surfactant molecules. Further weight loss occurred around 503°C, most probably due to the melting of excess Te metal

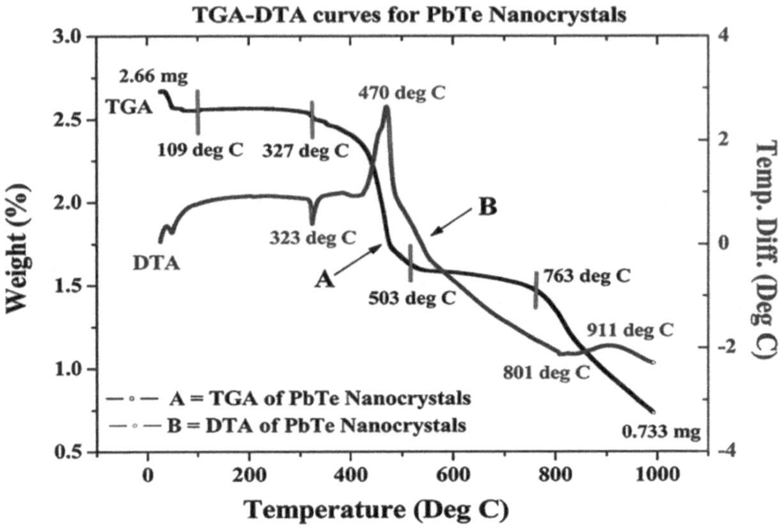

FIG. 5.21 Thermal analysis of as-synthesized PbTe nanocrystals, where curve A is the TGA curve and curve B is the DTA curve.

powder in the sample. The final weight loss started at 763°C and continued until the end, which might have been due to the melting of PbTe. Curve B shows the DTA curve, where there is an endothermic peak at 323°C and a strong exothermic peak at 470°C. This was probably due to the evaporation of excess cetyltrimethylammonium bromide (CTAB) or excess metal salt precursor. The final exotherm near 910°C was due to the melting of the synthesized PbTe nanocrystals. Similar types of thermoelectric analyses have been reported earlier by others. After heating 2.66 mg of sample up to 1000°C, the amount of sample remaining was 0.733 mg. Hence the percentage of weight loss during this process was 72.44%.

Another example, Bristow and his colleagues reported the formation of bismuth telluride-based thermoelectric material coatings for protection against thermal cycling effects [40]. They have studied the effect of heating on the composition and morphology by TGA, to locate temperatures of various thermal decomposition processes and to determine the amounts of volatiles for Bi_2Te_3 (Fig. 5.22). A slight mass loss by TGA up to 400°C was determined. There is a steeper slope (wt%/°C) in the mass loss up to ~160°C, likely due to the desorption and evaporation of absorbed residues such as moisture. The mass loss slope is more gradual from ~160°C to ~400°C likely due to the sublimation of tellurium. However, starting at approximately 400°C the diagram shows relatively sharp weight gain. Apparently, a chemical reaction with

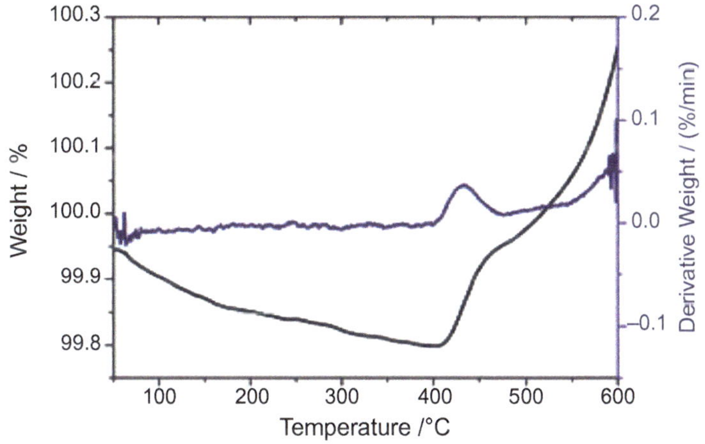

FIG. 5.22 TGA thermogram and its derivative for Bi_2Te_3.

something from the environment takes place. If tellurium is escaping, then oxidation of bismuth is a likely candidate for that reaction.

To select the proper processing temperatures of alloys, the phase change temperatures must be accurately determined [41a]. The traditional method for determining phase change temperatures in metallography is to microscopically examine a series of alloy samples which have been heated to below, near, and above the expected transition temperature. The accuracy of this method is ±5°C for titanium alloys and ±3°C for nickel-base alloys compared to ±0.5°C accuracy level on using DTA.

The accuracy in spotting phase transition temperatures can be improved substantially by applying the derivative function for determining the rate of temperature change with respect to time. Fig. 5.23 illustrates the DTA and its derivative curve for a titanium alloy of the composition Ti-6Al-4V. The phase change temperature of interest is β-transus, that is, the temperature at which the alloy's alpha phase, characterized by a hexagonal close packed structure, transforms to the beta phase, a body centered structure. It can be readily picked off the derivative trace at 994°C [41b].

5 Phase transformations in organic compounds and drugs

Phase transformations in organic compounds also affects their chemical and physical properties [42]. Fig. 5.24 shows a DSC calorigram of the monomerization of 9-cyanoanthracene photo-dimer (9 CNAD). Peak A is assigned to phase transformation in 9 CNAD with subsequent crystallization of the monomerization product (that is, 9 CNA). This peak occurs in crystalline samples but is absent in amorphous samples [43,44]. Peak B is assigned to decomposition

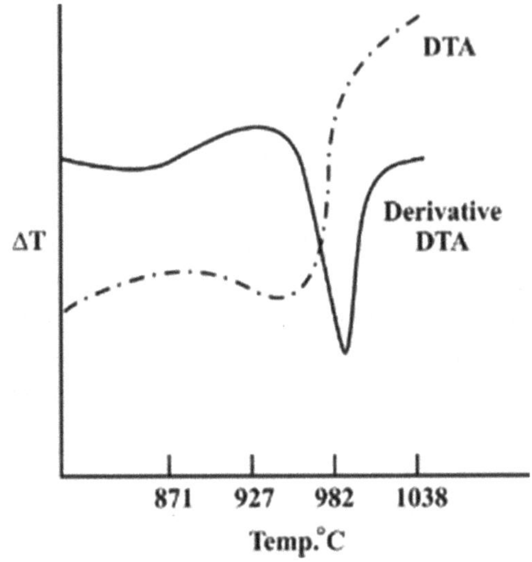

FIG. 5.23 DTA (–●–●–●–) and derivative DTA (——) traces for a Ti-6Al-4V sample.

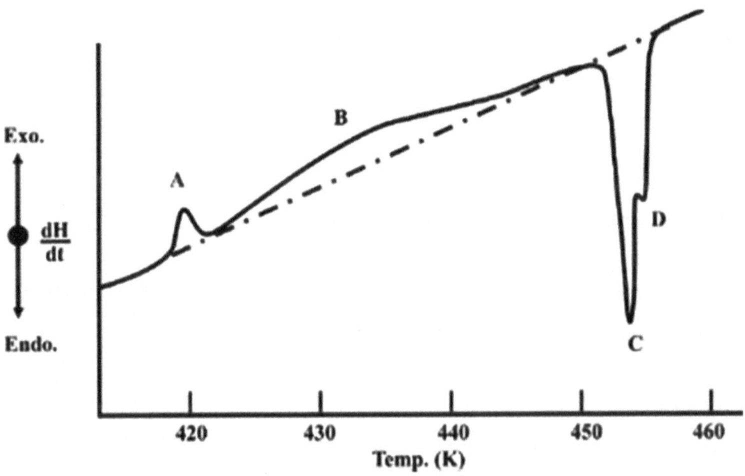

FIG. 5.24 DSC calorigram of crystalline 9 CNAD.

and peaks C and D are assigned to melting of 9 CNAD and 9CNA decomposition product, respectively [43]. Thermoanalytical data was supplemented by X-ray diffraction, microscopic as well as spectroscopic data to prove phase transformations.

Phase transformation in azodiisobutyronitrile (AIBN) crystals has also been detected by DSC at 64°C [45,46]. DSC has also been applied extensively to study

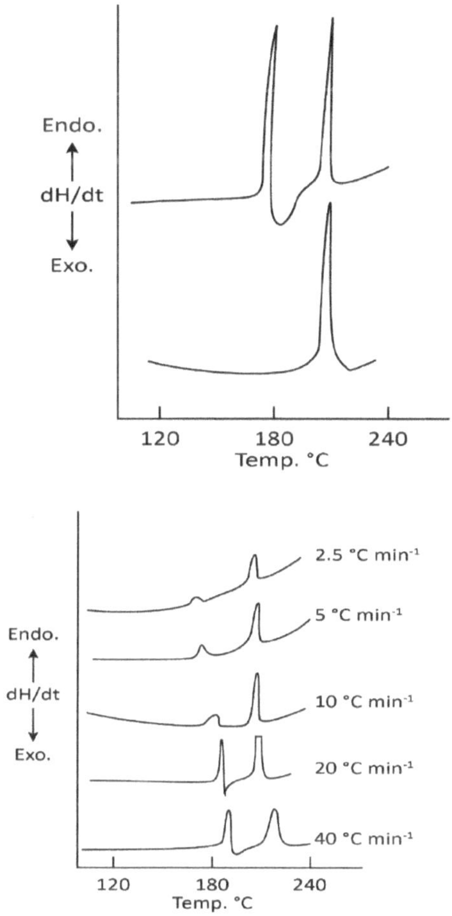

FIG. 5.25 DSC scans of form II gepirone before and after heat treatment for 3 h at 150°C, scan rate is 20°C min^{-1} *(top)* and form I recorded at different scanning rates *(bottom)*.

phase-changing materials for energy storage [47]. Several organic and inorganic salt hydrates are used in the so-called "heat batteries."

An example of enantiomorphic polymorphism in drugs is that of forms I and II of gepirone hydrochloride. Endotherms are obtained due to solid-solid transitions as the less stable polymorphs convert to the more stable form (Fig. 5.25A).

Typical DSC traces for the newly crystallized lower stable polymorph show the solid-solid transitions at low scanning rates (2.5–10°C min^{-1}) but at higher scanning rates, only the melting endotherms of the two polymorphs could be observed (Fig. 5.25B).

5.1 Polymorphism in ritonavir (or Norvir)

Ritonavir was introduced in 1996 as a drug to treat HIV infection by Abbott Drug Company. In 1998, a low energy, more stable, and less soluble form (form II) appeared, causing slow dissolution of the marketed dosage form and compromising the oral bioavailability of the drug. This event forced Abbott Drug Company to remove the oral capsule formulation from the market because polymorph II is the wrong crystalline form. HIV patients were offered the Norvir drug in liquid form until the company could reformulate the product to eliminate the wrong crystal structure.

The liquid form was formulated as a capsule filled with a hydroalcoholic solution containing the dissolved drug. Even these capsules began failing dissolution specifications after 2 years of their entry into the market. Evaluation of the failed drug products revealed that the insoluble crystalline form II has precipitated from the formulation [48].

The crystal form II is less than 50% as soluble as form I, resulting in the poor dissolution behavior and eventual withdrawal of the capsule from the market.

In 2003, Morissette et al. [49] reported the discovery of three additional crystalline forms of ritonavir. These three additional crystalline forms are as follows:

(i) Form III which is a crystalline solvate.
(ii) Form IV which is a true, unsolvated, metastable polymorph of ritonavir.
(iii) Form V which is the hydrated form of III and is obtained upon subjecting form III to an aqueous medium. Form V in turn converts spontaneously to needle-like form I crystals. The transformation V→I occurs within 30 min of exposure of form III to water or 0.1 M HCl by which time the phase change III→V→I occurs.

Growth conditions of the new three polymorphs are as follows:

Form	Solvent
III	Mixed solvent of formamide + toluene or butylacetate or acetone, which are immiscible formamide.
IV	Mixed solvent of acetointrile + butylacetate or isobutylacetate or isopropylacetate
V	Form III washed with deionized water.

Physical parameters of ritonavir (or Norvir) polymorphs are summarized as follows:

Form	m.p. °C	$\Delta H_{fus.}$ J/g	Crystal structure
I	122	78.2	Monoclinic
II	(most stable)122	87.8	Orthorhombic
III	78–82	60.3	Monoclinic
IV	116	59.8	Not assigned
V	97	32	Monoclinic

TA has been applied to study different polymorphs of Norvir drug.

Form I

Has a melting onset at $\sim$113 $\pm$ 0.5°C with a peak endotherm of 122 $\pm$ 0.5°C and shows no significant weight loss before the onset of decomposition at 150°C. Powder X-ray diffraction patterns (PXRD) correspond to form I.

Form II

Has a melting point onset at $\sim$116 $\pm$ 0.5°C with a peak endotherm of 122 $\pm$ 0.5° C. PXRD correspond to polymorph (II).

Form III

Crystalline solid that melts irreversibly in the temperature range 78–82°C (Fig. 5.26). This range is significantly lower than the m.p. of known forms of ritonavir and is accompanied by significant and variable weight loss of 30%–60% upon melting. The weight loss suggests that the sample is solvated, and given the variability of the weight loss observed, the solvate is likely a non-stoichiometric solvate that can accommodate various amounts of solvent molecules within the structure.

Despite low melting point, samples of form III show physical stability at room temperature in sealed vials for at least 3 months.

Form IV

Exhibits sharp melting at 116°C and no significant loss of volatile matter up to 150°C meaning that form IV is unsolvated (Fig. 5.27). It exhibits a unique PXRD pattern. It has a characteristic PXRD and both DSC and TGA curves show melting onset at $\sim$113°C with a peak endotherm at 122°C and a significant

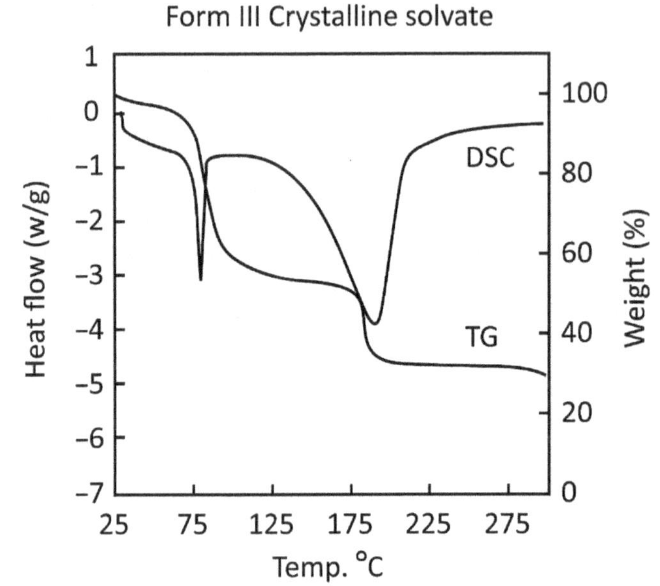

FIG. 5.26 Joint DSC/TG curves of polymorph III of ritonavir drug.

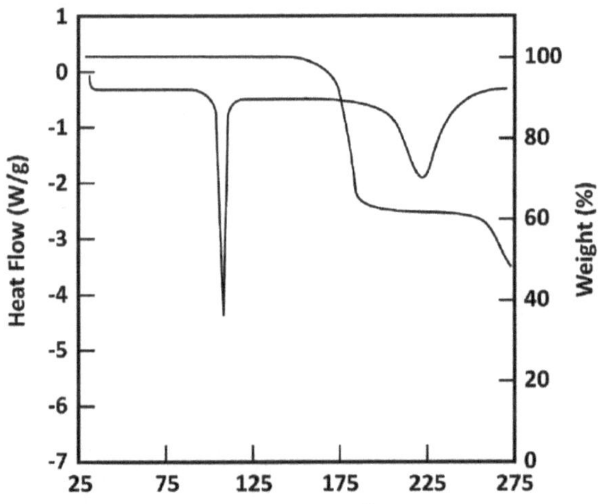

FIG. 5.27 Joint DSC/TG curves of ritonavir form IV.

weight loss at ~150°C. The absence of weight loss in the melting event indicates that form IV is unsolvated.

Form V

This form can be obtained by the hydration of form III. Form III before and after hydration giving Form V is shown in Fig. 5.28.

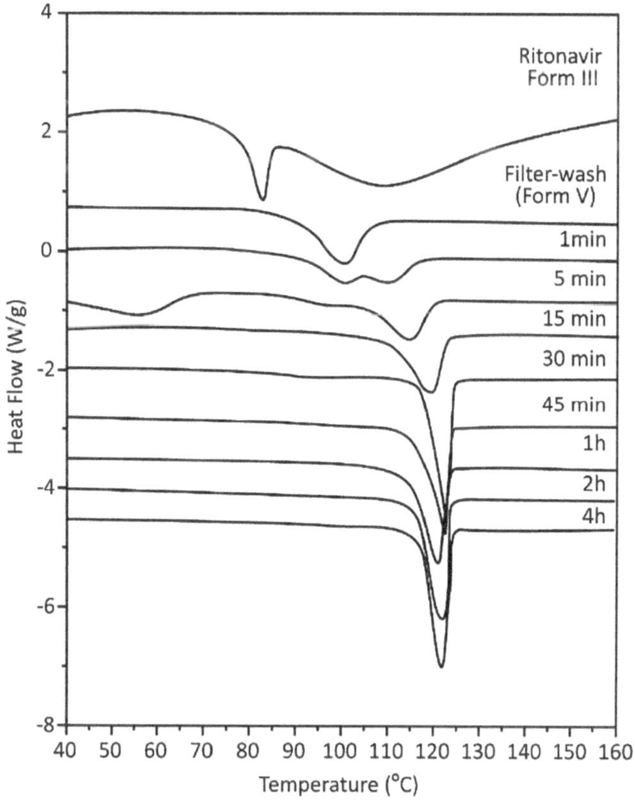

FIG. 5.28 DSC curves of ritonavir form III *(top)* and form V obtained by hydration of form III for different time intervals.

6 Summary

This chapter demonstrates the crystal structure and phase changes occurring during heating by TA. In this chapter we demonstrated the use of TA techniques for studying crystal structure transformations, topotactic thermal conversion, constructing phase diagrams of some materials, studying thermoelectric materials, and for studying phase transformations in organic compounds and drugs.

References

[1] M.B. Zakaria, M. Hu, M. Imura, R.R. Salunkhe, N. Umezawa, H. Hamoudi, A.A. Belik, Y. Yamauchi, Single-crystal-like nanoporous spinel oxides: a strategy for synthesis of nanoporous metal oxides utilizing metal-cyanide hybrid coordination polymers, Chem Eur J 20 (2014) 17375–17384.

[2] M.B. Zakaria, T. Chikyow, Recent advances in Prussian blue and Prussian blue analogues: synthesis and thermal treatments, Coord. Chem. Rev. 352 (2017) 328–345.

[3] M.B. Zakaria, E.M. Ebeid, M.M. Abdel-Galeil, T. Chikyow, Cyanide bridged coordination polymer nanoflakes thermally derived Ni_3C and fcc-Ni nanoparticles for electrocatalysts, New J. Chem. 41 (2017) 14890–14897.

[4] M.B. Zakaria, M.S.A. Hossain, M.J.A. Siddiky, M. Shahabuddin, E. Yanmaz, J.H. Kim, A.A. Belik, Y. Ide, M. Hu, S. Tominaka, Y. Yamauchi, Cyano-bridged trimetallic coordination polymer nanoparticles and their thermal decomposition into nanoporous spinel ferromagnetic oxides, Chem Eur J 22 (2016) 15042–15048.

[5] B.H. Toby, R.B. Von Dreele, GSAS-II: the genesis of a modern open-source all purpose crystallography software package, J. Appl. Crystallogr. 46 (2013) 544–549.

[6] D. Deng, S.T. Martin, S. Ramanathan, Synthesis of hollow porous nanospheres of hydroxyl titanium oxalate and their topotactic conversion to anatase titania, J. Mater. Res. 26 (2011) 1545–1551.

[7] A. Kolmakov, Y. Zhang, M. Moskovits, Topotactic thermal oxidation of Sn nanowires: intermediate suboxides and core-shell metastable structures, Nano Lett. 3 (2003) 1125–1129.

[8] Z. Trávníček, R. Zbořil, M. Matiková-Maľarová, B. Drahoš, J. Černák, Thermal decomposition of $[Co(en)_3][Fe(CN)_6]•2H_2O$: topotactic dehydration process, valence and spin exchange mechanism elucidation, Chem. Central J. 7 (2013) 1–28.

[9] S.G. Duyker, G.J. Halder, P.D. Southon, D.J. Price, A.J. Edwards, V.K. Peterson, C.J. Kepert, Topotactic structural conversion and hydration dependent thermal expansion in robust $LnM^{III}(CN)_6.nH_2O$ and flexible $ALnFe^{II}(CN)_6.nH_2O$ frameworks (A = Li, Na, K; Ln = La-Lu, Y; M = Co, Fe; $0 < n < 5$), Chem. Sci. 5 (2014) 3409–3417.

[10] X. Li, Z. Ao, J. Liu, H. Sun, A.I. Rykov, J. Wang, Topotactic transformation of metal-organic frameworks to graphene-encapsulated transition-metal nitrides as efficient Fenton-like catalysts, ACS Nano 10 (2016) 11532–11540.

[11] R. Ferro, A. Saccone, Thermal analysis and alloy phase diagrams, Thermochim. Acta 418 (2004) 23–32. C. Mailhé, M. Duquesne, A fast and low-cost dynamic calorimetric method for phase diagram estimation of binary systems, J. Therm. Anal. Calorim. 143 (2020) 587–598, https://doi.org/10.1007/s10973-020-09287-6. S. Pourbabak, B. Verlinden, J. Van Humbeeck, D. Schryvers, DSC cycling effects on phase transformation temperatures of micron and submicron grain Ni50.8Ti49.2 microwires, Shap. Mem. Superelast. 6 (2020) 232–241.

[12] W.L. Chiou, S. Riegelman, Pharmaceutical applications of solid dispersion systems, J. Pharm. Sci. 60 (1971) 1281–1302.

[13] C. Leuner, J. Dressman, Improving drug solubility for oral delivery using solid dispersions, Eur. J. Pharm. Biopharm. 50 (2000) 47–60.

[14] L. Yu, Amorphous pharmaceutical solids: preparation, characterization and stabilization, Adv. Drug Deliv. Rev. 48 (2001) 27–42.

[15] A.T.M. Serajuddin, Solid dispersion of poorly water-soluble drugs: early promises, subsequent problems, and recent breakthroughs, J. Pharm. Sci. 88 (1999) 1058–1066.

[16] D.T. Friesen, R. Shanker, M. Crew, D.T. Smithey, W.J. Curatolo, J.A.S. Nightingale, Hydroxypropyl methylcellulose acetate succinate-based spray-dried dispersions: an overview, Mol. Pharm. 5 (2008) 1003–1019.

[17] S. Janssens, G. Van den Mooter, Review: physical chemistry of solid dispersions, J. Pharm. Pharmacol. 61 (2009) 1571–1586.

[18] K. Six, J. Murphy, I. Weuts, D.Q.M. Craig, G. Verreck, J. Peeters, M. Brewster, G. Van den Mooter, Identification of phase separation in solid dispersions of Itraconazole and Eudragit E100 using microthermal analysis, Pharm. Res. 20 (2003) 135–138.

[19] M. Vasanthavada, W.Q. Tong, Y. Joshi, M.S. Kislalioglu, Phase behavior of amorphous molecular dispersions II: role of hydrogen bonding in solid solubility and phase separation kinetics, Pharm. Res. 22 (2005) 440–448.

[20] S.Z.D. Cheng, Phase Transitions in Polymers: The Role of Metastable States, Elsevier, Amsterdam, 2008.

[21] F. Qian, J. Huang, M.A. Hussain, Drug-polymer solubility and miscibility: stability consideration and practical challenge in amorphous solid dispersion development, J. Pharm. Sci. 99 (2010) 2941–2947.

[22] R. Koningsveld, W.H. Stockmayer, E. Nies, Polymer Phase Diagrams, Oxford University Press, New York, 2001.

[23] R. Mohan, H. Lorenz, A.S. Myerson, Solubility measurement using differential scanning calorimetry, Ind. Eng. Chem. Res. 41 (2002) 4854–4862.

[24] J. Tao, Y. Sun, G.G.Z. Zhang, L. Yu, Solubility of small-molecule crystals in polymers: d-mannitol in PVP, indomethacin in PVP/VA, and nifedipine in PVP/VA, Pharm. Res. 26 (2009) 855–864.

[25] M. Rubinstein, R.H. Colby, Polymer Physics, Oxford University Press, New York, 2003.

[26] D. Lin, Y. Huang, A thermal analysis method to predict the complete phase diagram of drug-polymer solid dispersions, Int. J. Pharm. 399 (2010) 109–115.

[27] E. Karavas, G. Ktistis, A. Xenakis, E. Georgarakis, Miscibility behavior and formation mechanism of stabilized felodipine-polyvinylpyrrolidone amorphous solid dispersions, Drug Dev. Ind. Pharm. 31 (2005) 473–489.

[28] G.H.G. Nakamura, D. Klimm, S.L. Baldochi, Thermal analysis and phase diagram of the LiF-BiF_3 system, Thermochim. Acta 551 (2013) 131–135.

[29] GTT Technologies, Kaiserstr. 100, 52134 Herzogenrath, Germany, FactSage 6.2, 2010. http://www.factsage.com/.

[30] S. Sagadevan, B. Janarthanan, A review on influence of thermal studies of nanomaterials, Int. J. Mater. Sci. Appl. 3 (2014) 370–377.

[31] K.F. Hsu, S. Loo, F. Guo, W. Chen, J.S. Dyck, Cubic AgPb(m)SbTe(2+m): bulk thermoelectric materials with high figure of merit, Science 303 (2004) 818–821.

[32] M.S. Dresselhaus, G. Chen, M.Y. Tang, R.G. Yang, R.H. Lee, D.Z. Wang, Z.F. Ren, J.P. Fleurial, P. Gogna, New directions for low-dimensional thermoelectric materials, Adv. Mater. 19 (2007) 1043–1053.

[33] B. Poudel, Q. Hao, Y. Ma, Y. Lan, A. Minnich, High thermoelectric performance of nanostructured bismuth antimony telluride bulk alloys, Science 320 (2008) 634–638.

[34] J.P. Heremans, V. Jovovic, E.S. Toberer, A. Saramat, K. Kurosaki, A. Charoenphakdee, S. Yamanaka, G.J. Snyder, Enhancement of thermoelectric efficiency in PbTe by distortion of the electronic density of states, Science 321 (2008) 554–557.

[35] T.M. Tritt, M.A. Subramanian, Thermoelectric materials, phenomena, and applications: a bird's eye view, MRS Bull. 31 (2006) 188–198.

[36] A. Minnich, M.S. Dresselhaus, Z.F. Ren, G. Chen, Bulk nanostructured thermoelectric materials: current research and future prospects, Energy Environ. Sci. 2 (2009) 466–479.

[37] H.J. Goldsmid, Thermoelectric Refrigeration, Plenum, New York, 1964.

[38] D.M. Rowe, Thermoelectrics Handbook Nano to Macro, CRC Taylor & Francis, Boca Raton, 2006.

[39] U. Nithiyanantham, M.F. Ozaydin, A.S. Tazebay, S. Kundu, Low temperature formation of rectangular PbTe nanocrystals and their thermoelectric properties, New J. Chem. 40 (2016) 265–277.

[40] W. Brostow, T. Datashvili, H.E.H. Lobland, T. Hilbig, L. Su, C. Vinado, J. White, Bismuth telluride-based thermoelectric materials: coatings as protection against thermal cycling effects, J. Mater. Res. 27 (2012) 2930–2936.

[41] (a) TA Instruments, Inc., Briefing number TA-131A, Detection of beta transus in titanium alloys by differential thermal analysis (DTA), thermal analysis & rheology a subsidiary of waters corporation, in: R.I. Jaffie, N.E. Promisel (Eds.), The Science, Technology and Applications of Titanium, Pergammon Press, New York, 1966. http://www.tainst.com. (b) W. Su, L. Gao, L. Wang, H. Zhi, Calibration of differential scanning calorimeter (DSC) for thermal properties analysis of phase change material, J. Therm. Anal. Calorim. (2020), https://doi.org/10.1007/s10973-020-09470-9.

[42] J.T. Carstensen, Solid-state Chemistry of Drugs. By Stephen R. Byrn. Academic Press, 111 Fifth Avenue, New York, NY 10003. 1982. pp. 1–346, Journal of Pharmaceutical Sciences 73 (1984) 573.

[43] E.M. Ebeid, S.E. Morsi, J.O. Williams, Calorimetric, spectroscopic and structural studies of the monomerization of crystalline 9-cyanoanthracene photodimer (9 CNAD), J. Chem. Soc. Faraday Trans. 1: Phys. Chem. Condensed Phases 76 (1980) 2170–2178.

[44] E.M. Ebeid, A.M. Donia, Structural and photochemical transformations in anthracenes, J. Phys. Chem. Solids 49 (1988) 263–266.

[45] H. Feilchenfeld, S. Sarig, Calcium chloride hexahydrate: a phase-changing material for energy storage, Ind. Eng. Chem. Prod. Res. Dev. 24 (1985) 130–133.

[46] E.M. Ebeid, M.M. Abou-Sekkina, Thermophysical and photochemical behavior of 2,5-distyrylpyrazine (DSP) and azodiisobutyronitrile (AIBN) crystals, Acta Chim. Hung. 124 (1987) 589–594.

[47] A.B. Jaffe, K.J. Skinner, J.M. McBride, Solvent steric effects. II. Free-radical chemistry of azobisisobutyronitrile and azobis-3-cyano-3-pentane in viscous and crystalline media, J. Am. Chem. Soc. 94 (1972) 8510–8515.

[48] J. Bauer, S. Sanpton, R. Henry, J. Quick, W. Dziki, W. Porter, J. Morris, Ritonavir: an extraordinary example of conformational polymorphism, Pharm. Res. 18 (2001) 859–866.

[49] S.L. Morissette, S. Soukasene, D. Levinson, M.J. Cima, O. Almarsson, Elucidation of crystal form diversity of the HIV protease inhibitor ritonavir by high-throughput crystallization, Proc. Natl. Acad. Sci. U. S. A. 100 (2003) 2180–2184.

Chapter 6

Thermal analysis in recycling and waste management

Waste management (or waste disposal) includes the activities and actions required to manage waste from its inception to its final disposal [1]. This includes the collection, transport, treatment, and disposal of waste, together with monitoring and regulation of the waste management process. Exploiting waste as a profitable resource would be a good way to resolve simultaneously several challenges and this inspires the aphorism that one's trash is another man's treasure. For instance, waste recycling could become the major source of metals in the second half of the 21st century [2]. Waste management methods include landfilling, incineration or combustion, recovery and recycling, composting, plasma gasification, and waste to energy conversion. There are certain waste types that are considered as hazardous and cannot be disposed of without special handling such as nuclear waste and biomedical waste that is found in health-care facilities and similar institutions.

TA techniques are frequently used to assess the optimum conditions in the areas of waste management and recycling. In this chapter, we discuss some of these applications with special reference to assess the identity of the mixed recycled plastics using DSC and TGA as a research method in the transformation processes of waste rubber and plastic products for energy carriers: waste-to-liquid (WtL) and waste-to-energy (WtE) processes, TGA characteristics and kinetics of scrap tire-biomass copyrolysis, capture of organic iodides from nuclear waste, Sandia Octahedral Molecular Sieves (SOMS) and their related perovskites in nuclear waste management, nuclear graphite using TGA, thermal characteristics of spent activated carbon (SAC) generated from air cleaning units (ACUs), cold-cap reactions in vitrification of nuclear waste glass, TA in vitrification of radioactive spent ion-exchange resins (SIERs) in borosilicate glass and TA in recycling electronic waste, and finally in waste electronic recycling.

1 Identity of the mixed recycled plastics using DSC

Plastics, which are made of hydrocarbons, are slowly biodegradable materials with polyethylene (PE) and polypropylene (PP) are the most used polymers in

Thermal Analysis. https://doi.org/10.1016/B978-0-323-90191-8.00002-6

our daily lives and in the industry. Products made from mixed recycled plastics with random volume ratios of PE and PP suffer from nonspecific weathering, thermal, mechanical and ageing properties, making them difficult to handle and finally increasing their production cost for the molding factories. Unfortunately, waste plastics floating in the world's oceans have entered the ocean's food cycle [3]. This turns our attention to the recycled plastic products. As a result, determining the concentration of each polymer in a mixed recycled plastic is of paramount importance and would contribute to the production of more efficient and low-cost recycled materials. Among other characterization techniques, the use of DSC has been proved an effective, reliable, and fast method for the identification and quantification of polymers such as PE and PP in waste and recycled plastics.

In his report, Rachellowe carried out studies revealing the identity of the mixed recycled plastics using DSC [4]. Plastics yield characteristic-unique melting, crystallization, and glass transition peaks in typical DSC dynamic or isothermal modes. The area below the melting curve (ΔH) is primarily used to determine the purity and the degree of crystallinity of the materials. In the case of polymer blends, the melting temperatures do not change with the percentage of the polymers in the mixed plastic sample. In addition, ΔH varies linearly with the percentage of the polymers in the blend. These would ultimately turn ΔH into an excellent tool for quantifying the polymers in the mixture.

A series of high-density polyethylene (HDPE):PP plastics were prepared and characterized. Plastic samples prepared from virgin resins of HDPE and PP were mixed to produce volume ratios of HDPE:PP 100:0, 70:30, 50:50, 30:70, and 0:100. The testing parameters included heating, cooling, and heating from −30°C to 240°C with heating rate 20.0K/min and nitrogen gas flow 40.0mL/min. Fig. 6.1 presents the melting peaks in the heat flow vs temperature DSC curves of HDPE:PP samples with various volume ratios. The HDPE:PP 100:0 and 0:100 samples showed one melting peak at approximately 136°C and 163°C, which compare well with the characteristic melting temperatures (T_m) of the HDPE and PP, accordingly. The HDPE:PP 70:30, 50:50, and 30:70 samples showed two melting peaks, which correspond to the HDPE and PP parts in the blend. The enthalpies (ΔH) below the melting peaks and the height of the melting peaks (showed with arrows in Fig. 6.1) vary with the percentage of the polymers in the blend.

Fig. 6.2A and B presents the plots of the melting enthalpies vs the %HDPE and PP, accordingly. Analysis of the melting enthalpies revealed that ΔH increases linearly (linear factors, $R^2 > 0.9$) with the concentration of the polymers in the plastics. Fig. 6.2C and D shows the relation between the melting peak height and the concentration of the polymers. The height of the melting peaks increases with the %HDPE/%PP exhibiting a much linear behavior ($R^2 > 0.97$) compared with the ΔH plots. The linear behavior of ΔH and melting peak heights vs polymer percentage plots offers an unprecedented technique for quantifying the polymer blends in the recycled plastics. The melting peak exhibited a nearly perfect linear increase with the increase of the polymer parts.

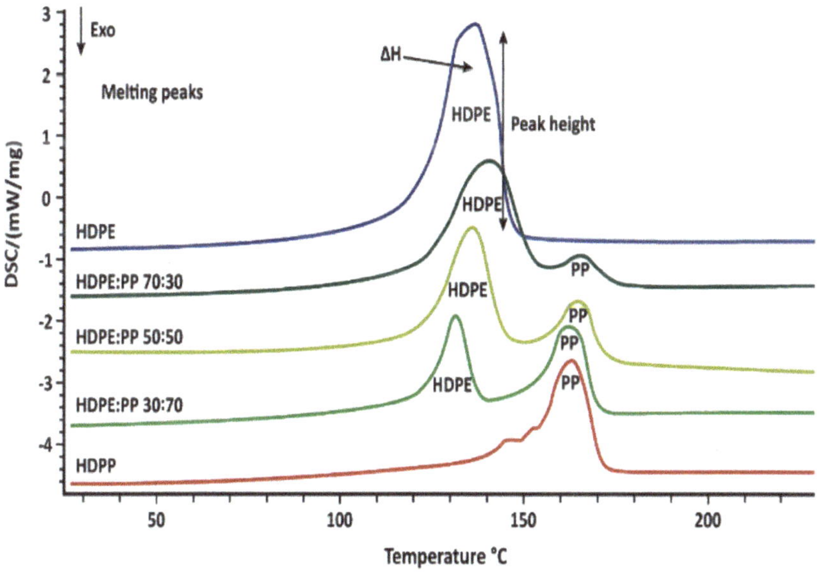

FIG. 6.1 DSC melting temperature peaks of the HDPE:PP 100:0, 70:30, 50:50, 30:70, and 0:100 mixed plastics. The areas below the melting peaks correspond to the melting enthalpies of each polymer in the plastic.

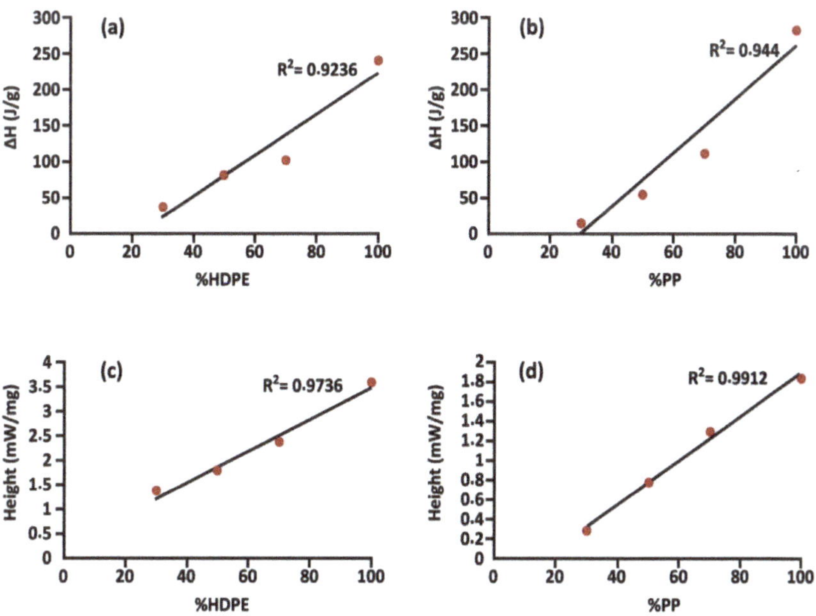

FIG. 6.2 (A, B) Melting enthalpy vs %HDPE and %PP, and (C, D) melting peak height vs % HDPE and %PP.

2 TGA in the transformation processes of waste rubber and plastic products for energy carriers

Nowadays there are three ways to utilize plastic and tire rubber waste: landfilling, incineration with or without energy recovery, and recycling. The largest amount of wastes is disposed of by landfilling (40%), followed by recycling about 39%, and incineration with energy recovery is only 20%. Traditionally the destruction of wastes via incineration and landfill deposition was regarded as the preferred treatment options. Incineration is a waste treatment process that involves the combustion of organic substances contained in waste materials [5].

Landfills emit by-products such as methane, dioxins, and leachate (a toxic liquid that is formed when waste breaks down in the landfill and filters through waste), which, when left untreated, can leach into the soil, contaminating water sources, plants, and even food. Suitable waste management is an important aspect of sustainable development. WtE technologies that process nonrenewable waste can reduce environmental and health damages, all the while generating sustainable energy. Plastic and rubber wastes represent a considerable part of municipal wastes, furthermore huge amounts of waste arise as a by-product or faulty product in industry and agriculture [6].

The concept "alternative fuels" was introduced as a means of using waste materials of industrial origin and household usage. In their study, Rostek and Biernat described the basic problems regarding the possibility of using waste plastics and tire rubber waste as raw materials for the preparation of liquid energy carriers for transport (WtL processes) or for direct energy use in a WtE process. The study was conducted for waste samples of different density, such as polyethylenes, polyethylene terephthalate, and rubber from waste tires. Gasification processes were studied using TG and DSC techniques to determine the kinetics of thermal decomposition in an atmosphere of argon/nitrogen as nonoxidizing atmospheres [7]. Heating runs showing melting behavior of ethylene-propylenediene rubber (EPDM) and polypropylene (PP) were reported by Schawe [8] at elevated temperatures, however, samples decomposition occurs.

It was found that the degradation is a suitable technique for converting waste polymers into liquid hydrocarbons. The volatile products could be used as feedstock components, e.g., in refineries. Their further utilization for petrochemical purposes has not yet been solved. One possibility is a fuel-like application or mixing in fuels as a blending component. Before blending the high-olefin content should be saturated with hydrogen, or hydro isomerized. These steps result in a high-quality synthetic diesel fuel, with high cetane number that refers to the ease with which diesel fuel ignites easily at a relatively low temperature. The product is theoretically free from sulfur, nitrogen, and metals. In practice, these fractions generally have a very low heteroatom content because most polymers contain dopes, e.g., sulfur containing antiflame or antioxidant additives, etc.

TGA combined with DSC was applied for the identification of thermal changes that proceed in examined specimen at elevated temperatures.

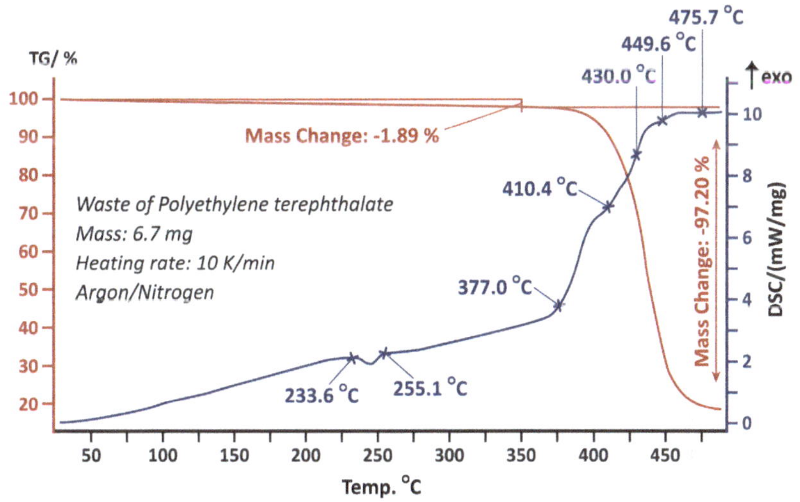

FIG. 6.3 TGA and DSC curves of PET sample conducted in a nitrogen atmosphere.

The decomposition products were identified in online mode. The kinetics of the thermal decomposition process were also determined for different atmospheres, with or without the participation of catalyst. Therefore, these methods not only used for the determination of technological parameters of thermal decomposition or the pyrolysis but may also be useful to estimate the effectiveness of gasification process. The significant feature of TGA processes is the possibility of precise definition of the enthalpy value and its sign for the exactly defined temperature range. Fig. 6.3 shows TGA and DSC curves for PET conducted in an atmosphere of nitrogen at elevated temperatures.

As a result of the process of thermal decomposition of PET samples, liquefaction of the material occurs in the temperature range from 233°C to 255°C. Two stages of exothermic nature are observed. The first occurs in the temperature range from 377°C to 410°C, and the other in the range 430–475°C. Initial thermal decomposition can be broadly seen as a two-stage thermal degradation process consisting of primary decomposition and secondary reactions. During the primary steps of decomposition, weak aliphatic bonds (nonaromatics) are broken. The fragments that do not condense at room temperature undergo further decomposition producing CO_2, light aliphatic gases (C_2–C_5), and H_2O. The secondary phase of pyrolytic thermal decomposition results in CH_4, CO, H_2, and light volatile nitrogenous species (NH_3, HCN, and H*i*NCO). Volatile nitrogen is expected to be present in all the waste samples [9–11].

Fig. 6.4 shows TGA and DSC curves for a granulated car radial tire conducted in a nitrogen atmosphere. The rubber sample was derived from waste tires. Thermal decomposition of rubber derived from waste tires reveals an initial plasticity of the native material, and then the gradual gasification staring at

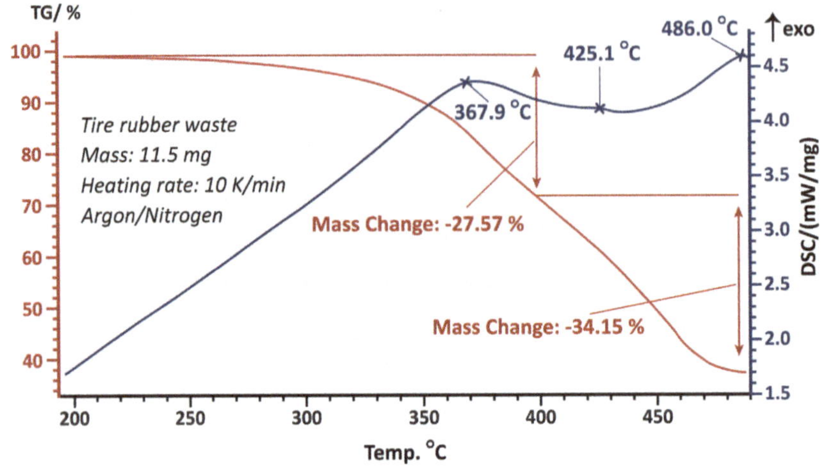

FIG. 6.4 TGA and DSC curves of granulated car radial tire samples conducted in a nitrogen atmosphere.

about 270°C. The endothermic peaks in the temperature range 368–486°C were assigned to the production of the gas phase such as carbon monoxide, carbon dioxide, and hydrogen sulfide (sulfone).

Fig. 6.5 shows DSC curves of HDPE and linear low-density polyethylene samples (LLDPE). The data is similar to earlier results [12]. The course of these curves confirms the radical nature of the distribution, consisting of a symmetrical break polymer chains, leading to the creation of two new radicals. For HDPE and LLDPE, the breakdown of polymer chains begins at 431°C and 428°C, respectively. For the temperatures over 428°C, LLDPE shows a small

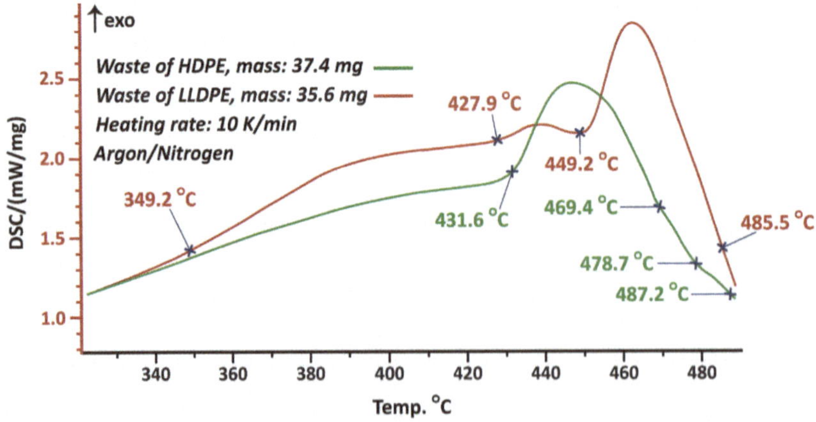

FIG. 6.5 DSC curves of polyethylene samples conducted in a nitrogen atmosphere.

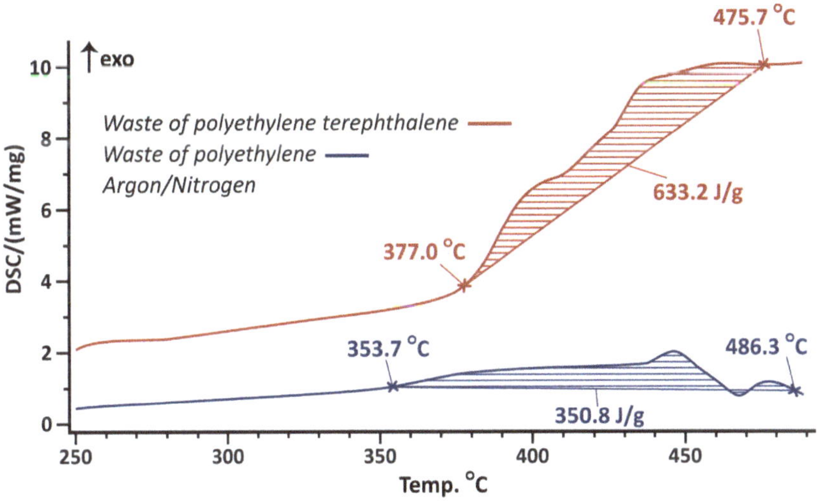

FIG. 6.6 Enthalpy changes in the processes of thermal decomposition of samples of waste from PET and PE.

exothermic peak, which may result from intramolecular or intermolecular hydrogen transport. An interpretation of the DSC curves was proposed following the identification of gaseous decomposition products.

The values of enthalpy of the exothermic processes of decomposition of waste polyethylene terephthalate and polyethylene are shown in Fig. 6.6. Enthalpy value for the degradation of PET ($\Delta H = 633.2\,J/g$) is almost two times higher than the enthalpy for the decomposition process of sample PE of $\Delta H = 350.8\,J/g$.

Fig. 6.7 shows the process of thermal decomposition of the sample of rubber derived from waste car tires. Meanwhile, the distribution of the abovementioned samples of rubber endothermic process was observed in the temperature range from 368°C to 486.0°C. The value of the enthalpy of the process is $\Delta H = -159.0\,J/g$. A negative value of enthalpy can be related to the energy needs associated with the formation of particulate methane assets.

TGA curves thus help understand the mechanisms of thermal decomposition of waste with the determination of the actual energy requirements in these processes. This permits the development of optimal manufacturing processes to enable energy use of waste polymer and rubber, both in WtE and WtL processes.

The carbon black content of an elastomer sample is often determined by heating it to 600°C in an inert atmosphere (usually nitrogen) since pyrolysis of the polymer is usually completed at this temperature. Carbon black was determined in a chloroprene rubber (CR) sample [8].

In TGA, specific gas changes are used to investigate particular constituents in a material. Fig. 6.8 displays TGA curves of an elastomer in which the carbon

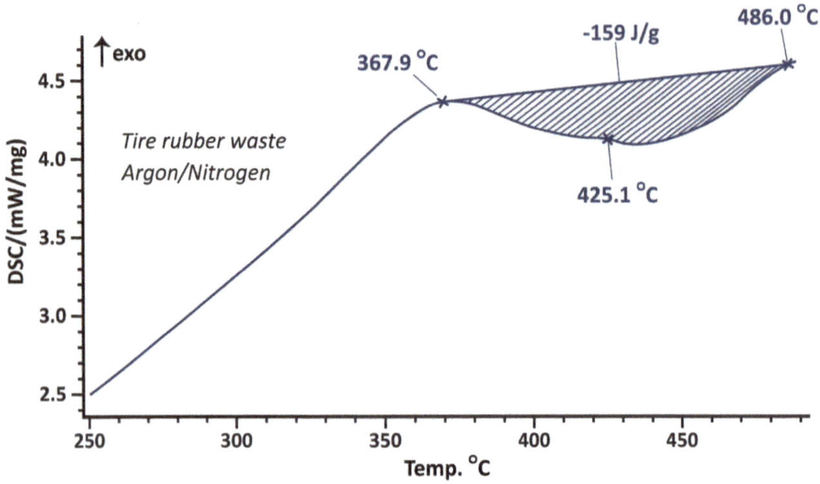

FIG. 6.7 Enthalpy change in the process of thermal decomposition of tire rubber waste.

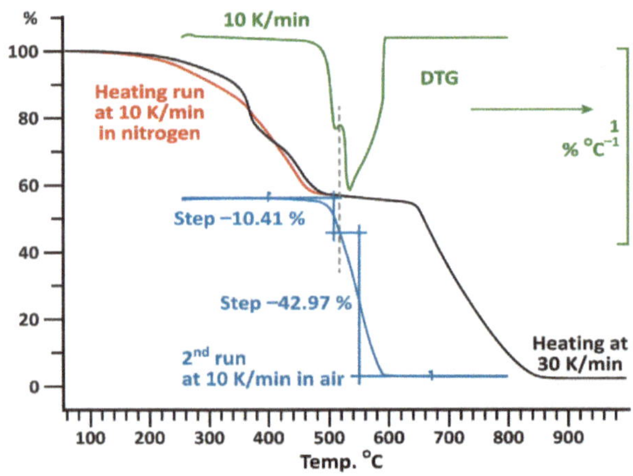

FIG. 6.8 Carbon black determination of a chloroprene rubber (CR) sample. Black curve: a heating run with gas switching at 600°C. Red curve: pyrolysis in nitrogen up to 650°C. Blue and green curves: TGA and DTG curves of the heating measurement in air.

black is a pyrolysis product. The gas is then switched, and the sample is heated to higher temperatures in an oxidative atmosphere (oxygen or air) whereby the carbon black additive burns. The carbon black content of the sample is determined from the step height. If, however, carbon black is also a pyrolysis product, the method cannot distinguish between the two types of carbon black, that is, the additive and the carbon black formed during pyrolysis. This is the case for chlorine-containing polymers (see black curve in Fig. 6.8) [8].

Separation of the two types of carbon black is however often possible if the sample is cooled in an inert atmosphere after the polymer has pyrolyzed (red curve). The atmosphere is then switched to air before starting the second heating run (lower blue curve). Under these conditions, both types of carbon black burn more slowly and at lower temperatures. Due to its larger surface area, the carbon black formed during pyrolysis burns more rapidly before the carbon black additive. The two steps can be clearly seen in the DTG curve (green curve, the DTG curve is the first derivative of the TGA curve). This allows the two types of carbon black to be separated.

Fig. 6.8 shows a measurement curve of a chloroprene rubber (CR) with a carbon black additive content of 43%. In the second heating run (blue and green curves) performed in air after pyrolysis of the polymer, the 10% pyrolysis carbon black burns first at a lower temperature. Variations of atmosphere include experiments in which the relative humidity of the air (or the sample) are set to defined values or changed. Generally, experiments of this type are performed in connection with TGA or DMA [13–15].

3 TGA characteristics and kinetics of scrap tire-biomass copyrolysis

Due to the increasing number of used tires, the valorization of this waste material could solve an important environmental problem [16]. Approximately 70%–85% of scrap tires are landfilled or stockpiled. Tires, on the other hand, cannot be remolded. They are made up of more than 100 different substances: rubber, fillers such as carbon black or silica gel, steel, sulfur, zinc oxide, and many other additives such as plasticizers. The most common rubbers used for tires are natural rubber (NR), styrene-butadiene rubber (SBR), and butadiene rubber (BR) or their blends. These components decompose at different temperature ranges and many researchers have emphasized that the thermal degradation behavior of scrap tires provide information about the type of rubber and its contents [17–19].

In the pyrolysis process, the organic volatile matter of tires is decomposed into low-molecular-weight products, which can be useful as fuels or chemicals. The inorganic components mainly from steel and the nonvolatile carbon black remain as 'solid residue,' relatively unchanged [20].

Synthetic rubber materials such as scrap tires have a high carbon content of about 70 wt% and a low oxygen content of about 17 wt%. Contrary to that, biomass has a low carbon content of 47–51 wt% and a high oxygen content of 42–46 wt%. Therefore, copyrolysis of scrap tire with biomass could balance the carbon and oxygen amount in the feedstock with a strong effect on the properties of degradation products. In copyrolysis, the yields and composition of the products strongly depend on the treatment method, processing conditions, type of biomass and synthetic polymers [21]. Copyrolysis of biomass and waste plastics constitutes a thermochemical conversion technology for high-grade biofuel production [22–24].

TABLE 6.1 Elemental analysis of *J. regia* shell and scrap tire.

Substance	C %	H %	N %	S %	O %
J. regia shell	47.50	6.39	0.46	-	47.65
Scrap tire	70.02	6.76	1.09	4.21	17.92

Different forms of biomasses have been studied in that respect including olive residue [25], spent coffee ground (SCG) [26], walnut shell [27–30], pine cone [24, 31, 32], waste wood chip [33, 34], cellulose [35], date palm [36], and corn stalks [37].

The elemental compositions were applied to calculate the calorific values by applying Du-Long's formula (Eq. 6.1). Weight fractions of carbon, hydrogen, and nitrogen were determined by elemental analysis for scrap tire and *Juglans regia* shell and is presented in Table 6.1:

$$\text{Calorific value} = 338.2\text{C} + 1442.8\left(\text{H} - \frac{\text{O}}{8}\right) + 94.2\text{S}\,(\text{kJ kg}) \tag{6.1}$$

where C, H, O, and S are the mass fractions of carbon, hydrogen, nitrogen, oxygen, and sulfur, respectively [38].

According to Uzun and Yaman [39], the calorific values of *J. regia* shell and scrap tire were calculated as 16.7 and 28.45 MJ kg^{-1}, respectively.

The kinetic analysis of pyrolysis processes has been the subject of many studies [39, 40]. Kinetic data are essential for designing any kind of device, in which the thermal decomposition takes place; on the other side, kinetics is intrinsically related to the decomposition mechanisms allowing the postulation of thermal decomposition mechanisms [39].

Uzun and Yaman investigated the degradation kinetics of *J. regia* shell, scrap tire, and their blends using TGA in nitrogen atmosphere in the range 293–973 K. Arrhenius and Coats-Redfern-based kinetic analysis methods have been applied to identify the pyrolysis reaction model [39]. According to their study, pyrolysis of *J. regia* shells shows three mass loss events assigned to loss of water, decomposition of both hemicelluloses and cellulose, and finally decomposition of lignin. The degradation curves of scrap tire showed one stage due to decomposition of SBR. The kinetic parameters were calculated using both Arrhenius and Coats-Redfern methods. By adopting the Arrhenius method, the average value of activation energies of *J. regia* shell, scrap tire, and their 1: 1 blend were found to be 69.22, 71.48, and 47.03 kJ mol^{-1}, respectively. By using the Coats-Redfern method, the corresponding average value of activation energies of *J. regia* shell, scrap tire, and their 1: 1 blend were determined as 99.85, 78.72, and 63.81 kJ mol^{-1}, respectively. The addition of *J. regia* shell to scrap tire caused a reduction in the activation energies. The difference of weight loss was measured to examine interactions between raw materials. The maximum difference between experimental and theoretical mass loss was 5% at about 648°C with a heating rate of 20°C min^{-1}. These results indicate a significant synergistic effect occurring during copyrolysis of *J. regia* shell and scrap tire in the high-temperature region.

3.1 Kinetic modeling

The decomposition rate of a solid is generally given as

$$\frac{dx}{dt} = k(T)f(x) \tag{6.2}$$

Conversion, x, is a normalized form of weight loss data of the decomposed sample and is defined as follows:

$$x = \frac{w_i - w_t}{w_i - w_f} \tag{6.3}$$

where w_i is the initial mass of the sample, w_t is the measured mass, and w_f is the mass after pyrolysis. Parameter $k(T)$ is defined as the rate constant of reaction whose temperature dependence is expressed by the following Arrhenius equation:

$$k(T) = Ae^{\frac{-E_a}{RT}} \tag{6.4}$$

where E_a is the activation energy (kJ mol^{-1}), T is the absolute temperature (K), R is the universal gas constant (8.314 J K^{-1} mol^{-1}), and A is the preexponential factor (min^{-1}) [34]. The expression of the function $f(x)$ and its derivative are used for describing the solid-state, first-order reaction; hence many authors restrict the mathematical function $f(x)$ to the following expression [36]:

$$f(x) = (1 - x)^n \tag{6.5}$$

For constant heating rate, $\beta = dT/dt$

$$\frac{dx}{dT} = \frac{A}{\beta} e^{-E_a/RT} (1 - x)^n \tag{6.6}$$

3.2 Arrhenius method

In the Arrhenius method, the final rate equation given below is obtained by linearization of Eq. (6.6) and making some rearrangements we obtain the following equation:

$$\ln\left(\frac{dx}{dT}\right) - n \ln(1 - x) = \ln\left(\frac{A}{\beta}\right) - \frac{E_a}{RT} \tag{6.7}$$

The plot of $\ln(dx/dT) - n\ln(1-x)$ vs. $(1/T)$ should give a straight line for the correct value of reaction order n. The activation energy and preexponential factors are calculated from slope $(-E_a/R)$ and intercept $(\ln(A/\beta))$, respectively [28].

3.3 Coats-Redfern method

The Coats-Redfern method is an integral method and involves the thermal degradation mechanism. Using an asymptotic approximation for the resolution of Eq. (6.6), we obtain

$$\ln\left[\frac{1-(1-x)}{T(1-n)}\right] = \ln\left[\frac{AR}{\beta E}\right] - \frac{E}{RT}(n \neq 1) \tag{6.8}$$

$$\ln\left[-\frac{\ln(1-x)}{T}\right] = \ln\left[\frac{AR}{\beta E}\right] - \frac{E}{RT}(n = 1) \tag{6.9}$$

Eqs. (6.8), (6.9) will result in a straight line with slope $-E_a/R$ and an intercept of $\ln[AR/\beta E_a]$. This was done by plotting a graph between the following terms [36]:

$$\ln\left[\frac{1-(1-x)^{1-n}}{T^2(1-n)}\right] \text{versus } 1/T\,(n \neq 1) \tag{6.10}$$

$$\ln\left[-\frac{\ln(1-x)}{T}\right] \text{versus } 1/T\,(n = 1) \tag{6.11}$$

4 Capture of organic iodides from nuclear waste

The number of studies on the capture of radioactive iodine compounds by porous sorbents has regained major importance in the last few years as it is related to operational safety and the treatment and safe disposal of generated radioactive waste. During nuclear accidents, such as that in Chernobyl (1986) and Fukushima (2011), gaseous radionuclides have been released in the off-gas stream. Among these released radionuclides are highly volatile and harmful long-lived ^{129}I, short-lived ^{131}I, and organic compounds such as methyl iodide (CH_3I). Immediate and effective means of capturing and storing these radionuclides are needed. Porous sorbents for the capture and storage of radioactive iodine compounds have been proposed.

The application of porous sorbents for the capture of radioactive iodine compounds has been recently reviewed by Huve et al. [41]. Porous sorbents have been in the forefront of the removal of radioactive contaminants due to their high removal efficiencies and adsorption capacities, high thermal stability, low maintenance cost, and the large range of structures and functionalizations that can be obtained [42–44]. One of the earliest compounds used for the treatment and capture of iodine compounds are triethylenediamine (TEDA) and potassium iodide (KI). To increase the iodine adsorption capacities of activated carbons, these materials are impregnated with TEDA with a concentration usually between 0 and 10 wt%. Silver-exchanged zeolites are also known candidates for the capture of iodine. These sorbents were highly praised for their good performance in iodine capture and were used in industrial applications for the trapping and treatment of iodine in the off-gas stream (in particular, activated carbons owing to their very low production cost). Activated carbons and zeolites were also investigated. In addition, new porous sorbents have been added to the list of potential sorbents for iodine trapping, including titanosilicates, mesoporous silicas, silver-impregnated silica and alumina ($Ag^0@SiO_2$ and $Ag^0@Al_2O_3$), silver-functionalized aerogels, chalcogels, macroreticular ion-exchange resins, metal- and covalent-organic frameworks (MOFs and COFs, respectively), and porous organic polymers (POPs) [45].

Among the MOFs known, MIL-101 emerged as one of the most promising porous materials. MIL-101 (MIL, Matérial Institut Lavoisier) with a chemical composition of $\{Cr_3F(H_2O)_2O(BDC)_3{\cdot}nH_2O\}$ ($n \sim 25$; 1,4-benzenedicarboxylate; BDC) and superior physicochemical properties emerged as potential adsorbent candidate [46]. However, effective capture of radioactive organic iodides from nuclear waste remained a significant challenge due to the drawbacks of current adsorbents such as low uptake capacity, high cost, and nonrecyclability.

In 2017, Li et al. [47] reported a general approach to overcome this challenge by creating radioactive organic iodide molecular traps through functionalization of metal-organic framework (MOF) materials with tertiary amine-binding sites. The molecular trap exhibited a high CH_3I saturation uptake capacity of 71 wt% at 150°C, which is more than 340% higher than the industrial adsorbent formed of metallic silver loaded on mordenite (Ag^0@MOR) under identical conditions. Mordenite's molecular structure is a framework containing chains of five-membered rings of linked silicate and aluminate tetrahedra (four oxygen atoms arranged at the points of a triangular pyramid about a central silicon or aluminum atom).

These functionalized MOFs also serve as good adsorbents at low temperatures. Furthermore, the resulting adsorbent can be recycled multiple times without loss of capacity, making recyclability a reality. In combination with its chemical and thermal stability, high capture efficiency, and low cost, the adsorbent demonstrates promise for industrial radioactive organic iodides capture from nuclear waste. The capture mechanism was investigated by experimental and theoretical methods. TGA (Fig. 6.9) shows that MIL-101-Cr-TED (triethylenediamine) and MIL-101-Cr-HMTA (hexamethylenetetramine) are stable up to 260°C and 220°C, respectively.

Isothermal TGA of amine grafted MIL-101-Cr samples clearly show that both TEDA and hexamethylenetetramine (HMTA) remain attached to the framework without losing mass upon prolonged heating at 150°C for 12 h (Fig. 6.10).

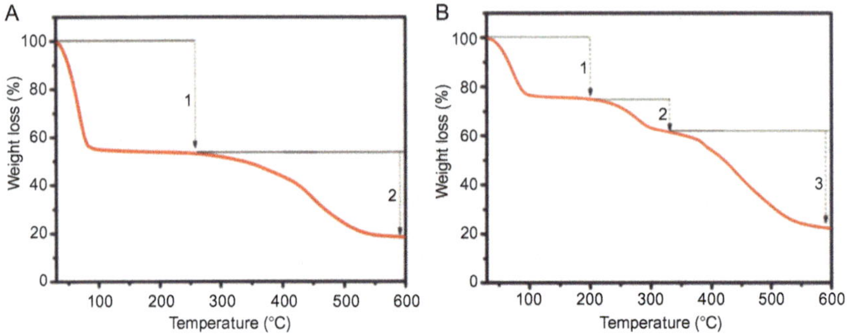

FIG. 6.9 The TG profiles of (A) MIL-101-Cr-TED and (B) MIL-101-Cr-HMTA.

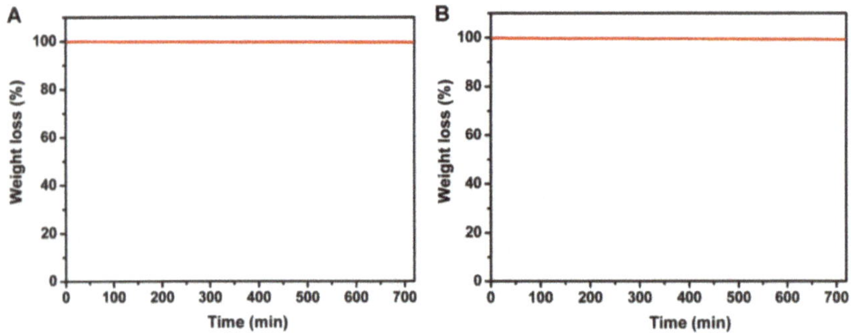

FIG. 6.10 The isothermal TG profiles of (A) MIL-101-Cr-TED and (B) MIL-101-Cr-HMTA at 150°C for 12h.

The high thermal stability of the two compounds enables their use at the elevated working temperature (such as 150°C) required for nuclear waste treatment. This result is consistent with the high binding energies of TED and HMTA to the open metal sites (OMSs) [47].

Compared to organic materials (e.g., macroreticular resins), silica-based substrates are generally more stable to radiolysis and can operate safely at elevated temperatures [48, 49]. Mnasri et al. [50] studied iodine uptake by mesoporous MCM-41 loaded by silver nanoparticles (AgNPs) using TGA/DTA together with other relevant techniques. It was demonstrated that the functionalized silica retained much gaseous iodine in an irreversible manner, mainly as an "interfacial" AgI. The AgNPs/MCM-41 system was recommended for the entrapment and immobilization of radioactive iodine in the nuclear industry since adsorbed radioactive iodine is primarily localized within the material pores and was thermally stable up to 800K in air. In their study, the thermal stability of the I_2-adsorbed phase in the C_{18}TAB templated silica material was investigated by TGA and DTA techniques. The corresponding TGA and DTA traces are presented in Fig. 6.11.

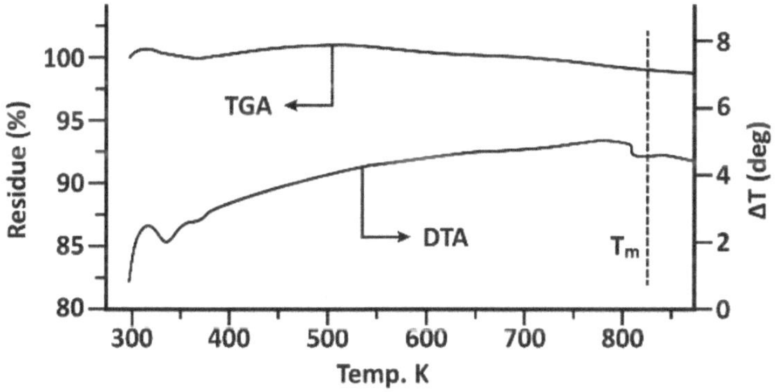

FIG. 6.11 TGA and DTA traces for the Ag-derivatized C_{18}TAB-templated silica after saturation with gaseous iodine. The vertical dashed line indicates the bulk melting temperature of AgI at 825 K.

Some mass change was observed before 373 K, giving rise to a small endothermic peak in the DTA curve that is ascribed to desorption of water vapor from the sample. In the temperature range between 373 and 800 K, little mass loss is observed in the TGA plot. This means that the adsorbed phase withstands heating up to 800 K in air. Another endothermic peak starting at about 806 K can be seen below the bulk melting temperature of AgI that is attributed to melting of AgI nanoparticles.

Two types of capture technology have been studied over the years: absorption of gaseous constituents by scrubbing into a liquid solvent and gas adsorption on an appropriate solid matrix [51]. The scrubbing technology, e.g., with an alkaline solution, generates a liquid residue which usually is to be transformed into a stable waste form suitable for disposal or eventual sequestration. This essentially means that the dissolved radioiodine should be converted to a form insoluble in ground water, e.g., Pb, Ag, and Hg iodides or hydrotalcite-like Bi-I-oxide compounds formed by iodide or iodate encapsulation in layered metal oxide-hydroxide anion scavengers [52].

The solid materials tested for the removal of gaseous iodine compounds by adsorption from the gas phase include macroreticular resins, silver-loaded zeolites, silica or alumina, and MOFs. The sorption technology offers some advantages over liquid scrubbing as being generally more simplistic in design and implementation, more efficient with respect to very low iodine concentrations in the off-gas and giving the possibility of sorbent regeneration.

5 Chalcogenide aerogels as sorbents for radioactive iodine

Chalcogens are the chemical elements in group 16 of the periodic table according to the modern IUPAC notation. This group is also known as the oxygen family. It consists of O, S, Se, Te, and the radioactive element polonium (Po). A chalcogel (or properly metal chalcogenide aerogel) is an aerogel made from chalcogens. An aerogel is a synthetic porous ultralight material derived from a gel, in which the liquid component for the gel has been replaced with a gas. The result is a solid with extremely low density.

In the context of finding a suitable matrix for capturing radioactive iodine, several sulfidic chalcogels were explored as iodine sorbents including $NiMoS_4$, $CoMoS_4$, $Sb_4Sn_3S_{12}$, $Zn_2Sn_2S_6$, and $K_{0.16}CoS_x$ ($x = 4-5$) by Subrahmanyam et al. [43].

The studied chalcogels showed high uptake, reaching up to 225 mass % (2.25 g/g) of the final mass owing to strong chemical and physical iodine-sulfide interactions. The analysis of the iodine-loaded specimens revealed that the iodine chemically reacted with $Sb_4Sn_3S_{12}$, $Zn_2Sn_2S_6$, and $K_{0.16}CoS_x$ to form the metal complexes SbI_3, SnI_4, and, KI, respectively. The $NiMoS_4$ and $CoMoS_4$ chalcogels did not appear to undergo a chemical reaction with iodine since iodide complexes were not observed with these samples. Once heated, the iodine-loaded chalcogels released iodine in the temperature range 75–220°C, depending on the nature of iodine speciation. In the case of $Sb_4Sn_3S_{12}$ and $Zn_2Sn_2S_6$, iodine release was observed around 150°C mainly in the form of SnI_4 and SbI_3, respectively. $NiMoS_4$, $CoMoS_4$, and $K_{0.16}CoS_x$ released elemental iodine at ~75°C, which is consistent with physisorption. Preliminary investigations on consolidation of iodine-loaded $Zn_2Sn_2S_6$ chalcogel with Sb_2S_3 as a glass forming additive produced glassy material whose iodine content was around 25 mass %.

TGA of iodine-loaded chalcogels was carried out to assess the nature of iodine binding and the thermal stability of chalcogels with and without iodine loading. Fig. 6.12 shows the TGA traces of both pristine and iodine-loaded samples of NiMoS, CoMoS, SbSnS, and ZnSnS. Among the pristine chalcogels, ZnSnS, and SbSnS were most stable with mass losses of ~10 mass % after heating to 600°C. The mass loss at the early stage in ZnSnS up to 200°C corresponds to the loss of solvent residue from the chalcogel synthesis process. The mass loss in both ZnSnS and SbSnS chalcogels from 200°C to 450°C is attributed to the removal of sulfur and thermal decomposition at ≥500°C.

The NiMoS and CoMoS chalcogels were less thermally stable relative to ZnSnS and SbSnS showing a loss of up to 20 mass %. Both chalcogels showed a sharp mass loss around 300°C followed by a gradual mass loss up to 600°C, which correspond to the thermal decomposition and loss of sulfur, respectively. Among the chalcogels used, KCoS was the least thermally stable and started decomposing even after 150°C with a total mass loss of about 40 mass % (Fig. 6.13). This behavior is consistent with the presence of polysulfide units in the structure which are prone to thermal decomposition.

Depending on the composition, the iodine-loaded chalcogels showed mass losses at various temperatures above 75°C. Iodine-loaded ZnSnS and SbSnS showed a mass drop at ~150°C, whereas iodine-loaded NiMoS, CoMoS, and KCoS started to lose mass starting at 75°C. The gradual mass losses at temperatures >200°C in iodine-loaded NiMoS, CoMoS, and KCoS follow a trend like

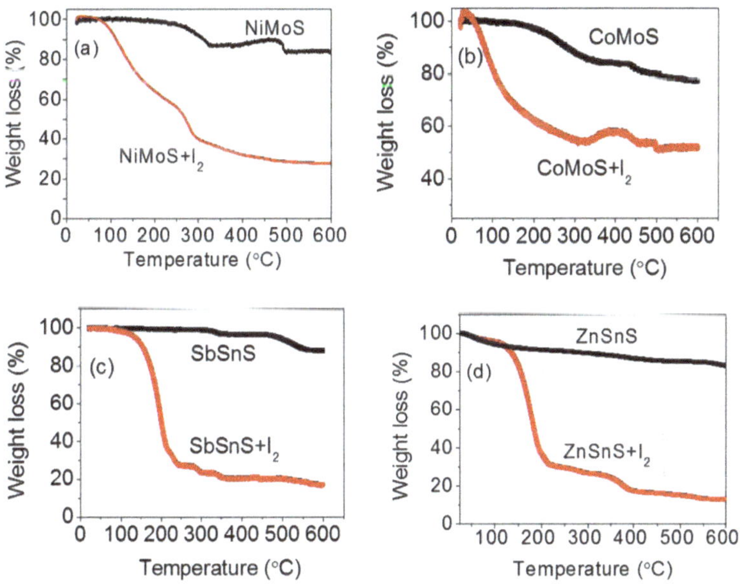

FIG. 6.12 TGA traces of pristine and iodine-loaded (A) NiMoS, (B) CoMoS, (C) SbSnS, and (D) ZnSnS.

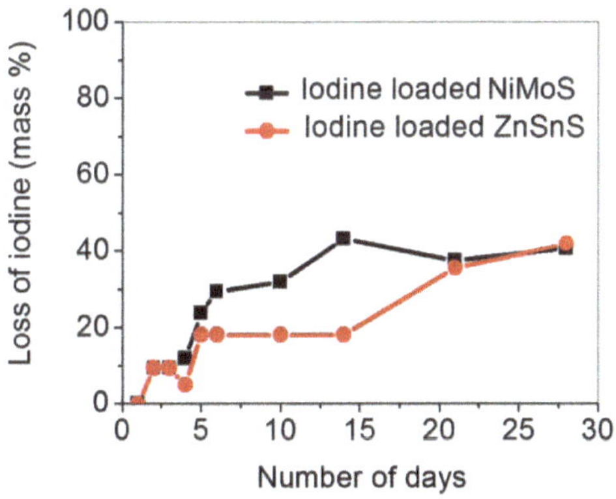

FIG. 6.13 Loss of iodine from iodine-loaded ZnSnS and NiMoS as a function of time.

that of their pristine counterparts. The mass losses at <200°C can be ascribed to the release of weakly physisorbed molecular iodine on the chalcogel surfaces as well as elemental sulfur. As the melting point of KI is above 600°C (~680°C), the iodine chemisorbed to form KI in KCoS was likely not released during the TGA experiment [43].

Iodine-loaded ZnSnS and SbSnS showed the major mass loss at ~150°C followed by a minor loss from 200°C to 400°C, which was not seen in the samples before iodine loading. The mass loss in the range 150–400°C in ZnSnS and SbSnS is attributed to the release of SnI_4 and SbI_3, respectively, as well as elemental sulfur. The amounts of iodine released during TGA in iodine-loaded NiMoS, CoMoS, and KCoS are lower than the total adsorbed iodine. As mass gain was measured after 2 days of completion of adsorption and the residence time was weeks, the observed discrepancy is due to escape of iodine vapor from the materials over time, which could be visually witnessed in the vials containing the materials as a yellowish hue on the interior of the vials [43].

After TGA, the Sn/S and Sn/Zn ratios in pristine ZnSnS were 2.6 and 1.3, respectively, whereas they decreased to 0.5 and 0.16 in iodine-loaded ZnSnS. This clearly confirms the volatilization of SnI_4 from the iodine loaded ZnSnS gel network upon heating. To probe the effect of chalcogel porosity on iodine uptake in addition to aerogels, iodine adsorption experiments on xerogels of NiMoS, CoMoS, SbSnS, and ZnSnS were carried out. The NiMoS and CoMoS xerogels showed very low mass uptake at ~10 mass % as revealed by mass differences. This is attributed to the significantly reduced fraction of exposed surfaces and pores in the xerogels as compared to the aerogel counterparts. Interestingly, the SbSnS and ZnSnS xerogels performed very well exhibiting up to 200 mass % uptake, like their aerogel counterparts. Since the I_2 absorption in these systems is due to a chemisorption redox process resulting in the formation of SbI_3 and SnI_4, the level of reactivity is the main factor, whereas porosity appears to be a minor factor [43].

The desorption of iodine from the chalcogel materials was visible in the vial, so it was monitored for ZnSnS and NiMoS as a function time. In these experiments, a new batch of ZnSnS and NiMoS chalcogels was used with specific surface areas of 302 and 213 $m^2 g^{-1}$, respectively. The iodine uptake of both samples was 160 mass % and was obtained by mass difference before and after iodine adsorption. The masses of both iodine-loaded samples were monitored over a period of 4 weeks at different time intervals. Both samples showed considerable mass loss during the first week but almost stabilized thereafter. The observed weight uptake values (160 mass %) were normalized to 100 mass %, and the mass loss from chalcogels as a function of time is depicted in Fig. 6.13. The observed total mass loss (40 mass %) is the same in both chalcogels over the period of study. The mass loss in NiMoS and ZnSnS can be attributed to sublimation of molecular iodine and SnI_4 from the respective chalcogels [43].

6 Sandia Octahedral Molecular Sieves and their related perovskites in nuclear waste management

Over 99% of the radioactivity in nuclear tank wastes is due to ^{137}Cs and ^{90}Sr that will be separated to concentrate and minimize final volumes of expensive high-level waste (HLW). Crystalline silicotitanate (CST), which was discovered by Sandia National Laboratories (SNL), Texas A&M, and Universal Oil Products

(UOP), is one of the best separation agents available for cesium [53]. It has a high selectivity for cesium over other alkali metals over a broad pH range. Additionally, it is stable in extreme radioactive and chemical environments [53].

These properties make the CST ion exchanger useful for the removal of ^{137}Cs from defense wastes. However, different from other ion exchanger technologies such as solvent extraction process, Cs is retained in CST and is nonelutable from CST. Su et al. [54] evaluated new metal niobate and silicotitanate ion exchangers for radioactive Cs and Sr nuclides removal. One class of the new phases $Na_2Nb_{2-x}M_x^{IV}O_{6-x}(OH)_x \cdot H_2O$ (M^{IV} = Ti, Zr, $x = 0.04 \approx 0.4$, SOMS) and their thermally converted Sr-loaded perovskites $Na_{2-x}Sr_xNb_{1.6}Ti_{0.4}O_{5.8+0.5x}$ ($0 \leq x \leq 0.4$) were synthesized and characterized. SOMS exhibits very high Sr selectivity with distribution coefficient K_d larger than 99,800 mL/g in the absence of competitive ions such as Na in solution, and about 10^3 mL/g when 0.1 M Na present in solution. This class of SOMS is easily converted to perovskites through low-temperature heat treatment (500–600°C). The thermally converted perovskites exhibit extremely low Sr leach rates, ranging from 2.4×10^{-7} to 1×10^{-6} g/m^2 day for 5% to 20% Sr loading over a 7-day leaching period. Fractional Sr release FR% is 0.001%–0.02% for 5%–20% Sr loading over the 7-day leaching period. These results indicate that SOMS could potentially be used for separating Sr from HLW and that their related perovskites could be used as potential ceramic waste forms [54].

The DTA-TGA analysis of SOMS-1 with 10% of Na^+ exchanged for Sr^{2+} is shown in Fig. 6.14. The weight loss between 100°C and 300°C corresponds to dehydration (calculated 7.6 wt%, observed 7.5 wt%) followed by structure change to a new crystalline phase, as observed by X-ray diffraction. The exothermic transition at 550°C is associated with conversion to a perovskite form. Perovskite (titanate-based) is a major component in the well-known SYNROC

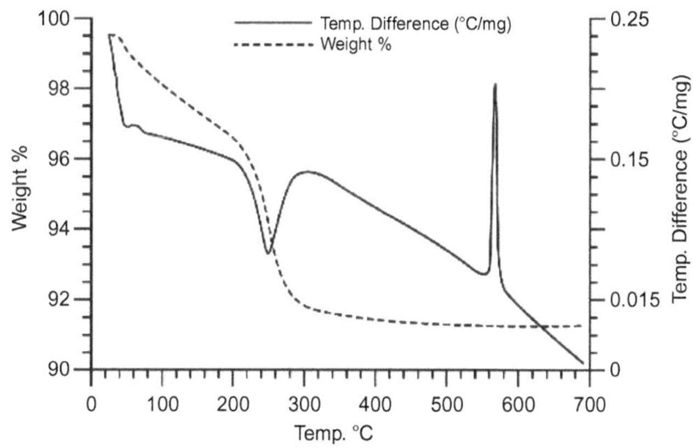

FIG. 6.14 DTA-TGA of Sr^{2+}-exchanged SOMS-1. Heating rate 10°C/min in argon atmosphere of flow rate 20 cm^3/min.

ceramic waste form for high-level radioactive waste (HLW) storage, and thus a reliable commodity for stability in radioactive fields and in repository conditions [55–58].

7 Study of nuclear graphite using TGA

The majority of graphite waste needing disposal because of decommissioning of the first generation of gas-cooled, graphite-moderated Magnox reactors is classified as intermediate level waste (ILW) [59]. It includes the long-lived radionuclide ^{14}C, which is a key radionuclide in the assessment of the safety of a geological disposal facility (GDF) for radioactive waste [60]. This key radionuclide is formed during the lifetime of the reactor as the graphite, which is used as a moderator and reflector material, is bombarded with thermal neutrons. Neutrons may be captured by ^{13}C, ^{14}N, and ^{17}O to form ^{14}C [61].

The thermal treatment of irradiated graphite has been studied previously aiming at achieving maximum ^{14}C release with minimum weight loss [62, 63]. Payne et al. studied the behavior of virgin Pile Grade A (PGA) graphite. PGA graphite is a pure petroleum coke graphite. Despite the irradiation conditions in the later advanced gas-cooled reactors (AGRs) precluded the use of PGA, the decommissioning of the first generation of gas-cooled, graphite-moderated Magnox reactors led to approximately huge amounts of graphite waste needing disposal [64].

Unlike fuels such as coal and charcoal, graphite itself is not readily combustible due to the lack of volatile hydrocarbons or oxygen [65]. However, when exposed to various gaseous environments at suitable elevated temperatures, this polycrystalline material will undergo thermal oxidation [66]. This behavior has been widely studied using various oxidizing gas species, different types of graphite, and different temperatures. Payne et al. [59] studied the oxidation behavior of this virgin PGA graphite in the temperature range 600–1150°C. Samples were run at various heating regimes in different gas environments as shown in Fig. 6.15.

Each run consisted of a dynamic heating stage followed by an isothermal stage and then a dynamic cooling stage. Of main interest was the stage where the sample was held at a constant temperature for a set period as the emerging data allowed for a rate of oxidation to be calculated. The oxidation rates were calculated during the 3600 s isothermal stage and calculated as a function of percentage mass loss per second because variations in initial masses between samples invalidated the use of mass units alone. The initial masses of the samples were measured using the XP5U, 5.0 g limit/0.1 μg resolution, balance (Mettler Toledo) incorporated in the TGA instrument. Samples were then heated at 10°C/min starting from room temperature (approximately 25°C) up to a final temperature of 600°C with air being introduced at 25 mL/min as a reactive gas. This was held for 3600 s with a continued flow of 25 mL/min of reactive gas. The samples were then cooled at −10°C/min until they returned

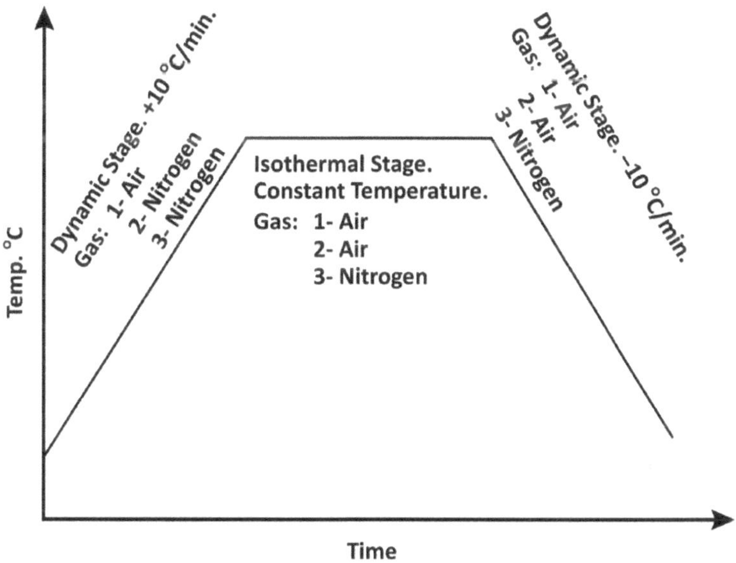

FIG. 6.15 The oxidation behavior of virgin Pile Grade A (PGA) graphite in the temperature range 600–1150°C at various heating regimes in different gas mixtures.

to room temperature. This was repeated for a fresh graphite sample four times each for the cubes and the discs, giving a total of eight analyses for each temperature and gas mixture. A new set of eight samples was then examined at a final temperature of 650°C, the final temperature was then increased in 50°C intervals for a fresh set of samples up to a maximum of 1050°C [59].

A control experiment following the same sequence was run on the cuboidal samples only using dry nitrogen instead of air (flow rate 25 mL/min) for the entirety of the experiment. Additionally, an intermediate experiment was performed by heating the graphite sample in a flow of dry nitrogen (flow rate 25 mL/min) and then switching the reactive gas to air once the final temperature had been achieved (flow rate 25 mL/min). Two larger disc samples were oxidized using the TGA instrument in several experimental runs in a flow of 25 mL/min air at 650°C and 900°C, these samples were weighed before and after using the balance. The duration of each run for each sample varied, due to the different oxidation rates, with each sample oxidized until it reached an approximate total weight loss of 18%, determined by recording the mass [59].

It was observed that with increasing oxidation temperature there was an associated increased loss of material from the outer geometrical surface of the graphite. This is in good agreement with the TGA results presented and the oxidation models discussed in previous literature. When oxidation is in the boundary-layer regime, it leads to material loss almost entirely from the external surface due to the diffusion of oxidizing species into the pore network being limited. Samples oxidized at 1050°C are characteristic of this phenomenon with the external surface being vastly different from both the virgin material and the samples oxidized at lower temperatures. Even though the internal pore structure is not visible, it is believed that oxidation occurred predominantly in the boundary-layer controlled regime as the sample lost approximately 92% of its original weight with the sample geometry changing significantly, indicating that the oxidation was from the external surface with material being lost progressively following a shrinking envelope regime [59].

At lower temperatures, 650°C, the chemical rate regime is anticipated as being the predominant method of oxidation. Samples oxidized at these low temperatures showed very little change in surface topography and size compared to virgin material however they still lost 6% of their initial mass suggesting that this weight loss had occurred evenly throughout the bulk of the material including the pore structure. Control by both boundary layer diffusion and chemical rate occurred at intermediate temperatures with samples oxidized at 850°C exhibiting characteristics of both regimes. With a total recorded weight loss at approximately 57%, the external surface of these samples displayed signs of oxidation. However, the sample geometry did not show such significant morphological change at higher temperatures suggesting that oxidation and associated weight loss occurred both on the surface and within the pore network, although not as deep into the pore network as at lower temperatures [59]. The oxidation of PGA graphite investigated with the use of TGA involves an oxidizing species (O_2, H_2O, CO_2) at these temperatures and will not proceed in an environment that does not contain such species, as demonstrated by the nitrogen-only control experiment [67]. When reacted with air, the predominant oxidizing species upon examining different types of graphite was found to be oxygen [66, 68–71]. Three regimes of oxidation of graphite are commonly discussed in the light of TGA results. At lower temperatures (<600°C) in the chemical rate-controlled regime, there is very little oxidation and the rate is slow. At temperatures exceeding approximately 900°C, the reaction rate is relatively fast and is boundary layer diffusion-controlled; with further increases in temperature there is minimal effect on the oxidation rate up to the final temperature studied (1050°C). Luo et al. [72] stated that in IG-11 (isostatic molded) graphite, there is a further increase in oxidation rate at higher temperatures. At temperatures between these two (600–900°C) is a thermal zone, known as the in-pore diffusion regime, where an increase in the temperature has a significant effect on the oxidation rate.

Electron micrographs and X-ray tomography taken of the PGA graphite correlate well with the TGA findings showing that at higher temperatures, the weight loss is quite significant and predominantly affects the external geometrical surface causing the sample to reduce in size considerably. At lower temperatures, there is very little evidence of oxidation on the surface and no change in sample geometry, however there is still appreciable weight loss of the samples, suggesting that this is occurring evenly throughout the graphite. For the in-pore diffusion regime a combination of the two regimes exists, giving oxidation on the surface and within the pore network.

The area of most interest for future work is the chemical rate-controlled regime, operating at temperatures <600°C, due to the relatively slow and uniform oxidation rate. This slow oxidation rate allows diffusion of gases throughout the graphite and deep into the pore network causing graphite oxidation not only in the external surface but also on the pore surfaces. This is critical for studying ^{14}C concentrations as this radioisotope may not be homogeneously distributed throughout the graphite [73]. Fachinger et al. [74] and Smith et al. [75] have both studied the possibility of using thermal treatment (selective oxidation) to selectively remove ^{14}C from the surface of irradiated graphite and even though this work was aimed at decontamination of graphite prior to disposal, it highlights that under suitable conditions ^{14}C can be removed and gases collected and subsequently analyzed.

8 Thermal oxidation of nuclear graphite

Nuclear waste management is a current problem facing many countries worldwide, with the disposal of significant amounts of irradiated graphite being prominent among these problems. The Magnox and the AGRs contain large amounts of nuclear grade graphite, principally in the core, where it is primarily used as a neutron moderator and as a reflector [76–78]. The global figure of irradiated graphite residing within reactors is believed to be >250,000 tons [79], all of which will require appropriate conditioning, treatment, and disposal. Graphite also contains a variety of radionuclides, some long-lived such as ^{14}C and ^{36}Cl [79], and others considered short-lived such as, ^{3}H and ^{60}Co [60]; there are many other radionuclides present in the graphite, generally formed via the neutron-induced activation of metallic impurities inherent to the origin and manufacture of the material [60].

Currently, the UK Nuclear Decommissioning Authority (NDA) baseline strategy is to suspend the abstraction of graphite cores, during a period of "care and maintenance" typically in the order of >85 years, after which the graphite will be removed and conditioned before being disposed of in a future GDF [80]. This approach, however, has several drawbacks and is likely to be both time consuming and expensive. Another possible approach for graphite could be large-scale gasification; under optimum conditions, graphite is known to readily oxidize to the gas phase, i.e., gasification, the topic of which has provided a wealth of literature [81, 82]. These authors have monitored the gasification process by measuring the oxidation-induced mass loss of samples in situ, TGA.

Theodosiou et al. investigated the laboratory-scale thermal oxidation of nuclear graphite as a proof of concept for the treatment and decommissioning of reactor cores on a larger industrial scale [83]. They based their study on the laboratory-scale gasification of nuclear grade graphite grades that were extensively used in the UK Magnox reactors as a neutron moderator. This graphite is known as PGA. PGA graphite bricks were used to construct the moderator cores for the Magnox fleet and therefore forms most of the UK graphite waste inventory currently ready for disposal. Further information on the origin and nature of PGA graphite was also reported [83–85].

Such treatment would lead to a very large volume reduction in irradiated graphite and provide an efficient, time reducing, and cost-effective decommissioning option that may decouple the need to store graphite in a GDF. However, the applicability of this method would depend largely on the available options to deal with the large amounts of gaseous ^{14}C being produced, such as potential aerial discharge (according to local regulations) or carbon-capture followed by liquefaction and storage.

The use of thermal treatment of such graphite waste is seen as advantageous since it will decouple the need for an operational GDF. Particulate samples of Magnox reactor PGA graphite were oxidized in both air and 60% O_2, over the temperature range 400–1200°C. Oxidation rates were found to increase with temperature, with a particular rise between 700°C and 800°C, suggesting a change in oxidation mechanism. A second increase in oxidation rate was observed between 1000°C and 1200°C and was found to correspond to a large increase in the CO/CO_2 ratio, as confirmed through gas analysis. Increasing the oxidant flow rate gave a linear increase in oxidation rate, up to a certain point, and maximum rates of 23.3 and 69.6 mg/min for air and 60% O_2, respectively, were achieved at a flow of 250 mL/min and temperature of 1000°C (Fig. 6.16).

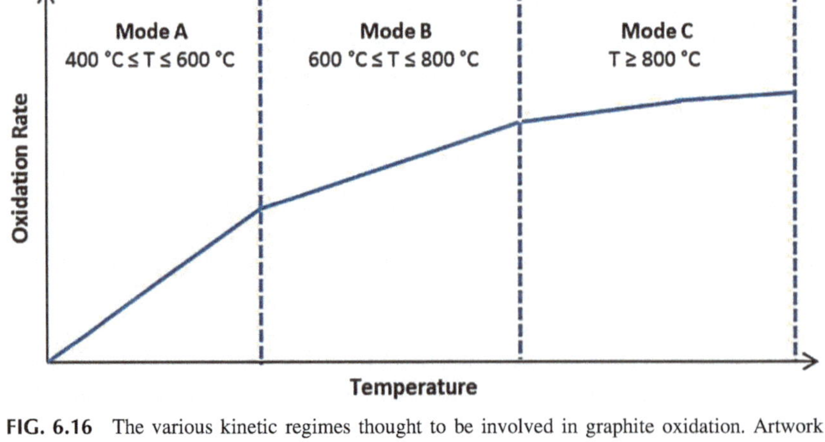

FIG. 6.16 The various kinetic regimes thought to be involved in graphite oxidation. Artwork inspired by Clark et al. [86]. The temperature ranges shown are an approximation and can vary depending on many factors.

These results show that large-scale thermal treatment could be a potential option for the decommissioning of graphite cores, although the design of the plant would need careful consideration in order to achieve optimum efficiency and throughput [76, 86].

Fig. 6.17 shows the CO and CO_2 release rates with increasing temperature. A significant increase in the amount of CO being produced at temperatures $\geq$1000°C is observed, reaching a maximum at 1200°C. In the case of air, CO/CO_2 $>$ 1 at $\approx$1150°C, reaching a maximum of 3.2 at 1200°C. The oxidation-induced weight loss of PGA both in air and 60% O_2 is shown in Fig. 6.18 and

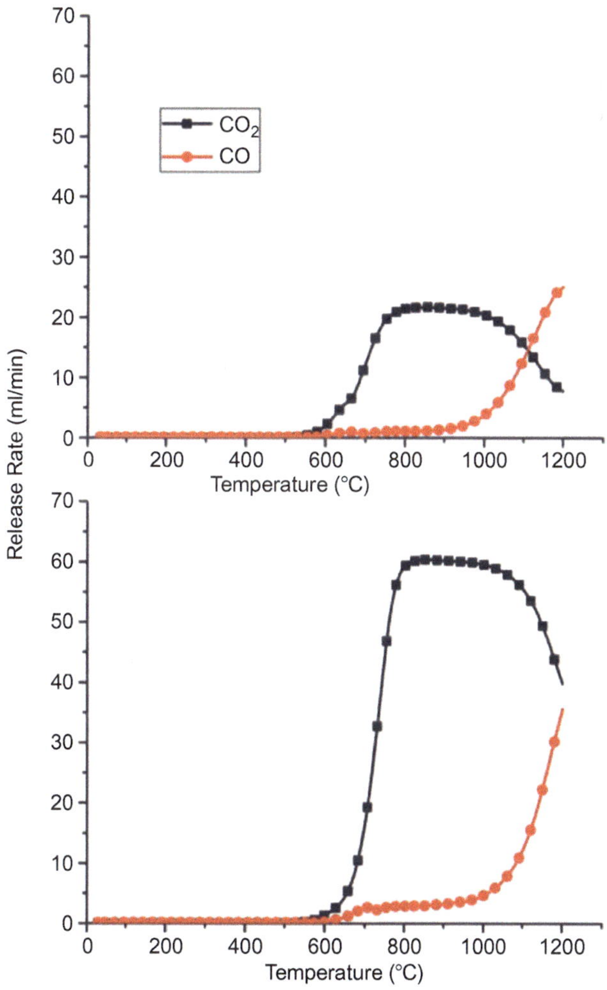

FIG. 6.17 The effect of temperature on the CO/CO_2 production ratio in air (top) and 60% O_2 (bottom) for PGA oxidation behavior.

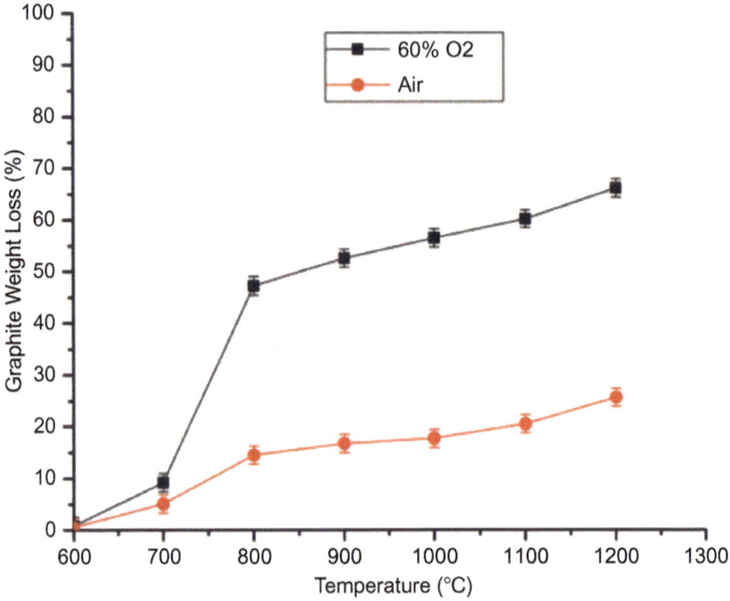

FIG. 6.18 Comparison of weight loss against temperature in air and 60% O_2 for PGA.

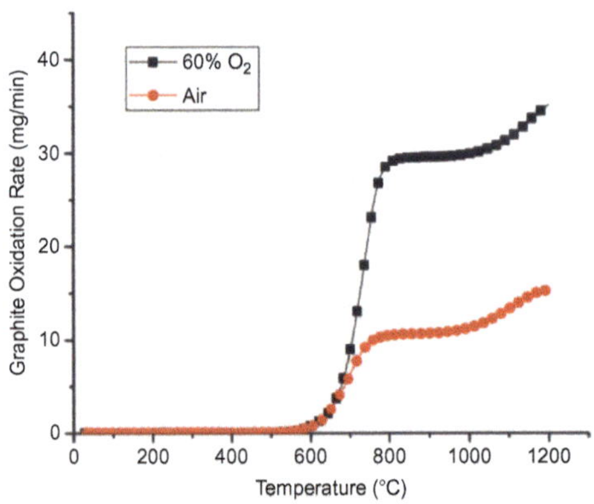

FIG. 6.19 The effect of increasing temperature on the graphite oxidation rate for both air and 60% O_2. The heating rate was set to 5°C/min and the oxidant flow rate was fixed at 100 mL/min.

the temperature dependence of graphite oxidation was studied over a wider temperature range of 30–1200°C [76]. The results are shown in Fig. 6.19.

Based on their study, Theodosiou et al. [76] concluded that thermal treatment of nuclear grade graphite is an effective method of gasification to CO/ CO_2 and therefore a potentially viable graphite waste management option, at industrial scale. Their work demonstrated that nuclear graphite particulate is readily oxidized in both air and 60% O_2 and the oxidation rate and, therefore, the throughput can be optimized through the control of several parameters which should be considered in the design of any potential treatment plant.

Primarily, the oxygen concentration will have the limiting effect on oxidation rate, with an almost linear relationship at suitably high temperature, e.g., 1000°C. However, from a commercial point of view any improvement in oxidation rate would need to be weighed against the additional cost of large volumes of oxygen-rich gas. Despite the lower oxidation rates air may still be the oxidant of choice, at pilot scale, due to the cost savings.

The flow rate of the oxidant also has a considerable effect, again exhibiting a linear relationship with oxidation rate at high temperature, although, it has been seen that this can deviate from linearity at high flow rates due to oxygen wastage. As a result, this would need to be minimized from a commercial viewpoint and the design of the pilot scale plant would need to incorporate methods to avert this.

Oxidation rates can also be optimized by increasing the process temperature. At $T \leq 700°C$, oxidation was found to proceed via the relatively slow diffusion-controlled mechanisms and at $T \leq 500°C$ the observed rates were considered negligible; hence such temperatures would not be suitable for an industrial scale process. A substantial increase in oxidation rate was observed at $T > 800°C$ due to the transition from a boundary layer diffusion-controlled mechanism. Maximum oxidation rates were achieved at $T > 1000°C$, due to a rapid and significant increase in the CO/CO_2 release ratio; this may be beneficial for a larger operation due to improved heat control arising from the CO production.

Fig. 6.20 shows Arrhenius plots showing the temperature dependence of graphite oxidation rates from 500°C to 1200°C. Data from $T < 500°C$ is omitted due to negligible oxidation rates. The plot provides the activation energy of the various kinetic regimes where the gradient is equal to $-E_a/R$ and the intercept is equal to the preexponential factor, $\ln A$, as described by the following equation [76]:

$$\ln k = -\frac{E_a}{R} \cdot \frac{1}{T} + \ln A \tag{6.12}$$

where k is the oxidation rate, T is the temperature in kelvin, and R is the universal gas constant. However, E_a values are tentative as the transition temperatures between different temperature regimes involve significant overlap and this empirical approach should be taken as a guideline only.

9 Thermal characteristics of spent activated carbon (SAC) generated from air cleaning units

SAC generated by the replacement of filters in air cleaning units (ACUs) in nuclear power plants (NPPs) is considered a nuclear waste since it contains ^{14}C as a problematic constituent. The ACUs in NPPs are classified into exhaust ACUs and makeup ACUs according to the function. Exhaust ACUs purify

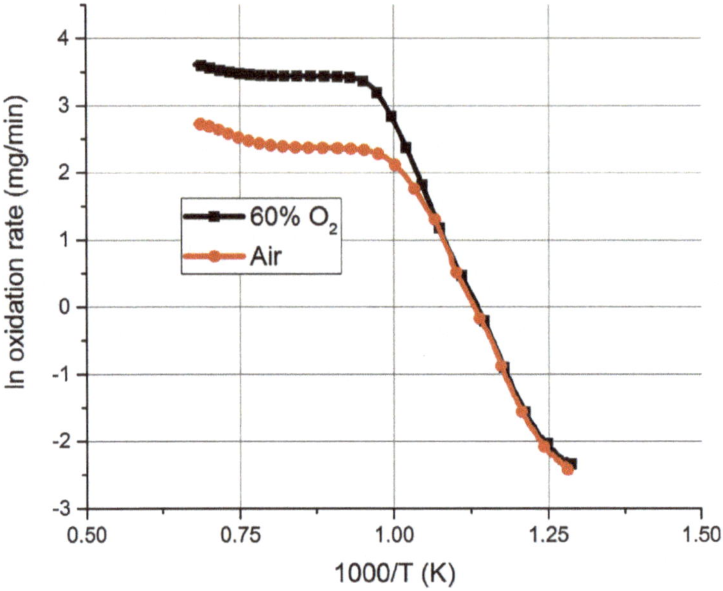

FIG. 6.20 Arrhenius plots showing the temperature dependence of graphite oxidation rates from 500°C to 1200°C. Data from $T < 500°C$ is omitted due to negligible oxidation rates.

radioactively contaminated air in the NPP buildings, while makeup ACUs provide the main control room and technical support room with fresh air by cleaning up outside air contaminated with radionuclides in emergency situations such as a severe accident. All these ACUs are equipped with activated carbon filters impregnated with TEDA. These units play a very important role in protecting power plant operators and populations from radiation exposure [87]. Air cleaning facilities in NPPs are equipped with high-efficiency particulate air (HEPA) filters and granular activated carbon filters. The former removes particulate radioactive air contaminants at the front end, while the latter remove remaining pollutants such as iodine compounds at the back end. SAC is generated by the replacement of filters according to performance degradation.

The radioactive waste classification and clearance level criteria (Nuclear Safety and Security Commission Notification No. 2014-3) limit the ^{14}C concentration for clearance to less than 1 Bq/g. However, the results of radioactivity analysis for the SAC from Korean NPPs, for example, showed that the ^{14}C concentration of about half of SAC exceeds this limit value [88].

One example of ^{14}C desorption techniques is a high-temperature treatment technology that heats the SAC to 700–900°C using a kerosene burner and separates it into activated carbon solid and $^{14}CO_2$ gas [89]. Another is a catalyst oxidation technology that grinds the SAC into granules of 1–1.5 mm diameter and heats it to 200–500°C under vacuum condition (10^{-2} to 10^{-5} Torr) in a nitrogen quantiler [90]. One chemical technology is also available that converts

^{14}C inside the SAC into inorganic $^{14}CO_2$ gas using strong acids, strong acidizing agents (4%–5% $K_2S_2O_8$) and catalysts (3%–5% $AgNO_3$) [91]. A recent ^{14}C removal technology of the Korean Atomic Energy Research Institute (KAERI) uses oxygen gas and maintains the ambient pressure at atmospheric level, unlike the old heating (up to 500°C) and decompressing techniques [92]. This technology uses microwaves to heat simply the absorbed organic materials to 100°C to vaporize them. The activated carbon waste is heated instantaneously indirectly or directly by microwaves, while the pressure is maintained at the normal atmospheric level to facilitate the carbonization of activated carbon waste absorbed with radioactive nuclides of various concentrations. Unlike the previously mentioned techniques, this technique focuses on the absorbed organic materials and carbonizes them without pretreatment such as grinding or chemical treatments. This is done to minimize the complexity of the process. The process is a closed system that enables desorption of only the radioactive materials vaporized under decompressed condition [93]. However, the patent owners' technologies only suggest measures for the SAC treatment process; they do not provide any field-proven results obtained using the abovementioned techniques. Moreover, a feasibility study of each technology for field application revealed that there are some difficulties due to interference with other systems in the NPP and the generation of secondary waste.

So and Cho reported the feasibility of disposing SAC as a clearance level waste [88]. They selected thermal treatment technology to remove ^{14}C and analyzed the moisture content and thermal characteristics using TGA in the temperature range from room temperature to 1000°C in a nitrogen quantiler. The results of TGA indicated that most of the SAC lost weight in three temperature ranges. Fig. 6.21 shows the differential TG (DTG) thermogram of one

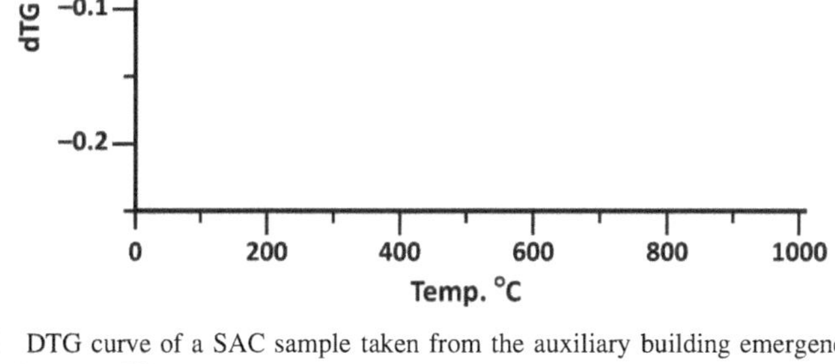

FIG. 6.21 DTG curve of a SAC sample taken from the auxiliary building emergency ACU in SHIN-KORI NPP 1 [88].

of the SAC samples with loss of weight in the three temperature ranges: room temperatures ~200°C, 200–500°C, and around 800°C. Pyrolysis-gas chromatography/mass spectrometry (Py-GC/MS) analysis was performed for the identification of gases generated in each temperature range. According to So and Cho assignment, the physically absorbed substances are desorbed under 200°C whereby moisture varies from 1.6 to 23.9 wt% according to SAC sample history. For the SAC sample shown in Fig. 6.21, the weight losses (in mg starting with 5 mg sample) in the temperature ranges 40–140, 140–520, and 520–1000 are, respectively, 0.2154, 0.60260, and 0.3771 mg. The chemical groups on the surface of the activated carbon are pyrolyzed at temperatures over 200°C with the desorption of organic and inorganic carbon compounds and TEDA. The gasification reaction terminates at 850°C. Therefore, So and Cho concluded that the gasification temperature and adequate moisture content are critical variables for ^{14}C desorption in each type of treatment for the three temperature ranges.

10 Cold-cap reactions in vitrification of nuclear waste glass

Research into reprocessing spent fuel from NPPs for civilian or military applications is actively going on by several international organizations. Glasses can integrate a wide variety of radionuclides (RN) because of their amorphous state and subsequent structural flexibility. Except the most volatile species, which are trapped in filters or released in the environment, vitrification provides a solid in which stable and radioactive elements are chemically bonded to the glass network formers. Vitrification is the transition from the liquid phase into an amorphous phase or glass, while avoiding the formation of crystalline state. Composition optimization has been the result of a compromise considering waste loading, technological feasibility, and the physical and chemical properties of glass in its liquid and solid states, as well as its containment properties [94]. In that respect, alkaline and alkaline-earth type borosilicate glasses, alumino-borosilicates or phosphates have been developed since the 1970s and are currently being used on an industrial scale (La Hague, Hanford, Savannah River, West Valley, Sellafield, Tokai, Rokkasho-Mura, Tarapur, Trombay, Karlsruhe, Mayak) [95].

Gin et al. reviewed three issues concerning radionuclides containment in nuclear glasses, namely, the RN incorporation in glass, the glass stability under self-irradiation, and the RN behavior during glass alteration by water [96].

Waste conditioning glasses are produced industrially in high-temperature melters, using glass frit as precursors and highly radioactive liquid or solid wastes containing fission products and actinides. These products are generally melted between 1000°C and 1200°C. Reaction between glass precursors and waste are mainly localized at the glass melt surface that is commonly called "cold cap" [96]. Cold cap is the floating layer of melter feed, or glass batch, on a pool of molten glass in a continuous electrical glass-melting furnace that is a melter.

In the cold-cap zone, a lower temperature driven by endothermic reactions between glass precursors can limit the volatilization. Consequently, within the cold cap and in the volume of the glass melt, the state (liquid, gas, solid, supercooled liquid) of the precursors and waste evolve with time and temperature until a high temperature homogeneous liquid is formed. The final glass package is then obtained after pouring the molten glass and cooling it into a metallic container. This reduces and manages the environmental risk since immobilized wastes in the form of glass constitute a long-term disposal [97].

Cold-cap reactions of a melter feed for vitrifying a high-alumina HLW were investigated by Xu et al. [94]. For vitrifying nuclear waste glass, the feed, a mixture of waste with glass-forming and modifying additives, is charged onto the cold cap that covers 90%–100% of the melt surface. As the feed moves through the cold cap, it undergoes chemical reactions and phase transitions through which it is converted to molten glass that moves from the cold cap into the melt pool. The process involves a series of reactions that generate multiple gases and subsequent mass loss and foaming, which significantly influence the mass and heat transfers. The rate of glass melting, which is greatly influenced by mass and heat transfers, affects the vitrification process and the efficiency of the immobilization of nuclear waste. The feed-to-glass conversion in the cold cap can then be summarized as follows (Fig. 6.22).

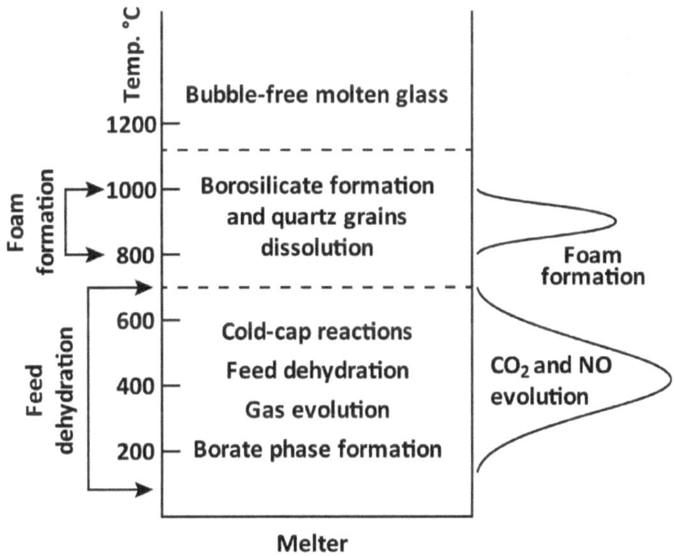

FIG. 6.22 Schematic diagram of feed-to-glass conversion in the cold cap. *(After K. Xu, P. Hrma, J. A. Rice, M.J. Schweiger, B.J. Riley, N.R. Overman, A.A. Kruger, Conversion of nuclear waste to molten glass: cold-cap reactions in crucible tests, J. Am. Ceram. Soc. 99 (2016) 2964–2971.)*

1. At 100°C $\leq T \leq$ 700°C, cold-cap reactions mainly consist of feed dehydration, gas evolution, and borate phase formation. The evolution of CO_2 and NO peaks at ~500°C, and ceases around 700°C. Amorphous, or unidentified, sodium borate phases form during feed preparation and subsequently dehydrate, melt, and/or form intermediate borate phases as the temperature increases to 700°C.
2. At 700°C $< T \leq$ 1100°C, the early glass-forming melt turns to borosilicate as quartz grains dissolve. The intermediate silicate phases form and then dissolve. As the melt connects above 800°C, it begins to expand and creates foam. Above ~1000°C, few solids remain, and foam starts to collapse into bubble-free molten glass.

During the different steps of the process, the RN can be integrated into the glass structure up to their incorporation limit. Beyond this limit, volatilization or heterogeneity formation processes may occur in the case of certain volatile RN (^{129}I, ^{36}Cl, ^{137}Cs, ^{99}Tc, ^{106}Ru) or for those with a low incorporation capacity (high oxidation state anions and cations). The waste loading must therefore be limited in certain cases to obtain a homogeneous glass at the microscopic scale. To contain larger quantities of waste, it is possible to increase the incorporation limits by using macroscopically homogeneous vitro-crystalline or vitro-ceramic matrices.

Gas-evolving cold-cap reactions release chemically bonded water and produce NO_x, O_2, and CO_x from reactions of nitrates with organics and reactions of nitrates, nitrites, and carbonates with solids. Understanding the cold-cap reactions over the temperature range of the conversion process helps formulate melter feeds for higher production rate, and hence an enhanced efficiency of the vitrification facility [98–109].

Fig. 6.23 shows TGA normalized weight and normalized weight derivative (DTG) of aluminum oxide (AO) feed heated at 10°C min^{-1} indicating release of various gases.

The TGA-GC-MS combination was used to perform qualitative and quantitative evolved gas analysis (EGA) for the cold-cap reactions [110–112]. Rodriguez et al. [113] carried out vitrification of nuclear waste glass. The authors studied the cold-cap reactions of a representative waste glass feed using DSC-TGA coupled with TGA-GC-MS as complementary tools to perform EGA. Analyses from DSC-TGA and EGA on the cold-cap reactions provide a key element for the development of an advanced cold-cap model. It also helps to formulate melter feeds for higher production rate.

Pokorný et al. [114] modeled the kinetics of the gas-evolving cold cap reactions using data from nonisothermal TGA. Their model describes the overall reaction rate as a sum of mutually independent *n*th-order reaction kinetics with the Arrhenius rate coefficients. For simplification, they neglected interactions between consecutive reactions and the complex responses of multicomponent molten salts and reactants. For the DSC-TGA study, the run/rerun method was utilized to separate the reaction heat from both the heat associated with the heat capacity of the feed and experimental artifacts. Calculating the degree of

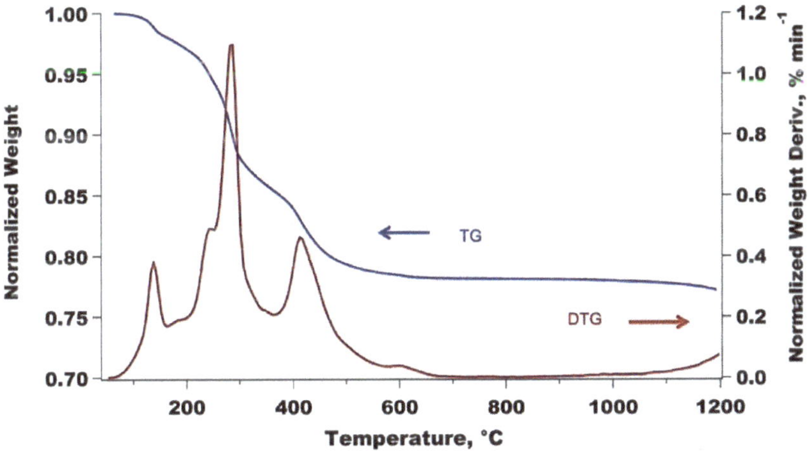

FIG. 6.23 Normalized TGA and DTG of aluminum oxide (AO) feed heated at 10°C min^{-1}. *(After C. Rodriguez, J. Chun, D. Pierce, M. Schweiger, P. Hrma, Kinetics of cold-cap reactions for vitrification of nuclear waste glass based on simultaneous differential scanning calorimetry-thermogravimetry (DSC-TGA) and evolved gas analysis (EGA), in: Prepared for the U.S. Department of Energy Assistant Secretary for Environmental Management, Published in Proceedings of WM March Conference, Phoenix, Arizona, USA, March 2–6, 2014.)*

conversion based on the reaction heat, an *n*th-order kinetic model was applied, where the kinetic parameters were obtained using the Kissinger method [115], combined with least-squares analysis. Furthermore, the heat capacity of the reacting feed from this analysis was estimated.

Rodriguez et al. [113] illustrated the correlation between the gases detected by the GC-MS and the mass loss rate from TGA to obtain a quantitative analysis of contributions of individual gases to mass losses associated with the feed-to-glass conversion and its correspondence to the DSC-TGA results.

Although details in cold-cap reactions are rather complicated, Rodriguez et al. [113] represented individual reactions with an *n*th-order kinetic model along with the Arrhenius rate coefficient [114]:

$$\frac{d\alpha_i}{dt} = A_i(1-\alpha_i)^{n_i}\exp\left(-\frac{E_i}{RT}\right) \tag{6.13}$$

where α_i is the *i*th reaction conversion degree, A_i is the *i*th reaction preexponential factor, E_i is the *i*th reaction activation energy, n_i is the *i*th reaction apparent order, T is the absolute temperature, and R is the gas constant. Assuming that the reactions are mutually independent, the overall reaction rate is represented as a weighted sum of the rates of individual reactions:

$$\frac{d\alpha}{dt} = w\sum_{i=1}^{N}\frac{d\alpha}{dt} = \sum_{i=1}^{N} wA(1-\alpha)^{n}\exp\left(-\frac{E}{RT}\right) \tag{6.14}$$

where α is the overall degree of conversion, N is the number of reactions, and w_i denotes the ith reaction fraction such that

$$\sum\nolimits_{i=1}^{N} w_i = 1 \tag{6.15}$$

According to Kissinger [115], E_i can be estimated by determining the temperature of the ith peak maximum, T_{im}, for experiments carried out at different heating rates, $\beta = dT/dt$, using the formula:

$$\frac{E_i}{R} = \frac{d\left(\ln\left(\beta/T_{im}^2\right)\right)}{d(1/T_{im})} \tag{6.16}$$

Pokorný et al. [114] showed that this formula can be applied to gas-evolving cold-cap reactions. The remaining coefficients (A_i, n_i, and w_i) were determined using the least-squares optimization. With ongoing reactions, the DSC essentially measures the total heat flow to the sample from two contributions concurrently: the heat flow associated with the heat capacity of the sample and the heat flow associated with heat of reactions. The simultaneous DSC-TGA measures the total specific heat flow, Q, which when divided by the rate of heating attains the heat capacity unit, i.e., heat per unit mass and temperature. Thus, it becomes an "apparent" heat capacity, $c_p^{app} = Q/\beta$ which comprises heat capacity of the sample, c_p, and the heat generated/consumed by the reactions:

$$c_p^{app} = c_p + \Delta H \partial_T \alpha \tag{6.17}$$

where ΔH is the overall specific reaction enthalpy [116]. Note that, although DSC (or simultaneous DSC-TGA) is not the best technique to measure heat capacities accurately, it can provide an estimate for heat capacity under appropriate experimental conditions (e.g., $\beta \geq 10\,\mathrm{K\,min^{-1}}$) [116, 117]. The total heat flows from the simultaneous DSC-TGA were used to obtain, by Eqs. (6.12)–(6.17), the kinetic parameters for individual peaks, ΔH, and estimated c_p. The run/rerun technique and the least-squares methods were used.

Assuming that the mass loss of the batch is only associated with gas evolution reactions [114], the mass of the feed batch at time t is

$$m(t) = m_i - m_{\text{loss}}(\text{t}) = m_i - \sum\nolimits_{j=1}^{N_g} m_j(t) \tag{6.18}$$

where m_i is the initial mass, m_{loss} is the mass loss, m_j is the mass loss associated with the jth off-gas, and N_g is the number of gas species. Differentiating Eq. (6.17) with respect to time yields

$$-\frac{dm(t)}{dt} = \sum_{j=1}^{N_g} \frac{dm_j(t)}{dt} \tag{6.19}$$

For analysis with GC-MS, the flux of each gas species, represented via abundance, is proportional to the number of molecular ions (or instrumentally induced charges or produced current). Provided that instrumental artifacts are negligible, quantitative analysis can be established by assuming that the mass change rate of the jth off gas is linearly proportional to the jth off-gas abundance or intensity, I_j. Thus,

$$\frac{dm_j(t)}{dt} = F_j I_j(t + \Delta t) \tag{6.20}$$

where F_j is the jth off-gas calibration coefficient, F_j is dimensionless such that $F_j\ I_j$ has an appropriate mass/(time)^2 unit. Here, Δt represents a time lag, assumed to be invariant over time or temperature, between the TGA signal and the MS detector reading, due to the off-gas transfer to the MS detector. By inserting Eq. (6.19) into Eq. (6.18), one obtains

$$-\frac{dm(t)}{dt} = \sum_{j=1}^{N_g} F_j I_j(t + \Delta t) \tag{6.21}$$

Eq. (6.20) correlates the mass loss rate, $dm(t)/dt$, from TGA, with I_j from MS via N_g calibration coefficients, which can be obtained using the least-squares method. Under ideal experimental conditions, similar to calibrations performed for quantitative and semiquantitative MS analyses shown elsewhere [118], a jth calibration coefficient can be established by comparing the integrated intensity with the mass change of a solid sample that releases the jth gas in a single reaction. Accordingly,

$$\Delta m_j = C_j \int I_j(t)dt \tag{6.22}$$

where C_j is the jth gas calibration coefficient and Δm_j is the mass change of the solid sample during the gas-evolving process. The range of integration is determined by the start and end of the gas-evolving reaction that generates the gas.

A simulated high-aluminum oxide (AO) melter feed was used by Rodriguez et al. [110]. The feed was formulated to vitrify a high-alumina HLW to produce glass that was designed for the Hanford Tank Waste Treatment and Immobilization Plant, at the Hanford Site in Washington State, USA [119]. The batch was heated and stirred at a constant rate in the temperature range 60–80°C, and dried overnight at 105°C, as described elsewhere [120].

The previously reported TGA analysis [114] was solely focused on reaction kinetics. Identifying the gases by EGA provides a step toward the identification of the gas-evolving reactions for individual TGA peaks with the goal of understanding the reaction mechanisms. The DSC-TGA and TGA-GC-MS have been shown to be good experimental tools for understanding EGA. A detailed EGA is one of the main aspects that could provide a key element of a more complete understanding of the cold cap reactions. The TGA-GC-MS analysis identifies gases evolved from melter feeds during their conversion to molten glass. A simple calibration approach correlates the mass loss rate from TGA to the production rate of individual gases from EGA on a quantitative basis. This identifies where specific mass loss occurs as melter feed is heated and goes through various reactions and releases CO_2, H_2O, NO, and O_2.

11 TA in vitrification of radioactive spent ion-exchange resins (SIERs) in borosilicate glass

The underground disposal of waste arising from the nuclear industry needs improvements regarding minimizing the volume and cost without compromising the safety of any leakage from the radioactive waste form. The selection of waste form materials for higher activity nuclear waste containment was reviewed recently by Jantzen and Ojovan [121]. The utilization of materials such as glasses, ceramics, glass composite materials, and cements was discussed as practiced in different countries.

SIERs are released during the operation of nuclear facilities, including the treatment of radioactive waste, demineralization of the wastewater of NPP, decontamination of nuclear installation, and decommissioning of nuclear facilities [122].

SIERs are a problematic type of waste, which in many cases require a special approach to reduce their volume (via thermal treatment) and reduce their carbon and sulfur contents [122]. For instance, divinyl benzene (DVB) exchangers are rich in carbon (49.9 wt%) and sulfur (15.5 wt%), which restricts the waste loading in the glass matrix [123].

An organic resin stored in an aqueous phase requires thermal treatment to be demoisturized and reduced in carbon and sulfur contents. The storage of spent resins in water and wet media poses a major challenge in nuclear waste management. In the case of wet media storage, the generation of hydrogen gas from SIERs is an understood phenomenon and is based on the radiolysis decomposition of water by alpha, beta, and gamma radiation. Water decomposition results in the formation of hydrogen gas and hydrogen peroxide. The potential hazard, therefore, exists from the buildup of flammable hydrogen concentrations (4 vol% H_2) in air if the facility is not properly ventilated. The experimental G values of the hydrogen gas evolved from other spent resin slurry (Argentine resin samples, for example) were in the range of 0.3–0.8 molecules/100 eV radiation exposure. There are other potential hazards relating to the isotopes migrating from the stored resin to the ground water in the case of a tank leakage. All these identified hazards must be isolated, reduced, controlled, and eliminated throughout safe disposal of the spent resin [124].

Hamodi et al. [125] investigated the immobilization of the nuclear spent resin (NRW-40) in borosilicate glass. The analysis of surrogate resin doped with radioactive and nonradioactive cesium (Cs) and cobalt (Co) was carried out to investigate their thermal and chemical properties and their compatibility with an alkaline borosilicate glass. The thermal analysis indicates that the structural damage caused by 1 mSv gamma radiation to the radioactive resin has altered its properties in comparison to the nonradioactive resin. The same amount of cesium (8.88 wt%) and cobalt (1.88 wt%) were used in both resins. This is attributed to the role of radiation in breaking down the carbon-hydrogen-sulfur bonds within the radioactive resin matrix.

Accordingly, TGA combined with X-ray fluorescence (XRF) results revealed that the volatilization of Cs and Co in IEX-type resin started at 1050°C, while their volatilization in IER-type resin started at 950°C. The immobilization of residue shows that the excess sulfur in the residue caused phase crystallization in the final glass matrix. It was found that the volatilization of Cs-137 and Co-60 from the successful radioactive resin-glass matrix (HG-3-IER-500) were more than that in the nonradioactive resin-glass matrix (HG-3-IEX-500). The TGA results revealed that radiation damage in the IER residue at 1000°C (7.7 wt%) was less (≈12 wt%) than IEX residue at the same temperature, despite the minor difference (3.8 wt%) in Cs doped in both the resins. The authors studied vitrification of thermally treated nuclear spent resin (NRW-40) to assess the ability of glass matrix to integrate the resulting ash. The major radioactivity in the spent resin used in the nuclear facilities originate from cobalt (Co-58 and Co-60) and cesium (Cs-134 and Cs-137); these contributions accounted for 90% of the radioactivity measured in the cationic ion-exchange resins involved in the pressurized water reactors (PWR) operation [123, 126, 127].

Thermo-analytical techniques such as TGA and DTA combined with carbon-hydrogen-nitrogen-sulfur (CHNS) elemental analysis were applied by Hamodi et al. [125] to study the thermal behavior of NRW-40 nuclear spent resin. NRW-40 resin is a gel/macroporous polystyrene cross-linked with divinylbenzene (DVB). NRW-40 cartridges consist of mixed bed nuclear grade resin containing a combination of 40% of strong acid cations ($—SO_3\ H^+$) that is highly selective for cesium-137, cobalt-60 and 60% strong base anions (hydroxyl groups) highly selective for cations in the water [126, 127].

The weight reductions in the nonradiative resin forms were analyzed and compared with the radioactive ones (Fig. 6.24). DTA graphs in Fig. 6.25 show five degradation regions of resin forms:

1) Zone (25–150°C) represents the evaporation of water from the resin. The TG graph is like the indion-223 and indion-225 thermal treatments, which were reported earlier by Singare et al. [128].
2) Zone (150–490°C) represents the liberation of SO and SO_2 gases. At this intermediate temperature range, the degradation of the functional group ($—SO_3H^+$) attached to the benzene group corresponds to the first peak labeled "a" and the decomposition of sulfonate group between the phenol rings to the second peak labeled as "b" in Fig. 6.25. Similar results were reported by Bothe et al. [129].
3) Zone (490–650°C) represents the decomposition of hydrocarbon matrix into CO and CO2 gases [129].

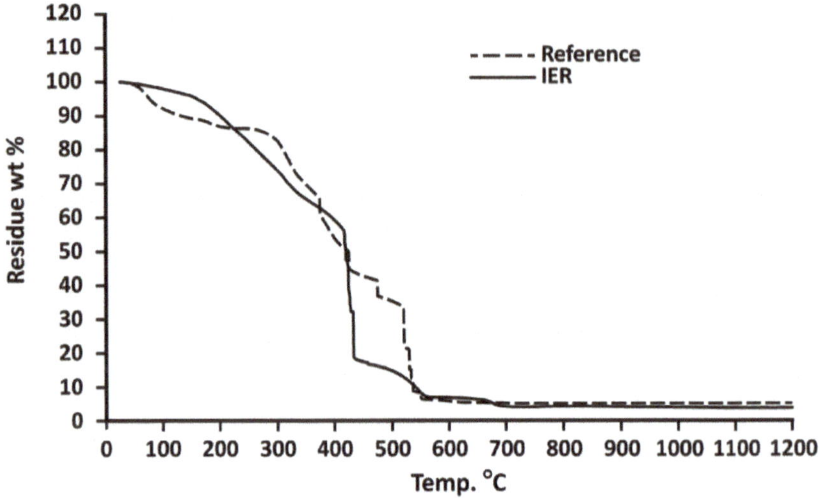

FIG. 6.24 TGA of IER and reference fresh resin forms.

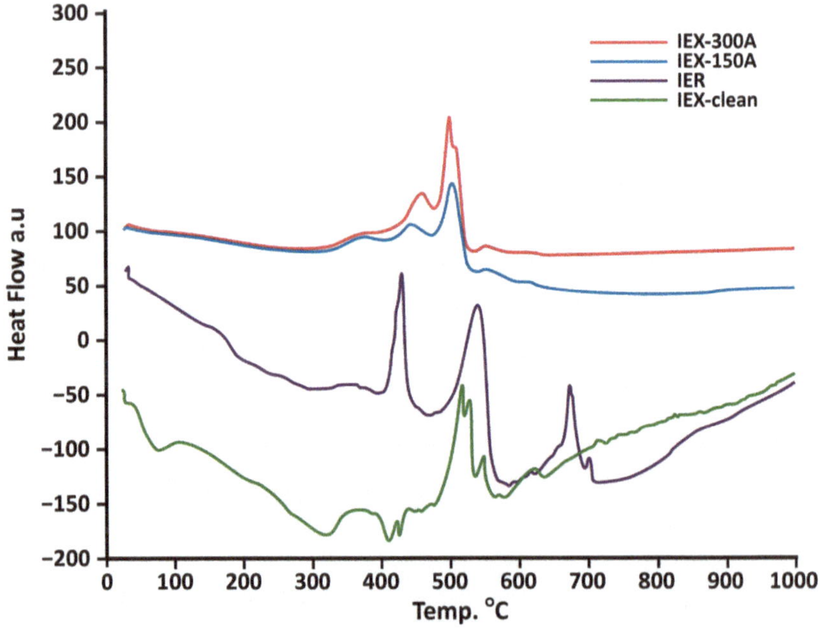

FIG. 6.25 DTA plots of resin forms. Code: nonirradiated resins treated at 150°C and 300°C (IEX-150 and IEX-300, respectively), irradiated resin (IER), and fresh unloaded resin (IEX-clean) as a reference.

4) Zone (650–950°C) in which the observed weight increase might be due to oxidation of the doped Cs and Co oxides in the resin to higher oxidizing valence such as $Cs_{11}O_3$, Cs_4O and Cs_7O, and Co_2O_3 and Co_3O_4. The oxidation zone may result in oxygen-rich compounds, which help in the reduction of the redox state of the glass during later vitrification, adding to the quality of the final glass [130].
5) Zone (1050–1200°C) represents the volatilization of Cs and Co as evidenced by elemental analysis [131].

12 TA in electronics waste recycling

Environmental management of noxious wastes across the world has attracted an increasing interest of researchers and industry due to the unbridled dumping into the ecosystem in a reckless way leading to possible risks. In the recent years, nearly 1.3 billion tons of wastes are generated every year in the world, which are anticipated to increase to 4.3 billion tons per annum by 2025. Especially, electronic waste (E-waste) generation is growing very fast all over the world, with the rate of generation being threefold higher than any other waste. It is one of the principal contributors to environmental contamination, the reason why it is strongly required to develop effective recycling techniques for its mitigation. It is predicted that E-waste generation from old computers would rise by 500% and that from the discarded mobile phones was estimated to be nearly 18 times higher in 2020 compared with 2007 [132].

E-waste contains both valuable and hazardous metals that require special handling and recycling methods to diminish environmental contamination and hazardous effects on human health. Knowledge and understanding the compositions of E-waste components help to recycle the components effectively. Several existing recycling processes and techniques are convenient for the management of E-waste such as mechanical, chemical, and thermal treatments, but the selection and adoption of an appropriate process need a cautious approach, which depends on the material composition, and thus it becomes quite important to know about its components. The conventional thermal process for the materials recovery from E-waste is currently analyzed in its critical aspects and an advanced solution is proposed to improve its economic sustainability and environmental impact. Particular attention has been paid on the removal of the polymeric materials thermally contained in the module's structure. A thermal treatment for the complete degradation of residual polymeric encapsulants is proposed in many E-waste recyclers, which require TA methods [133].

12.1 Silicon photovoltaic modules at end-of-life: Removal of polymeric layers and separation of materials

The considerable volume of photovoltaic (PV) panels at end-of-life that is expected to be generated in the next years imposes the need to adopt appropriate recycling processes. Improper disposal practices of PV waste may cause considerable environmental issues. Instead, correct recycling methods can play a key role in protecting the environment by reducing the risks associated with many hazardous materials such as heavy metals, polybrominated flame retardants, ozone depleting chemicals such as chlorofluorocarbons (CFCs), etc. On the other hand, recycling is indispensable to avoid the loss of valuable materials and reduce the total demand of new materials production, contributing to meet the target for the reduction of the greenhouse gas emissions. In this context, Fiandra et al. [34] reported an eco-friendly process to recover valuable materials deriving from silicon-based PV panels at end-of-life. Particularly, a new two-step process to separate and recover glass, Si, and metals has been investigated and discussed. A preliminary mechanical treatment to remove fluorinated polymers allows to exclude dangerous emissions of hydrofluoric acid and fluorinated compounds coming out from conventional heat treatments. A subsequent thermal treatment allows the complete removal of the residual polymers and the separation of valuable materials (Fig. 6.26). The results have shown that the combination of the two mechanical/thermal processes allows energy efficiency and environmental sustainability with respect to conventional recovery treatments. The optimal operating conditions for the thermal treatment have turned out to be at 500°C for 1 h in oxidizing atmosphere [133].

The TGA conducted on the ethylene vinyl acetate (EVA) copolymer encapsulating polymer and on the whole rear polymeric layer provides indications about the operating conditions for driving the subsequent thermal step of the material separation process. The TGA/DSC graphics are shown for EVA and

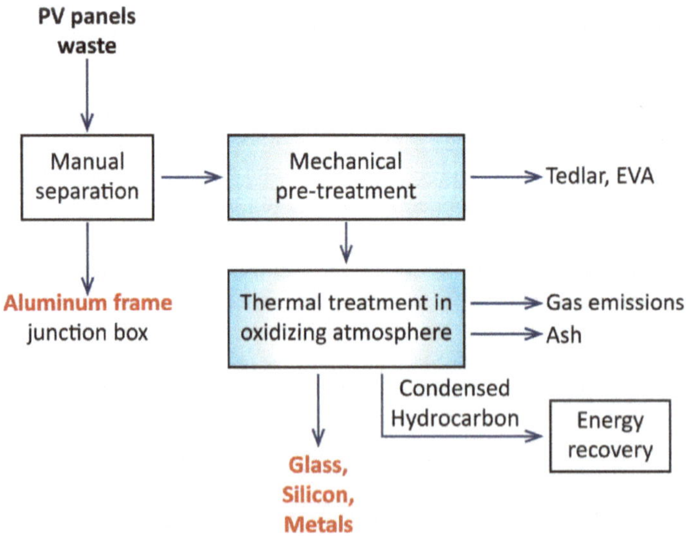

FIG. 6.26 Schematic diagram of the materials recovery process from PV panels waste.

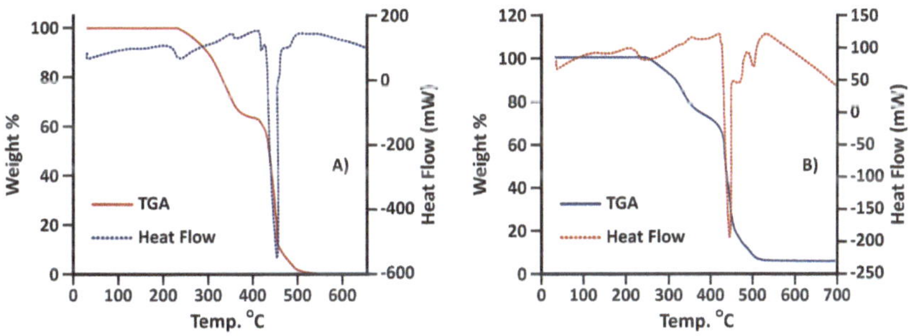

FIG. 6.27 TGA and DSC results for the (A) EVA layer and (B) back sheet of the PV module in air.

backsheet (EVA/Tedlar/PET/Tedlar) sampled from the PV panel used for the experimentation (Fig. 6.27A and B). The curves of TGA and heat flow of EVA and back sheet are similar indicating that EVA is the main plastic component of the panel. TGA analysis has detected two main weight losses during the polymer degradation in air. The first weight loss is related to the evolution of the acetic acid at about 300–350°C, due to the decomposition of vinyl acetate groups. The second one corresponds to the main-chain decomposition and to the evolution of aliphatic/ethylenic volatiles at 420–480°C. The data agree with the literature which reports that the thermal degradation of EVA occurring in two steps [134–136].

The third step may be assigned to the thermo-oxidative degradation of the carbonaceous residue at 480–530°C, formed during the degradation of the polyethylene chains. The residual mass is attributable to the presence in the polymers of flame-retardant materials such as aluminum hydroxide and to the formation of Al_2O_3 [134] or compounds used to promote adhesion between EVA and inorganic surfaces such as trialkoxy silane and then SiO_2 [135]. DSC analysis detects chemical reactions occurring by the consumption or release of heat. After the melting of the polymer, different overlapped and exothermic peaks indicate that the complete thermal decomposition took place in different steps.

Fig. 6.28 shows EVA weight losses vs temperature in inert atmosphere and in air, at the 5°C/min heating rate, in the range 200–550°C, where the decomposition occurs. As it can be observed, the presence of oxygen in the atmosphere can accelerate the process. The exothermic character of the combustion is clearly appreciated, causing an increase of the temperature. The exothermal peak occurring in oxidizing atmosphere may indicate burning reactions of the partial degradation products. The pyrolysis curve presents a higher slope than the combustion one. The peak temperatures of the two processes are different and refer to the two main weight losses occurring at 335°C and 473°C for pyrolysis and 316°C and 434°C for combustion. The results of the TGA indicate that the total decomposition of the residual polymers after the mechanical

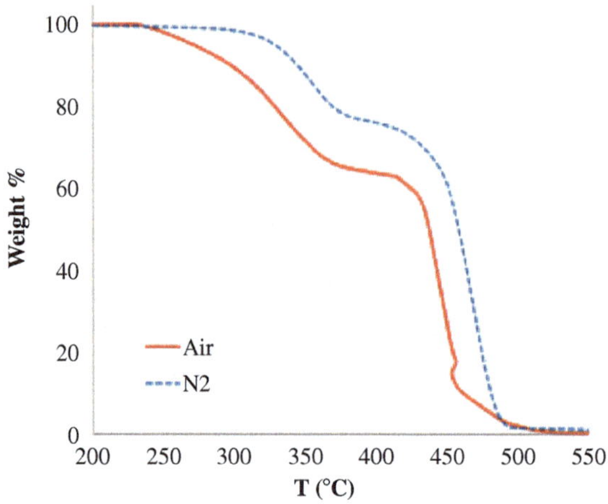

FIG. 6.28 Influence of the atmosphere on the EVA weight loss in the range 200–550°C.

treatment for the back sheet removal occurring in the temperature range between 450°C and 550°C. Therefore, the PV panels without back sheet were subjected to a successive thermal treatment to completely degrade the residual EVA in oxidizing atmosphere.

12.2 Kinetics study of thermal decomposition of electronic packaging material

A high-purity silica was obtained from thermal decomposition of molding resin used as electronic packaging materials. The reaction was performed at high temperature and in oxidizing atmosphere. The product was suitable for reuse as inorganic filler in the production of semiconductor devices, as well as the starting materials in the manufacture of silicon materials. Thermal decomposition kinetics of the electronic packaging material were investigated under various reaction parameters including gas flow rate, sample loading, grain size, oxygen concentration, and heating rate by using TGA. The results indicate that thermal degradation of electronic packaging material consisted of two distinct reaction stages. The corresponding kinetic parameters including the activation energy, preexponential factor, and reaction order in the chemical reaction-controlled region are presented. A mechanism of thermal decomposition was proposed, which is in good agreement with the experimental results [137].

When the gas flow rate exceeds 40 mL/min and the sample loading is smaller than 40 mg, the resistance due to mass transfer is negligible, and the overall rate of thermal decomposition reaction-controlled by chemical kinetics can be obtained. The degree of conversion of electronic packaging material, X, is defined as

$$X = (W_0 - W)/(W_0 - W_\infty) \tag{6.23}$$

where W_0, W, and W_∞ represent initial, instantaneous, and final masses of sample, respectively.

Fig. 6.29 presents plots of conversion vs temperature of electronic packaging material decomposed in air at a heating rate of 2, 5, or 10 K min^{-1}. The total temperature range of thermal decomposition is within 500–825 K. The conversion increased because of the loss in mass during thermal decomposition, indicating that the mass of volatile matter has been removed. For all three heating rates, a single inflection point is observed during the reaction, indicating that there are two principal stages of reaction distinguished by two significant and distinct variations of conversion. For a given conversion, the corresponding temperature for thermal decomposition at a small rate of heating is less than that at a large rate of heating. The reason for this phenomenon is that as the heating rate increases, the time interval during which the sample is exposed at a given temperature decreases. The same tendency was also observed by Iji and Ikuta [138]. During the thermal decomposition, it can be clearly observed that the volatilization of gas products is accompanied by the formation of tar. From the variation of conversion shown in Fig. 6.29, about 55% of conversion is observed in

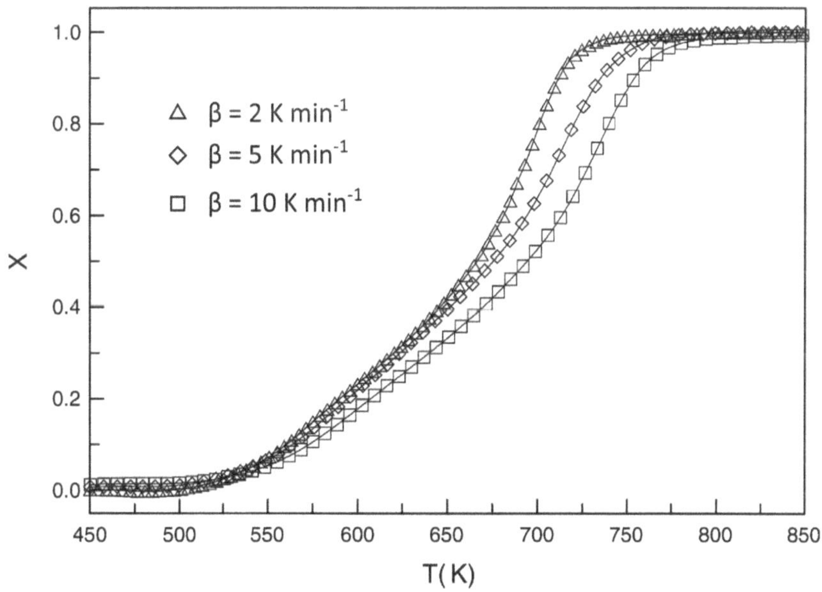

FIG. 6.29 Effect of heating rate on the thermal decomposition of electronic packaging material specimen. Initial conditions: gas flow rate 60 mL/min, sample loading 10 mg, initial grain size 38–45 μm, and oxygen concentration 21 vol.%.

first stage, when the temperature varied from 500 to 675 K for all three heating rates. The increase of conversion in the first stage may be due to evolution of volatile matter. This phenomenon may result from breakdown or depolymerization of the original resin constituents that are further decomposed to yield volatiles and char. The second stage, when the temperature varied from 675 to 825 K, is mainly attributed to dehydration or oxidation of residual resin components (or char) that decompose to yield gases and the other char. The slope of the heating curves in the second stage is slightly larger than that in the first stage, indicating that part of the reactant is not easily decomposed in the beginning stage. When the temperature exceeded 825 K, the electronic packaging materials were almost completely decomposed, and a higher reaction temperature did not affect the loss of mass.

12.3 Carbon elimination from silicon kerf

About 40% of precious ultrapure silicon is lost as kerf during slicing to produce wafers. Kerf is currently not being recycled due to engineering challenges and costs associated with removing its abundant impurities. Carbon left behind from the lubricant remains as one of the most difficult contaminants to remove in kerf without significant silicon oxidation. Pufleau et al. [139] studied the kinetics of carbon elimination from silicon kerf in air and N_2 atmospheres under a regime of no-diffusion limitation. A better understanding of the mechanism of carbon elimination in kerf can aid design of better processes for kerf recycling and subsequent low-cost PVs. The authors reported the apparent activation energies in air and N_2 atmospheres using three methods: Kissinger and two isoconversional approaches. In both atmospheres, a bimodal apparent activation energy was observed, suggesting a two-stage process. A reaction mechanism is proposed in which (a) C—C and C—O bond cleavage reactions occur in parallel with polymer formation and (b) at higher temperatures, this polymer fully degrades in air but leaves a tarry residue in N_2 that accounts for about 12% of the initial total carbon.

The Kerf sample used by Pufleau et al. was characterized by BET and had a surface area of 9.6 m^2/g, a pore volume of 0.052 cm^3/g, and a pore radius of 15.6 Å, proving its large surface area and flake-like morphology.

The conversion of carbon calculated using traditional TGA weight loss method was compared with a TGA-TC method. The TGA-TC method implies using TGA and stopping the run to determine its remaining total carbon (TC) content at various temperatures. The heating rate chosen of 20°C/min was sufficiently high to emulate real aerosol reactor conditions and at the same time, low enough to guarantee mass and heat transfer, thus, providing accurate TG readings. Fig. 6.30 shows that the conversion using the two methods differs.

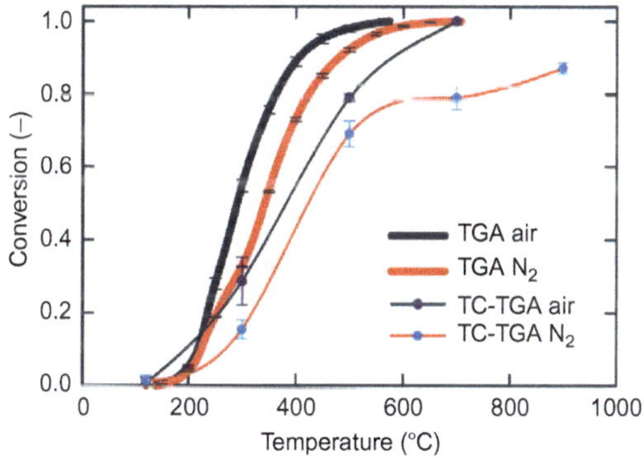

FIG. 6.30 Volatiles evolution from kerf using a heating rate of 20°C/min. Conversion is calculated by weight loss from TGA and total carbon (TC) of the remaining sample in air and in N_2 atmospheres. Error bars indicate the standard deviation.

The kinetics of carbon elimination in kerf were evaluated using three TG analysis methods: Ozawa-Flynn-Wall (OFW), Kissinger-Akahira-Sunose (KAS), and Kissinger. The three methods are used in conjunction as their combined results to provide more certainty on the accuracy of the obtained activation energy [140].

The raw quadruplet data at four different heating rates and its derivatives are shown for air in Fig. 6.31A and for N_2 in Fig. 6.31B. The EA determined with the Kissinger method was obtained by computing the peak values of the derivatives from insets of Fig. 6.31.

13 Summary

In this chapter, we discuss some applications of TA techniques in assessing the optimum conditions in the vital areas of recycling and waste management. This includes assessing the identity of the mixed recycled plastics using DSC and TGA as a research method in the transformation processes of waste rubber and plastic products for energy carriers: WtL and WtE processes, TGA characteristics, and kinetics of scrap tire-biomass copyrolysis. Some applications in the field of nuclear industry were given including the capture of organic iodides from nuclear waste, SOMS and their related perovskites in nuclear waste management, nuclear graphite treatment using TGA, thermal characteristics of SAC generated from ACUs, cold-cap reactions in vitrification of nuclear waste glass, TA in vitrification of radioactive SIERs in borosilicate glass, and finally TA in waste electronic recycling.

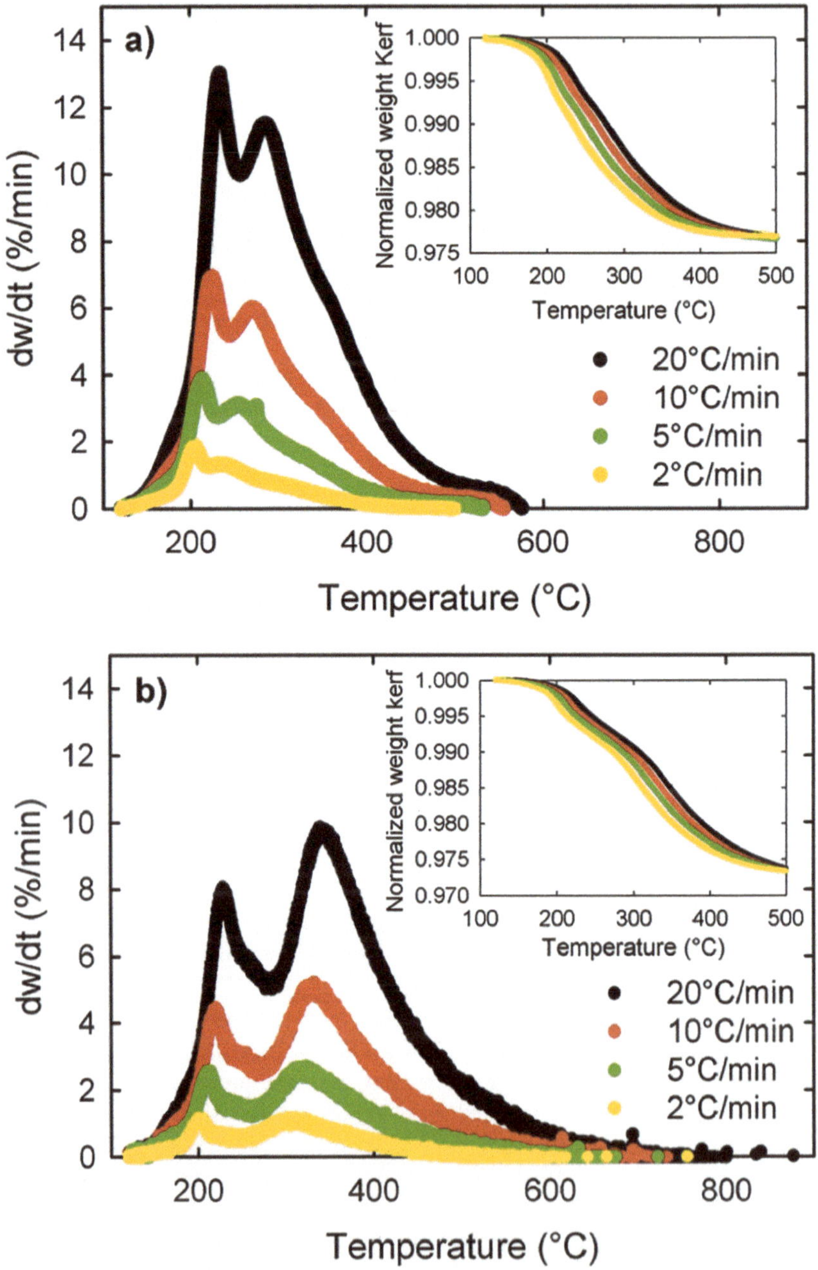

FIG. 6.31 Rate of volatiles elimination for kerf at various heating rates (A) in air and (B) in N_2. Insets show carbon elimination from kerf in the given atmosphere.

References

[1] United Nations Statistics Division—Environment Statistics, https://unstats.un.org/unsd/envstats/index.cshtml, Retrieved 3 March, 2017.

[2] O. Vidal, Mineral Resources and Energy: Future Stakes in Energy Transition, Elsevier Ltd., 2018.

[3] A. Cózar, F. Echevarría, J.I. González-Gordillo, X. Irigoien, B. Úbeda, S. Hernández-León, Á.T. Palma, S. Navarro, J. García-de-Lomas, A. Ruiz, M.L. Fernández-de-Puelles, C.M. Duarte, Plastic debris in the open ocean, Proc. Nat. Acad. Sci. U. S. A. (PNAS) 111 (2014) 10239–10244.

[4] Rachellowe, Revealing the Identity of The Mixed Recycled Plastics Using Differential Scanning Calorimetry (DSC), Impact Solutions: Independent Plastic Experts, UK, 2014.

[5] X. Li, C. Zhanga, Y. Lia, Q. Zhia, The status of Municipal Solid Waste Incineration (MSWI) in China and its clean development, Energy Procedia 104 (2016) 498–503.

[6] B.B. Garcia, G. Lourinho, P. Brito, P. Romano, in: K. Biernat (Ed.), Review of Biofuel Technologies in WtL and WtE, Elements of Bioeconomy, IntechOpen, 2019.

[7] E. Rostek, K. Biernat, TGA as a research method in the transformation processes of waste rubber and plastic products for energy carriers (WtE and WtL processes), J. Sust. Dev. Energy Water Environ. Syst. 1 (2013) 163–171.

[8] J.E.K. Schawe, Remarks regarding the determination of the initial crystallinity by temperature modulated DSC, Thermochim. Acta 657 (2017) 151–155.

[9] P.R. Solomon, M.A. Serio, E.M. Suuberg, Coal pyrolysis: experiments kinetic rates and mechanisms, Prog. Energy Combust. Sci. 18 (1992) 133–220.

[10] P. Glarborg, A.D. Jenses, J.E. Johansson, Fuel nitrogen conversion in solid fuel fired systems, Prog. Energy Combust. Sci. 29 (2003) 89–113.

[11] M. Zhao, N.H. Florin, A.T. Harris, Mesoporous supports cobalt catalysts for enhanced hydrogen production during cellulose decomposition, Appl. Catal. B Environ. 97 (2010) 142–150.

[12] S. Kumar, A.K. Panda, R.K. Singh, A review on tertiary recycling of high-density polyethylene to fuel, Resour. Conserv. Recycl. 55 (2011) 893–910.

[13] Determination on the Adsorption and Desorption of Moisture in Pharmaceutical Substances, Thermal Analysis Application No. UC 213, METTLER TOLEDO Thermal Analysis UserCom 21.

[14] Influence of Absorbed Moisture on The Mechanical Properties of Polyamide, Thermal Analysis Application No. UC 241, METTLER TOLEDO Thermal Analysis UserCom 24.

[15] DMA Measurements in Humid Atmospheres, Thermal Analysis Application No. UC 366, METTLER TOLEDO Thermal Analysis UserCom 36.

[16] R. Aguado, M. Olazar, D. Vélez, M. Arabiourrutia, J. Bilbao, Kinetics of scrap tyre pyrolysis under fast heating conditions, J. Anal. Appl. Pyrolysis 73 (2005) 290–298.

[17] M.R. Islam, H. Haniu, J. Fardoush, Pyrolysis kinetics behavior of solid tire wastes available in Bangladesh, Waste Manag. 29 (2009) 668–677.

[18] D.Y.C. Leung, C.L. Wang, Kinetic study of scrap tyre pyrolysis and combustion, J. Anal. Appl. Pyrolysis 45 (1998) 153–169.

[19] S. Seidelt, M. Muller-Hagedorn, H. Bockhorn, Description of tire pyrolysis by thermal degradation behavior of main components, J. Anal. Appl. Pyrolysis 75 (2006) 11–18.

[20] I.M. Rodriguez, M.F. Laresgoiti, M.A. Cabrero, Pyrolysis of scrap tyres, Fuel Process. Technol. 72 (2001) 9–22.

[21] C. Vasile, M. Brebu, Thermal valorisation of biomass and synthetic polymer waste. Upgrading of the pyrolysis oils, Cell. Chem. Technol. 40 (2006) 489–512.

[22] B.B. Uzoejinwa, X. He, S. Wang, A.E.-F. Abomohra, Y. Hu, Q. Wang, Co-pyrolysis of biomass and waste plastics as a thermochemical conversion technology for high-grade biofuel production: recent progress and future directions elsewhere worldwide, Energy Convers. Manag. 163 (2018) 468–492.

[23] L. Zhou, Y. Wang, Q. Huang, Thermogravimetric characteristics and kinetics of plastic and biomass blends co-pyrolysis, Fuel Process. Technol. 87 (2006) 963–969.

[24] M. Brebu, S. Ucar, C. Vasile, Co-pyrolysis of pinecone with synthetic polymers, Fuel 89 (2010) 1911–1918.

[25] A. Aboulkas, K. El Harfi, M. Nadifiyine, A. Aboulkas, K. El Harfi, A. El Bouadili, Thermogravimetric characteristics and kinetic of co-pyrolysis of olive residue with high density polyethylene, J. Therm. Anal. Calorim. 91 (2008) 737–743. Pyrolysis kinetics of olive residue/plastic mixtures by non-isothermal thermogravimetry, Fuel Process. Technol. 90 (2009) 722–728.

[26] E. Atabani, O.K. Al-Rubaye, Valorization of spent coffee grounds for biodiesel production: blending with higher alcohols, FT-IR, TGA, DSC, and NMR characterizations, Biomass Conver. Bioref. (2020), https://doi.org/10.1007/s13399-020-00866-z.

[27] E. Yaman, Investigation of Thermal Degradation Kinetics of Walnut Shell With Waste Plastics via Thermogravimetric Analysis, M.Sc., Thesis, Anadolu University, Institute of Applied Science, 2013.

[28] K. Acıkalın, Thermogravimetric analysis of walnut shell as pyrolysis feedstock, J. Therm. Anal. Calorim. 105 (2011) 145–150.

[29] Y. Kar, Co-pyrolysis of walnut shell, and tar sand in a fixed-bed reactor, Bioresour. Technol. 102 (2011) 9800–9805.

[30] H. Pirayesh, A. Khazaeian, T. Tabarsa, The potential for using walnut (*Juglans regia* L.) shell as a raw material for wood-based particleboard manufacturing, Compos. Part B 43 (2012) 3276–3280.

[31] S. Kumagai, K. Fujita, Y. Takahashi, Y. Nakai, T. Kameda, Y. Saito, T. Yoshioka, Beech wood pyrolysis in polyethylene melt as a means of enhancing levoglucosan and methoxyphenol production, Sci. Report. 9 (2019) 1955.

[32] C. Cardunculus, T. Damartzis, D. Vamvuka, S. Sfakiotakis, Thermal degradation studies and kinetic modelling of cardoon pyrolysis using thermogravimetric analysis (TGA), Bioresour. Technol. 102 (2011) 6230–6238.

[33] M.J. Jeon, S.J. Choi, K.S. Yoo, C. Ryu, S.H. Park, J.M. Lee, J.K. Jeon, Y.K. Park, S. Kim, Co-pyrolysis of block polypropylene with waste wood chip, Korean J. Chem. Eng. 28 (2011) 497–501.

[34] K. Slopiecka, P. Bartocci, F. Fantozzi, thermogravimetric analysis and kinetic study of poplar wood pyrolysis, Appl. Energy 97 (2011) 491–497.

[35] Y. Matsuzawa, M. Ayabe, J. Nishino, Acceleration of cellulose co-pyrolysis with polymer, Polym. Degrad. Stab. 71 (2001) 453–454.

[36] H.H. Sait, A. Hussain, A.A. Salema, Pyrolysis and combustion kinetics of date palm biomass using thermogravimetric analysis, Bioresour. Technol. 118 (2012) 382–389.

[37] B.B. Uzun, N. Sarioğlu, Rapid and catalytic pyrolysis of corn stalks, Fuel Process. Technol. 90 (2009) 705–716.

[38] J.F. Nzihou, S. Hamidou, M. Bouda, J. Koulidiati, B.G. Segda, Using Dulong and Vandralek formulas to estimate the calorific heating value of a household waste model, Int. J. Sci. Eng. Res. 5 (2014) 1878–1883.

[39] B.B. Uzun, E. Yaman, Thermogravimetric characteristics and kinetics of scrap tyre and *Juglans regia*: shell co-pyrolysis, Waste Manag. Res. 32 (2014) 961–970.

[40] M.V. Navarro, J.M. López, A. Veses, M.S. Callén, T. García, Kinetic study for the co-pyrolysis of lignocellulosic biomass and plastics using the distributed activation energy model, Energy 165 (Part A) (2018) 731–742.
[41] J. Huve, A. Ryzhikov, H. Nouali, V. Lalia, G. Augéd, T.J. Daou, Porous sorbents for the capture of radioactive iodine compounds: a review, RSC Adv. 8 (2018) 29248–29273.
[42] S.U. Nandanwar, K. Coldsnow, V. Utgikara, P. Sabharwall, D.E. Astona, Capture of harmful radioactive contaminants from off-gas stream using porous solid sorbents for clean environment—a review, Chem. Eng. J. 306 (2016) 369–381.
[43] K.S. Subrahmanyam, D. Sarma, C.D. Malliakas, K. Polychronopoulou, B.J. Riley, D.A. Pierce, J. Chun, M.G. Kanatzidis, Chalcogenide aerogels as sorbents for radioactive iodine, Chem. Mater. 27 (2015) 2619–2626.
[44] Y. Kim, I. Kim, T.S. Lee, E. Lee, K.J. Lee, Y. Kamachi, M.B. Zakaria, N.L. Torad, T. Nakato, T. Ahamad, S.M. Alshehri, V. Malgras, Victor, Y. Yamauchi, Yusuke, Porous hydrogel containing Prussian blue nanoparticles for effective cesium ion adsorption in aqueous media, J. Indust. Engin. Chem. 60 (2018) 465–474. Hydrogels containing Prussian blue nanoparticles toward removal of radioactive cesium ions, J. Nanosc. Nanotechnol. 16 (2016) 4200–4204.
[45] W. Xie, D. Cui, S.-R. Zhang, Y.-H. Xu, D.-L. Jiang, Iodine capture in porous organic polymers and metal-organic frameworks materials, Mater. Horizons 6 (2019) 1571–1595.
[46] G. Férey, C. Mellot Draznieks, C. Serre, F. Millange, J. Dutour, S. Surblé, I. Margiolaki, A chromium terephthalate-based solid with unusually large pore volumes and surface area, Science 309 (2005) 2040–2042.
[47] B. Li, X. Dong, H. Wang, D. Ma, K. Tan, S. Jensen, B.J. Deibert, J. Butler, J. Cure, Z. Shi, T. Thonhauser, Y.J. Chabal, Y. Han, J. Li, Capture of organic iodides from nuclear waste by metal-organic framework-based molecular traps, Nat. Commun. 8 (2017) 485.
[48] Y. Wang, H. Gao, C.J. Brinker, Y. Xiong, K. Holt, A. Miller, P. Pohl, N. Ockwig, M.A. Rodriguez, D.N. Bencoe, H. Tellez, J.N. Kruichak, R. Turnham, A.W. Murphy, Development of A New Generation of Waste Form for Entrapment and Immobilization of Highly Volatile and Soluble Radionuclides, SANDIA Reports, Sandia National Laboratories, Albuquerque, 2010, pp. 1–56.
[49] K.W. Chapman, P.J. Chupas, T.M. Nenoff, Radioactive iodine capture in silver-containing mordenites through nanoscale silver iodide formation, J. Am. Chem. Soc. 132 (2010) 8897–8899.
[50] N. Mnasri, C. Charnay, L.-C. de Ménorvala, Y. Moussaoui, E. Elaloui, J. Zajac, Silver nanoparticle-containing submicron-in-size mesoporous silica-based systems for iodine entrapment and immobilization from gas phase, Microporous Mesoporous Mater. 196 (2014) 305–313.
[51] D.R. Haefner, T.J. Tranter, Methods of Gas Phase Capture of Iodine From Fuel Reprocessing Off-Gas: A Literature Survey, Idaho National Laboratory, Idaho Falls, 2007.
[52] B.J. Riley, J.D. Vienna, D.M. Strachan, J.S. McCloy, J.L. Jerden Jr., Materials and processes for the effective capture and immobilization of radioiodine: a review, J. Nucl. Mater. 470 (2016) 307–326.
[53] R.G. Anthony, C.V. Phillip, R.G. Dosch, Selective adsorption and ion exchange of metal cations and anions with silico-titanates and layered titanates, Waste Manag. 13 (1993) 503–512.
[54] Y. Su, L. Li, T.M. Nenoff, M.D. Nyman, A. Navrotsky, H. Xu, Investigation of Sandia Octahedral Molecular Sieves (SOMS) and their related perovskites, in: P.W. Wang, T. Zachry (Eds.), Nuclear Waste Management, Accomplishments of the Environmental Management Science Program, American Chemical Society, Oxford University Press, 2006 (Chapter 15).

[55] R.G. Dosch, C.J. Northrup, T.J. Headley, Crystalline titanate ceramic nuclear waste forms: leaching and radiation damage, J. Am. Ceram. Soc. 68 (1985) 330–337.

[56] L. Li, S. Luo, B. Tang, D. Wang, Immobilization of sodium-bearing high-level radioactive waste in synroc containing ($Na_{0.5}Nd_{0.5}$) TiO_3-type perovskite, J. Am. Ceram. Soc. 80 (1997) 250–252.

[57] S. Luo, L. Li, B. Tang, D. Wang, Synroc immobilization of high level waste (HLW) bearing a high content of sodium, Waste Manag. 18 (1998) 55–59.

[58] S. Donze, L. Montagne, G. Palavit, Thermal conversion of heavy metal chlorides ($PbCl_2$, $CdCl_2$) and alkaline chlorides (NaCl, KCl) into phosphate glasses, Chem. Mater. 12 (2000) 1921–1925.

[59] L. Payne, P.J. Heard, T.B. Scott, A study of the oxidation behaviour of pile grade A (PGA) nuclear graphite using thermogravimetric analysis (TGA), scanning electron microscopy (SEM) and X-ray tomography (XRT), PLoS One 10 (2015), e0143041.

[60] E. Atherton, Review of Baseline Assumptions Regarding Disposal of Core Graphite in a Geological Disposal Facility, Nuclear Decommissioning Authority, Harwell Oxford, Didcot, UK, 2012.

[61] M.S. Yim, F. Caron, Life cycle and management of Carbon-14 from nuclear power generation, Prog. Nucl. Energy 48 (2006) 2–36.

[62] D. Vulpius, K. Baginski, B. Kraus, B. Thomauske, Thermal treatment of neutron-irradiated nuclear graphite, Nucl. Eng. Des. 265 (2013) 294–309.

[63] M.L. Dunzik-Gougar, T.E. Smith, Removal of Carbon-14 from irradiated graphite, J. Nucl. Mater. 451 (2014) 328–335.

[64] The Long-term Management of Reactor Core Graphite Waste Credible Options (Gate A), Higher Activity Waste, Nuclear Decommissioning Authority, SMS/TS/D1-HAW-6/002/A, 2013.

[65] S. Mohanty, S. Majumdar, HTGR Graphite Core Component Stress Analysis Research Program-Task 1 Technical Letter Report, Argonne National Laboratory, Argonne, IL, 2011.

[66] W.K. Choi, B.J. Kim, E.S. Kim, S.H. Chi, S.J. Park, Oxidation behavior of IG and NBG nuclear graphites, Nucl. Eng. Des. 241 (2011) 82–87.

[67] C.F. Cullis, J.G. Yates, Reaction of carbon with nitrogen, Trans. Faraday Soc. 60 (1964) 141–148.

[68] F.E. Loren, J.M. Okoh, Kinetics and mechanisms of the reaction of air with nuclear grade graphites: IG-110, J. Nucl. Mater. 240 (1997) 241–250.

[69] X. Luo, X. Yu, S. Yu, Oxidation performance of graphite material in reactors, Front. Energy Power Eng. China 2 (2008) 471–474.

[70] Y.S. Lim, S.H. Chi, K.Y. Cho, Change of properties after oxidation of IG-11 graphite by air and CO_2 gas, J. Nucl. Mater. 374 (2008) 123–128.

[71] C.I. Contescu, S. Azad, D. Miller, M.J. Lance, F.S. Baker, T.D. Burchell, Practical aspects for characterizing air oxidation of graphite, J. Nucl. Mater. 381 (2008) 15–24.

[72] X. Luo, R. Jean-Charles, S. Yu, Effect of temperature on graphite oxidation behavior, Nucl. Eng. Des. 227 (2004) 273–280.

[73] D. LaBrier, M.L. Dunzik-Gougar, Characterization of ^{14}C in neutron irradiated NBG-25 nuclear graphite, J. Nucl. Mater. 448 (2014) 113–120.

[74] J. Fachinger, W. von Lensa, T. Podruhzina, Decontamination of nuclear graphite, Nucl. Eng. Des. 238 (2008) 3086–3091.

[75] T.E. Smith, S. McCrory, M.L. Dunzik-Gougar, Limited oxidation of irradiated graphite waste to remove surface carbon-14, Nucl. Eng. Technol. 45 (2012) 211–218.

[76] A. Theodosiou, N. Jones, B.J. Marsden, Thermal oxidation of nuclear graphite: a large-scale waste treatment option, PLoS One 12 (2017), https://doi.org/10.1371/journal.pone.0182860, e0182860.

[77] M.E. Pick, Magnox Graphite Core Decommissioning and Disposal Issues in IAEA TECDOC-1647, IAEA, 2010, pp. 1–15.

[78] R.E. Nightingale, B.J. Masrden, G.N. Hall, Early use of nuclear graphite, in: R.E. Nightingale (Ed.), Nuclear Graphite, Academic Press, New York and London, 1962, pp. 1–17. Graphite in gas-cooled reactors, Comprehen. Nucl. Mater. 4 (2012) 325–390.

[79] Characterization, Treatment and Conditioning of Radioactive Graphite from Decommissioning of Nuclear Reactors, (IAEA) International Atomic Energy Agency, Vienna International Centre, PO Box 100, A-1400 Vienn, Austria, 2006.

[80] Strategic Position Paper on the Management of Waste Graphite, Higher Activity Waste, Nuclear Decommissioning Authority, SMS/TS/D1-HAW-6/003/PP, 2014.

[81] H.K. Hinsenn, K. Kuhn, R. Moormann, B. Schlogl, M. Fechter, Oxidation experiments and theoretical examinations on graphite materials relevant for the PBMR, Nucl. Engin. Des. 238 (2008) 3018–3025.

[82] E.F. Loren, J.M. Okoh, P.J. Walker Jr., F.R. Rusinko Jr., L.G. Austin, Kinetics and mechanisms of the reaction of air with nuclear grade graphites: IG-110, J. Nucl. Mater. 240 (1997) 241–250. Gas reactions of carbon, Adv. Catal. 11 (1959) 133–221.

[83] M. Inagaki, F. Kang, M. Toyoda, H. Konno, Isotropic High-Density Graphite and Nuclear Applications, Advanced Materials Science and Engineering of Carbon, Elsevier, Butterworth Heinemann, 2014, pp. 387–410.

[84] H.M. Freeman, A.N. Jones, M.B. Ward, F.S. Hage, N. Tzelepi, On the nature of cracks and voids in nuclear graphite, Carbon 103 (2016) 45–55.

[85] M. Mostafavi, M.J.J. Schmidt, B.J. Marsden, T.J. Marrow, Fracture behavior of an anisotropic polygranular graphite (PGA), Mater. Sci. Eng. A 558 (2012) 265–277.

[86] T.J. Clark, R.E. Woodley, D.R. De Halas, Gas-graphite systems, in: R.E. Nightingale (Ed.), Nuclear Graphite, Academic Press, New York and London, 1962, pp. 387–445.

[87] Design, Inspection, and Testing Criteria for Air Filtration and Adsorption Units of Post-accident Engineered-Safety Feature Atmosphere Cleanup System in Light-Water-Cooled Nuclear Power Plants, Regulatory Guide, United States Nuclear Regulatory Commission, 2012.

[88] J.Y. So, H.-R. Cho, Thermal characteristics of spent activated carbon generated from air cleaning units in Korean nuclear power plants, Nucl. Eng. Technol. 49 (2017) 873–880.

[89] D.L. Brier, M.L. Dunzik-Gougar, Characterization of ^{14}C in neutron irradiated NBG-25 nuclear graphite, J. Nucl. Mater. 448 (2014) 113–120.

[90] J.G. Kim, Treatment Methods and Process for Waste Activated Carbon Contaminated by Radiocarbon and Tritium Compounds, Vacuum Plant Co., Ltd., Korea Hydro Nuclear Power (KHNP) Co., Ltd., 2011. Korean, Patent 1020110034364, 13 April.

[91] Method and Apparatus for the Decontamination of Spent Activated Carbon, HUVISWATER Co., Ltd., Korean, Patent 1020120151155, 21, Dec., 2013.

[92] Treatment Methods for Waste Decontaminant, Korea Atomic Energy Research Institute (KAERI), Korean, Patent 1020140030542, 14, Mar., 2015.

[93] Y. Fernandez, A. Arenillas, J.A. Menendez, Microwave Heating Applied to Pyrolysis, Instituto Nacional del carbon (CSIC), Advances in Induction and Microwave Heating of Mineral and Organic Materials, Stanisław Grundas, IntechOpen, 2011.

[94] K. Xu, P. Hrma, J.A. Rice, M.J. Schweiger, B.J. Riley, N.R. Overman, A.A. Kruger, Conversion of nuclear waste to molten glass: cold-cap reactions in crucible tests, J. Am. Ceram. Soc. 99 (2016) 2964–2971.

[95] M.I. Ojovan, W.E. Lee, Glassy waste forms for nuclear waste immobilization, Metallurg. Mater. Trans. A 42 (2011) 837–851.

[96] S. Gin, P. Jollivet, M. Tribet, S. Peuget, S. Schuller, Radionuclides containment in nuclear glasses: an overview, Radiochim. Acta 105 (2017) 2016–2658.

[97] R.H. Gray, C.D. Becker, K.D. Crowley, J.F. Ahearne, Environmental cleanup: the challenge at the Hanford site, Washington, USA, Environ. Manage. 17 (1993) 461–475. Managing the environmental legacy of U.S. Nuclear-Weapons Production, Am. Scientist 90 (2002) 514–523.

[98] F.W. Wilburn, C.V. Thomasson, The application of differential thermal analysis and thermogravimetric analysis to the study of reactions between glass-making materials. Part 1. The sodium carbonate-silica system, J. Soc. Glas. Technol. 42 (1958) 158–175.

[99] C.V. Thomasson, F.W. Wilburn, The application of differential thermal analysis and thermogravimetric analysis to the study of reactions between glass-making materials. Part 2. The calcium carbonate-silica system with minor batch additions, Phys. Chem. Glasses 1 (1960) 52–69.

[100] F.W. Wilburn, C.V. Thomason, The application of differential thermal analysis and thermogravimetric analysis to the study of reactions between glass-making materials. Part 3. The sodium carbonate-silica system, Phys. Chem. Glasses 2 (1961) 126–131.

[101] F.W. Wilburn, C.V. Thomason, The application of differential thermal analysis and differential thermogravimetric analysis to the study of reactions between glass-making materials. Part 4. The sodium carbonate-silica-alumina system, Phys. Chem. Glasses 4 (1963) 91–98.

[102] F.W. Wilburn, S.A. Metcalf, R.S. Warburton, Differential thermal analysis, differential thermogravimetric analysis, and high temperature microscopy of reactions between major components of sheet glass batch, Glass Technol. 6 (1965) 107–114.

[103] E. Bader, Thermoanalytical investigation of melting and fining of thüringen laboratory glassware, Silikattechnik 29 (1978) 84–87.

[104] J. Mukerji, A.K. Nandi, K.D. Sharma, O. Abe, T. Utsunomiya, Y. Hoshino, Reaction in container glass batch, Ceram. Bull. 22 (1979) 790–793. The reaction of sodium nitrate with silica, Bull. Chem. Soc. Jpn. 56 (1983) 428–433.

[105] T.D. Taylor, K.C. Rowan, Melting reactions of soda-lime-silicate glasses containing sodium sulfate, J. Am. Ceram. Soc. 66 (1983) C227–C228.

[106] M. Lindig, E. Gehrmann, G.H. Frischat, Melting behavior in the system SiO_2-K_2CO_3-CaMg $(CO_3)_2$ and SiO_2-K_2CO_3-PbO, Glastech. Berich. 58 (1985) 27–32.

[107] C.A. Sheckler, D.R. Dinger, Effect of particle size distribution on the melting of sodalime-silica glass, J. Am. Ceram. Soc. 73 (1990) 24–30.

[108] K.S. Hong, R.E. Speyer, Thermal analysis of reactions in soda-lime-silicate glass batches containing melting accelerants. I. One- and two-component systems, J. Am. Ceram. Soc. 76 (1993) 598–604.

[109] K.S. Hong, S.W. Lee, R.E. Speyer, Thermal analysis of reactions in soda-lime silicate glass batches containing melting accelerants. II. Multicomponent systems, J. Am. Ceram. Soc. 76 (1993) 605–608.

[110] J. Chun, D.A. Pierce, R. Pokorný, P. Hrma, Cold-cap reactions in vitrification of nuclear waste glass: experiments and modeling, Thermochim. Acta 559 (2013) 32–39.

[111] P.A. Smith, J.D. Vienna, P. Hrma, The effects of melting reactions on laboratory scale waste vitrification, J. Mater. Res. 10 (1995) 2137–2149.

[112] J. Matyáš, P. Hrma, D.-S. Kim, Melt Rate Improvement for High-Level Waste Glass, PNNL-14003, Pacific Northwest National Laboratory, Richland, Washington, 2002.

[113] C. Rodriguez, J. Chun, D. Pierce, M. Schweiger, P. Hrma, Kinetics of cold-cap reactions for vitrification of nuclear waste glass based on simultaneous differential scanning calorimetry-thermogravimetry (DSC-TGA) and evolved gas analysis (EGA), in: Prepared for the U.S.

Department of Energy Assistant Secretary for Environmental Management, Published in Proceedings of WM March Conference, Phoenix, Arizona, USA, March 2-6, 2014.
[114] R. Pokorný, D.A. Pierce, P. Hrma, Melting of glass batch: model for multiple overlapping gas-evolving reactions, Thermochim. Acta 541 (2012) 8–14.
[115] H.E. Kissinger, Reaction kinetics in differential thermal analysis, Anal. Chem. 29 (1957) 1702–1706.
[116] M. Reading, A. Luget, R. Wilson, Modulated differential scanning calorimetry, Thermochim. Acta 238 (1994) 295–307.
[117] M.J. O'Neill, Measurement of specific heat functions by differential scanning calorimetry, Anal. Chem. 38 (1966) 1331–1336.
[118] P.R. Hrma, M.J. Schweiger, B.M. Arrigoni, C.J. Humrickhouse, J. Marcial, A. Moody, C. Rodriguez, R.M. Tate, B. Tincher, Effect of Melter-Feed-Makeup on Vitrification Process, PNNL-18374, Pacific Northwest National Laboratory, Richland, Washington, 2009.
[119] M. Maciejewski, C.A. Müller, R. Tschan, W.D. Emmerich, A. Baiker, in: E. Kaiserberger (Ed.), Special issue of Thermochim. Acta, vol. 295, 1997, p. 167.
[120] M.J. Schweiger, P. Hrma, C.J. Humrickhouse, J. Marcial, B.J. Riley, N.E. TeGrotenhuis, Cluster formation of silica particles in glass batches during melting, J. Non-Crystal. Solids 356 (2010) 1359–1367.
[121] C.M. Jantzen, M.I. Ojovan, On selection of matrix (waste form) material for higher activity nuclear waste immobilization (review), Russ. J. Inorg. Chem. 64 (2019) 1611–1624.
[122] J. Wang, Z. Wan, Treatment and disposal of spent radioactive ion-exchange resins produced in the nuclear industry, Prog. Nucl. Energy 78 (2015) 47–55.
[123] P. Antonetti, Y. Claire, H. Massit, P. Lessart, C. Pham Van Cang, A. Perichaud, Pyrolysis of cobalt and cesium doped cationic ion-exchange resin, J. Anal. Appl. Pyrolysis 55 (2000) 89–91.
[124] N. Hamodi, Immobilisation of Spent IEX in Borosilicate Glass, M. Sc. Thesis, The University of Manchester, Manchester, 2008.
[125] N. Hamodi, K. Papadopoulou, T. Lowe, T. Abram, Thermal analysis and immobilization of spent ion exchange resin in borosilicate glass, New J. Glass Ceram. 2 (2012) 111–120.
[126] N.H. Hamodi, Y. Iqbal, Immobilization of spent ion exchange resin arising from nuclear power plants: an introduction, J. Pak. Mater. Soc. 3 (2009) 7–18.
[127] D. Vermeeren, Method for Purifying the Cooling Circuit of an Alternator Stator Operating in Ventilated Circuit and Implementing Device, US Patent 6572777 B1, 2003.
[128] P.U. Singare, R.S. Lokhande, R.S. Madyal, Thermal degradation studies of some strongly acidic cation exchange resins, Open J. Phys. Chem. 1 (2011) 45–54.
[129] N. Bothe, F. Doscher, J. Klein, H. Widdecke, Thermal stability of sulphonated styrene-divinylbenzene resins, Polymer 20 (1979) 850–854.
[130] R.S. Juang, T.S. Lee, Oxidative pyrolysis of organic ion exchange resins in the presence of metal oxide catalysts, J. Hazard. Mater. 92 (2002) 301–314.
[131] M.A. Dubois, J.F. Dozol, C. Nicotra, J. Serose, C. Massiani, Pyrolysis and incineration of cationic ion exchange resins-identification of volatile degradation compounds, J. Anal. Appl. Pyrolysis 31 (1995) 129–1340.
[132] S.M. Abdelbasir, C.T. El-Sheltawy, D.M. Abdo, Green processes for electronic waste recycling: a review, J. Sustain. Metall. 4 (2018) 295–311.
[133] V. Fiandra, L. Sannino, C. Andreozzi, F. Corcelli, G. Graditi, Silicon photovoltaic modules at end-of-life: removal of polymeric layers and separation of materials, Waste Manag. 87 (2019) 97–107.

[134] B. Girardin, G. Fontaine, S. Duquesne, M. Försth, S. Bourbigot, Characterization of thermo-physical properties of EVA/ATH: application to gasification experiments and pyrolysis modeling, Materials 8 (2015) 7837–7863.

[135] M.C.C. de Oliveira, A.S.A.D. Cardoso, M.M. Viana, V.F.C. Lins, The causes and effects of degradation of encapsulant ethylene vinyl acetate copolymer (EVA) in crystalline silicon photovoltaic modules: a review, Renew. Sustain. Energy Rev. 81 (2018) 2299–2317.

[136] L. Gasnot, V. Decottignies, J.F. Pauwels, Kinetics modelling of ethyl acetate oxidation in flame conditions, Fuel 84 (2005) 505–518.

[137] T.-H. Liou, Kinetics study of thermal decomposition of electronic packaging material, Chem. Eng. J. 98 (2004) 39–51.

[138] M. Iji, Y. Ikuta, Pyrolysis-based material recovery from molding resin for electronic parts, J. Environ. Eng. 124 (1998) 821.

[139] M.V. Pufleau, T.S. Chadha, G. Yablonsky, P. Biswas, Carbon elimination from silicon kerf: thermogravimetric analysis and mechanistic considerations, Sci. Rep. 7 (2017) 40535, https://doi.org/10.1038/srep40535.

[140] S. Vyazovkin, A.K. Burnham, J.M. Criado, L.A. P-Maqueda, C. Popescu, N. Sbirrazzuoli, ICTAC Kinetics Committee recommendations for performing kinetic computations on thermal analysis data, Thermochim. Acta 520 (2011) 1–19.

Index

Note: Page numbers followed by *f* indicate figures, and *t* indicate tables.

O

P

Q

R

S

T

Kolibri 360 GmbH
Ferdinand-Jühlke-Straße 7
99095 Erfurt, Deutschland
produktsicherheit@kolibri360.de